A new legal framework for managing the world's shared groundwaters

Water Law and Policy Series

The role of law in the development and management of the world's water resources is often overlooked by water resource experts. This series aims to make "water law" - international, national and transnational - more accessible to the global water community.

While focusing primarily on the legal component of water resources management, the series will integrate other water-related disciplines in order to ensure a truly comprehensive and interdisciplinary approach to the problem.

SERIES EDITORS

Patricia Wouters
International Water Law Research Institute, University of Dundee, UK.

Sergei Vinogradov
Centre for Energy, Petroleum and Mineral Law and Policy,
University of Dundee, UK.

Forthcoming titles in the series include:

International Law and Sustainable Development: *Lessons from the Law of International Watercourses*

Hydrology and Water Law – Bridging the Gap

Water Rights: *An International and Comparative Study*

A new legal framework for managing the world's shared groundwaters

A case study from the Middle East

Fadia Daibes-Murad

IWA Publishing
LONDON • SEATTLE

IWLRI
Water law • Water for all
international • national • transnational

Published by IWA Publishing, Alliance House, 12 Caxton Street, London SW1H 0QS, UK
Telephone: +44 (0) 20 7654 5500; Fax: +44 (0) 20 7654 5555; Email: publications@iwap.co.uk
Web: www.iwapublishing.com

First published 2005

British Library Cataloguing in Publication Data
A CIP catalogue record for this book is available from the British Library

Library of Congress Cataloging- in-Publication Data
A catalog record for this book is available from the Library of Congress

ISBN 1843390760

Contents

Foreword

In 1997 the United Nations adopted, with the support of 103 States, the United Nations Convention on the Non-Navigational Uses of International Watercourses, following close to 30 years study of the topic.[1] Attached to the working Draft Rules considered in New York by the Legal Committee (convened as a "Working Group of the Whole) was a resolution recommending that the legal obligations and entitlements set forth in the Draft Rules should apply also to shared confined aquifers. Despite this suggestion, however, the proposed resolution was not adopted, leaving open the question of what rules of international law determine the lawful use of shared groundwaters.

The UN International Law Commission now works to prepare Draft Articles on Transboundary Groundwaters -- an exercise that will address the very important matter of how sovereign states should be permitted to manage and develop an international shared resource needed by close to one-half of the earth's population.[2] Dr Daibes-Murad has worked with the UNESCO ISARM programme,[3] which provided expert assistance to Ambassador Yamada in his role as Special Rapporteur on this new UN

1 See Wouters, P., "The Legal Response to Water Conflicts: The UN Watercourses Convention and Beyond", 42 German Yearbook of International Law 293 (2000).

2 See Preliminary Draft Articles on Transboundary Groundwaters presented to the ILC by Special Rapporteur Chusei Yamada at the 57th Session of the International Law Commission in Geneva, May-July 2005.

3 UNESCO, ISARM – International Shared Aquifer Resource Management Project, a collaborative project with the goal of developing and championing best practice for the management of groundwater resources shared between neighbouring countries., see http://www.iah.org/isarm/.

ILC project.[4] Nonetheless, the area remains controversial, with huge national interests at stake and only the future will show whether or not this work in progress will result in an agreed and enforceable global legal instrument.

Clearly, this gap in the law, now being closely examined and progressively developed, provides the ideal opportunity for consideration of Dr Daibes-Murad new book, which sets forth a "new legal framework" for managing shared aquifers. Of particular interest is the case study that she has selected for this work – the Middle East, surely one of the most difficult arenas within which to negotiate and agree upon shared scarce waters. This study examines the rights and obligations of States related to the use of their shared groundwaters – an area that is still evolving and remain unclear in many respects. Fadia, following a thorough survey of the relevant issues, literature and state practice, concludes that the governing rule of "equitable and reasonable utilization", which governs shared surface and connected surface/groundwater, should apply also shared aquifers. This recommendation is based upon a well-researched and argued position, with a rigorous examination of comparative state activity in this field.

Can Israel and Palestine benefit from an agreed legal framework that is forward-looking and different from the current regime? Dr Daibes-Murad is convinced (and convincing on this point) that a step-by-step approach to operationalising a cooperative legal instrument is essential to the future peaceful management of the shared water resource. Water can be a catalyst for peace, and not an instrument for conflict.

Dr Daibes-Murad, with her background as an engineer, and her doctoral work in international law combines two disciplinary approaches to considering this problem. As a result there is a depth of hydrological and legal information considered and presented carefully and thoroughly in mounting her case. The work is a provocative read and should be of interest to a wide range of water experts.

It is a real pleasure to have this contribution to our book series. Fadia continues to work to promote regional cooperation in the Middle East and uses her expertise in water law and water science to achieve this. Dr. Daibes-Murad won the Edberg Prize in 2005 (Swedish Environmental Award), recognising her important work on water related issues in the Middle East, including this book that introduces innovative solutions for using international water law to assist in resolving the Palestinian/Israeli conflict over shared transboundary resources.

Dr Patricia Wouters

Series Editor

Dundee 16 September 2005

4 See http://www.un.org/law/ilc/sessions/56/56sess.htm#shared_natural_resources.

Preface

The rights and obligations of States concerning their shared transboundary groundwaters are not fully understood. The primary reasons for this are the complex characteristics and behaviours of groundwater, and the lack of full understanding of its interconnection with the surrounding environment. Consequently, States appear to be reluctant in concluding legally binding arrangements that specifically address this topic. This complexity is further intensified under difficult political situations, whereby the Parties sharing groundwater are unable to foresee the benefits of cooperation concerning this resource. This book represents an inquiry into the rules and principles governing the utilization of shared transboundary groundwater, and how cooperative frameworks can be initiated and implemented, in the complex political context of the Mountain Aquifer shared between Israel and Palestine. The main objective of this research is to achieve an integrative mutually beneficial arrangement in the case of the Mountain Aquifer, based on lessons learnt from international law. The proposal for approaching the problem within the Palestinian-Israeli context is to progressively establish a solid basis for long-term sustainable arrangements. The progressive approach starts with a non-binding flexible arrangement that can serve the short-term need, which gradually builds up into a final legally binding arrangement based on international law.

Abbreviations

ADPILC	Annual Digest of Public International Law Cases
AJIL	American Journal of International Law
ASDI	Annuaire Suisse de Droit international
ASILP	Proceedings of the American Society of International Law
CalWILJ	California Western International Law Journal
CBR	Canadian Bar Review
CoJIELP	Colorado Journal of International Environmental Law and Policy
CYIL	Canadian Yearbook of International Law
DenvJILP	Denver Journal of International Law and Policy
EMAB	Eastern Mountain Aquifer Basin
EC	European Commission
EEB	The European Environmental Bureau
EJIL	European Journal of International Law
ELQ	Environmental Law Quarterly
Emory Int'l L Rev	Emory International Law Review
EU	European Union
FAO	United Nation Food and Agriculture Organization
GeoIELR	Georgetown International Environmental Law Review
GW-MATE	Groundwater Management Advisory Team (The World Bank Group)
GYIL	German Yearbook of International Law

HarvILJ	Harvard International Law Journal
ICJ	International Court of Justice
IDI	Institute of International Law
IGL	International Groundwater Law
ILA	International Law Association
ILC	International Law Commission
ILM	International Legal Materials
ILR	International Law Reports
IPCC	Inter-governmental Panel on Climate Change
ISARM	International Shared Water Resources Management
IWLRI	International Water Law Research Institute
IWRM	Integrated Water Resources Management
J. Hydrol	Journal of Hydrology
NCJIL	North Colorado Journal of International Law
NEMAB	North Eastern Mountain Aquifer Basin
NethILR	Netherlands International Law Review
NRF	Natural Resources Forum
NRJ	Natural Resources Journal
PaceELR	Pace Environmental Law Review
PASSIA	Palestinian Academic Society for the Study of International Affairs
PCIJ	Permanent Court of International Justice
PYIL	Palestine Yearbook of International Law
RIAA	Reports of International Arbitral Awards (United Nations)
TAR	Transboundary Aquifer Resources Management
TFD	Transboundary Freshwater Dispute Database
UBCLR	University of British Columbia Law Review
UN	United Nations
UNGA	United Nations General Assembly
UNCED	United Nation Conference on Environment and Development
UNECA	United Nations Economic Commission for Africa
UNECE	United Nations Economic Commission for Europe
UNEP	United Nations Environmental Programme
UNDP	United Nations Development Programme
UNTS	United Nations Treaty Series
UTLJ	University of Toronto Law Journal
WMAB	Western Mountain Aquifer Basin
WFD	Water Framework Directive
YBIEL	Yearbook of International Environmental Law
YILC	Yearbook of International Law Commission

1

Introduction

We Know the Value of Water When the Well Runs Dry

(Benjamin Franklin)

1.1 WHY TRANSBOUNDARY GROUNDWATER?[1]

Solutions to partly resolve the world's water crisis may lie hidden underground.[2] More than half of the world's population already depends on groundwater that is pumped from the layers of rock or other geological strata, known as aquifers,[3]

---

[1] "Groundwater means"
"All water, which is found below the surface of the ground, in the saturation zone, aquifers, and the structures containing deep "fossil waters", *See generally* C.W. Fetter, Applied Hydrogeology, (Prentice-Hall: New Jersey, 1994), H. Bouwer, Groundwater Hydrology (McGraw-Hill: New York, 1978) and M. Price Introducing Groundwater (second edition) (Chapman and Hall: London, 1996).

[2] UNESCOPRESS Press Release No.2002-35, *found at* http://www.unesco.org/bpi/eng/unescopress/2002/02-39e.shtml (*last visited* May 2003).

[3] "Aquifer" means:

which lie hidden below the Earth's surface. These formations can span thousands of kilometres and contain enough water to satisfy much of humanity's demands for water for many decades.[4] Groundwater is the largest source of fresh water available in reserve on Earth.[5] Of 37 million cubic kilometres of freshwater estimated to be present on this planet, about 22% occurs below the land surface in the form of groundwater resources. Excluding water locked in polar icecaps, this constitutes 97% of all freshwater potentially available for human use.[6] Groundwater is in general a high-value resource[7] and is especially important as a source for drinking water. In Europe for instance, 75% of drinking water supplies come from groundwater sources, with reliance up to 98% in Denmark[8] and 100% in Lithuania.[9] In arid or semi-arid regimes, groundwater is almost the sole source of water supply for human and other consumption. Human dependence on groundwater as a major source of water supply has been recently acknowledged and the risks associated with depleting this valuable resource are widely recognized.[10]

Transboundary groundwater,[11] sometimes referred to as international or cross-border groundwater, is used virtually in every continent of the world[12] and is

---

"Any subsurface geological formation, such as a layer or layers of rock or other geological strata of sufficient porosity and permeability, which contains water and from which it may be extracted in appreciable quantities. Water that is in the upper zone of the soil is not included", *See generally supra* note 1.

[4] *Ibid.*

[5] S. Burchi, *National Regulations for Groundwater: Options, Issues and Best Practices,* found in: S. Salman (ed.) Groundwater: Legal and Policy Perspectives (World Bank Technical Paper No. 456, Washington, 1999).

[6] S. Foster, *Essential Concepts for Groundwater Regulators,* found in S. Salman (ed.) (World Bank Technical Paper No. 456, Washington, 1999).

[7] *See infra* Section 1.3.1 in Chapter Two, which explains what is meant by a high value resource through highlighting the overall functions of groundwater with respect to human beings and the ecology.

[8] S. Burchi, *supra* note 5.

[9] UNECE Task Force on Monitoring and Assessment, under the Convention on the Protection and Use of Transboundary Watercourses and International Lakes (Helsinki 1992): *Inventory of transboundary ground waters,* Vol. 1, ISBN 9036952743, Lelystad, 1991.

[10] United Nations Publication, *Groundwater storage and Artificial Recharge*, Natural Resources/Water Series No. 2 (Sales No. E.74.IIA.11, p.1).

[11] J. Barberis, *The Development of International Transboundary Groundwater Law,* 31 Natural Resources Journal 167-186 (1991). Barberis explains the various situations where an aquifer is international or transboundary: (i) Where a confined aquifer is intersected by an international boundary, this aquifer is not linked hydrologically with other groundwater or surface water, and consequently, only the aquifer itself can be considered a shared natural resources (ii) Where an aquifer lies entirely within the territory of one state but has interconnections and interdependence with an international river; (iii) where an aquifer is

subject to the same competition for its use as indigenous water resources. Many transboundary aquifers are found in Africa,[13] North America,[14] Europe,[15] and the Middle East.[16] Because of the special problems that are present when more than one national jurisdiction is involved, extraordinary concern has been shown for waters that are international in character.[17] The research addresses two primary questions; the first is what are the rules and principles of international law that govern the utilization, development and management of transboundary groundwater resources? while the second is what lessons could be learnt from this law to help achieve an integrative mutually beneficial arrangement for the Palestinian and Israelis sharing the Mountain Aquifer? These questions are relevant for two main reasons: (i) the global importance of transboundary groundwater and the need for a legal regime that governs its utilization, development and protection and (ii) the critical role transboundary groundwater plays in the area of the case study, whereby the existing water conflict is likely to grow in the coming few years.

---

entirely situated within the territory on one state but is getting recharge in another State and (iv) Aquifers found entirely in the territory of one State but is hydrogeologically connected with another aquifer in a neighbouring State.

[12] International Law Association, Report of the Sixty second Conference, p. 238, (1986).

[13] Examples on aquifers in Africa: the Northern Sahara Aquifer shared between Algeria Tunisia and Libyan Arab Jamahiriya, Chad Aquifer shared between Niger, Chad, Sudan, Central African Republic, Nigeria and Cameron, Aquifer on the Taoudeni Basin shared between Chad, Egypt, Libyan Arab Jamahiriya and Sudan and the Kalahari/Karoo multi-layered aquifer that is shared by Namibia Botswana and South Africa. See generally S. McCaffrey, *International Groundwater Law: Evolution and Context*, found in: Groundwater: Legal and Policy Perspective, World Bank Technical Paper No. 456 (S. Salman (ed.),Washington D.C. , 1999) and S. Puri et al, Internationally Shared Aquifer Resources Management: Their Significance and Sustainable Management: Framework Document (International the United Nations Educational Scientific and Cultural Organization, Paris 2001*)* *also* *found* *at* http://www.unece.org/env/water/documents/isarm.pdf (*last visited* May 2003).

[14] Eighteen transboundary aquifers in the Mexico-United States border area. See UNESCO Bulletin, International, *found at* www.unesco.org/ubb/Forum24/HTML/000002.html (*last visited* May 2003).

[15] The UNECE Report, *supra* note 9, has documented over 100 aquifers that are shared by two or more states. (Examples are: the aquifer system of the Vechte Catchment is shared by Germany and the Netherlands, The alluvial aquifer system is shared by the Czech Republic and Poland).

[16] Examples are: the Mountain Aquifer shared between Palestine and Israel, the Saq aquifer shared between Saudi Arabia and Jordan, Egypt and Israel sharing the Nubian Sandstone aquifer in Sinai and Negev. See S. Puri et al *supra* note 13.

[17] UN/Natural Resources/Water Series No. 1, Management of International Water Resources: Institutional and Legal Aspects (United Nations: New York, 1975).

The question of transboundary confined[18] aquifers will be addressed in a limited manner. The reason for doing this is the fact that the case study only involves connected aquifers,[19] which will be the main focus of the study. However, the issues, challenges and recent development of international law in the area of confined aquifers will be addressed and discussed.

1.2 TRANSBOUNDARY GROUNDWATER LAW: THE NEED FOR MODERN DEVELOPMENT

The unique characteristics of a watercourse and the need to use its waters in a rational way require legal solutions that are different from those applicable in other branches of international law.[20] This explains the fact that international water law is widely recognised as a distinct branch of international law.[21] Furthermore, the special requirements of optimal and sustainable development of international

[18] "Confined Aquifers" are often described in the legal literature as "fossil" aquifers which contain Non-renewable groundwaters. These are aquifers with no appreciable recent recharge and which cannot discharge naturally. As a result, water in these aquifers is stagnant and has little if any flow. By definition, they cannot be utilized sustainably as any withdrawal eventually will exhaust the resource. Such aquifers generally contain groundwater that is trapped in a geologic formation, either because of physical isolation of the aquifer from sources of recharge, impermeability of overlying strata, or paucity of recharge in an arid region. Typically, water in confined aquifers is hundreds if not thousands (or millions) of years old, [hereinafter referred to as confined aquifers].To understand the basic concept of renewable and non-renewable aquifers it is recommended to *see generally* C.W. Fetter, Applied Hydrogeology, (Prentice-Hall: New Jersey, 1994), H. Bouwer, Groundwater Hydrology (McGraw-Hill: New York, 1978) and M. Price Introducing Groundwater (second edition) (Chapman and Hall: London, 1996), *see also* http://www.unesco.org/ubb/Forum24/HTML/000003.html (*last visited* May 2003).

[19] "Connected Aquifer" is :"an aquifer that is connected to the surface water system and annually replenished by water which finds its way to the saturated rock formations that can hold and transmit water to wells and springs". Such aquifers occur in the following situations: (i) when the aquifer is overlain by a low permeability or a leaky confining layer that permits water to slowly flow through it; (ii) when the aquifers are close to the land surface, with continuous layers of materials of high intrinsic permeability extending from the land surface to the base of the aquifer; and (iii) when a layer of low permeability material is found as a lens in more permeable materials where water moving downwards is intercepted by this layer and will accumulate on the top of the lens. The definition is derived from Fetter, *supra* note 1, at 635 but the classification is by this researcher.

[20] L. Caflisch, found in: P.K. Wouters (ed.) International Water Law (Kluwer Law International: the Netherlands, 1997).

[21] *See* generally P.K. Wouters (ed.), *ibid*, also see R.D. Hayton, and A.H. Garretson, C.J. Olmstead (ed.), The Law of International Drainage Basins (1967) (GARRETSON).

groundwater aquifers[22] and the alarming rate at which water resources are being overexploited and polluted suggest that groundwater is increasingly becoming a high-value resource.[23] This fact is recently prompting the development of more appropriate legal, economic social and institutional frameworks. These frameworks are being supplemented by the establishment of co-operative arrangements and mechanisms for the utilization, development and protection of groundwater resources.

Whilst the international law relating to the use of international surface watercourses has recently been subjected to remarkable development,[24] the law governing transboundary groundwater appears to remain rather uncertain.[25] International law governing the non-navigational uses of international watercourses has been developing during the last 50 years. At present this body of law provides suitable guidance and adequate legal instruments to settle conflicts and prevent disputes over shared water resources.[26] Hundreds of treaties are signed on major rivers of the world including the Niger, Nile, Danube and the Mekong. Only few regulate both international surface and groundwater and rarely deal with transboundary groundwater as a distinct subject of the law.[27]

At the commencement of the 21st century water scarcity has the potential to end in conflict, especially in the Middle East[28] and these risks have a global dimension. The concern is becoming most evident in the arid and semi-arid areas of Africa where shortage and lack of access to water form the main causes for poverty, dwindling public health and food insecurity.[29] The scarcity and rapid

---

[22] *See infra* Chapter 1 section 1.2 "Key Features of Groundwater".

[23] S. McCaffrey, *supra* note 13.

[24] At this stage it suffices to mention the work of some international organizations on the codification of customary international law that governs the non-navigational uses of international watercourses, see *infra* chapter 3. Also *see* generally R.D. Hayton, *supra* note 19, D.A. Caponera, Principles of Water Law and Administration (Balkema: Rotterdam, 1992); W.L. Griffin, *The Use of Waters of International Drainage Basins under Customary International Law*, 53 AJIL 1959, 50-80; S. McCaffrey, *The Evolution of the Law of International Watercourses*, 45 AJPIL 1993, 87-111.

[25] For a review of international groundwater law evolution, see the Introduction to the Law of International Groundwater Resources, Part II, International Law Association (ILA), Seoul Conference (1986). Committee on International Water Resources Law, Report of the Committee 8-20 (R.D. Hayton Rapporteur), reproduced in ILA, Report of the Sixty Second Conference held at Seoul 1987. See *infra* discussion in Chapter 4.

[26] W. Sheumann and A. Klaphake, *The Convention on the Law of Non-Navigational Uses of International Watercourse*, Deutsches Institute fur Entwicklungpolitilk, Bonn, 2001.

[27] See *infra* Chapter 4.

[28] See *infra,* Chapter 3.

[29] International Workshop on Managing Shared Aquifers Resources in Africa: 2nd To 4th June 2002, Tripoli - Workshop Report And Recommendations, *found at* http://www.iah.org/isarm/TripoliWorkshopRecommendations.doc (*last visited* May 2003).

depletion of water in one of the world's driest regions constitutes a major determinant of the domestic and external policies of the major actors in the Middle East.[30] The large variations in rainfall and limited surface resources have led to widespread scarcity of the fresh water resources in the region, resulting in a heavy reliance on groundwater as the major source for various uses. The contribution of surface water to the overall water balance is limited and marginal.[31] Finally, due to the increasing dependence on groundwater, conflicts over the use and development of transboundary groundwater aquifers are expected to grow in the coming few decades and there is an increasing need to achieve a more optimal, equitable and sustainable utilisation of these resources through better co-operation between those sharing the waters. In the view of this research, cooperation is considered essential for the maintenance of peace and the avoidance of conflicts.

1.3 THE MOUNTAIN AQUIFER[32] CASE STUDY

A water scarce situation coupled with political instability and inequality in power structures between two nations sharing the same fixed amounts of waters, is regarded by this research, as the most complex setting for resolving water conflicts. This is true in the region of the Middle East where the political factors inevitably play a decisive role in the settlement of an international water dispute. The hostilities and conflicts arising over the utilization and development of the Mountain Aquifer shared between Israel and Palestine, which has been selected as the case study of this research, are one of the most contentious in the Middle East region. As of 1967[33] up to 1995[34], the Palestinian utilization of transboundary

---

[30] T. Naff and R. Matson (ed.), Water in the Middle East: Conflict or Cooperation, (Westview Press: USA, 1984).

[31] F. Daibes, *Water the Blue Gold of the Middle East*, PASSIA Water Special Bulletin, June 2002, *found at* www.passia .org (*last visited* May 2003).

[32] According to the definition *supra* note 1, the Mountain Aquifer is considered to be a "connected aquifer" that is shared between Israel and Palestine, more details on the physical and natural characteristics of this Aquifer is found *infra* Chapter 3.

[33] Israel took control of the West Bank in 1967, including the recharge areas for the Mountain Aquifer that flow west and northwest into Israel (at about 360 mcm/yr and 145 mcm/yr, respectively) and east to the Jordan Valley (about 125 mcm/yr). The entire renewable recharge of these first two aquifers is already being exploited by Israel and the recharge of the third is close to being depleted as well, See *infra* Chapter 3.

[34] In September 1995, Israel signed an agreement with the Palestinian Liberation Organization PLO regarding the interim arrangement on the West Bank and Gaza Strip; whereby 80 mcm of additional water was allocated for the Palestinians. See Israeli - Palestinian Interim Agreement On the West Bank and the Gaza Strip Washington DC, December 1995 *found at* http://www.ariga.com/treaties/interim.html (*last visited* May 2003).

groundwater has been governed by the Israeli legislations and military orders.[35] The Palestinians' access to the only major international surface watercourse –the Jordan River- has been denied since 1967. Accordingly, groundwater in Palestine, most of which is transboundary with Israel, constitutes the major source of supply for all uses. For many years, and in the context of shared water resources between Israel and Palestine, it appears that there has been a violation of the rights of Palestinians to equitable and reasonable utilization of these resources. This fact, in the opinion of this research, has for long time been one of the major obstacles for co-operation and the achievement of peace between the Palestinians and Israelis.[36]

---

[35] Prior 1967, Jordan issued rules and regulations governing the use, development and other water related matters. These laws remained in effect even after the 5th June 1967, whereby an Israeli proclamation stated that these could be subject to changes introduced by the military Commander. After 1967, successive Military Orders were issued to amend parts of these laws. Significant Military Order 92 of 7 June 1967, Order 158 (19 November 1967) Military Order 291 (19 December 1968) and many others. World Foundation for Environment and Development (WFED): *Water Institutions and Water Laws in Palestine (Gaza Strip and the West Bank)*, March 1995 (on file with author).

[36] See *infra* discussion on applicable law in Chapter 2.

2

Groundwater: Hydrogeological, Management and Legal Perspectives

2.1 INTRODUCTION

This chapter presents the first step in the proposed research methodology, as it sets the context for transboundary groundwaters. It considers issues related to groundwater management from two different perspectives: the first concerns the hydrogeological and physical aspects of groundwater, focusing on the basic scientific knowledge needed as a prerequisite for effective groundwater management and regulation. It presents in a nutshell the key features, concepts, and behaviours that demonstrate the need for a special set of rules dealing with the subject of groundwater. The second focuses on the issues, concerns and challenges requiring responses by the various actors at the national and international political levels. This type of understanding is necessary in order to decide what responses are called for by policy makers, regulators and water managers in order to effectively manage and regulate groundwater. The present chapter also includes an illustrative analysis of modern trends in national groundwater regulations with a view to identify lessons that might be relevant at the international level. The findings of this chapter lay groundwork for later discussions in chapters four and five, which focus on the law that applies to groundwater in transboundary contexts.

Groundwater recharge results from direct percolation of precipitation through the unsaturated zone after evapotranspiration and runoff have taken place, and when the soil moisture deficit of the soil is satisfied. This type of recharge is referred to as direct or precipitation-recharge and is more common in humid and sub-humid areas where rainfall is regular.[1] Recharge may as well originate from surface and subsurface sources and this form is more common in arid and semi-arid areas. Groundwater recharge processes are therefore very complex and often require modern and sophisticated techniques for their estimation.[2] Groundwater is stored in reservoirs known as aquifers. In physical terms aquifers are of two major types connected aquifers[3] and confined aquifers.[4] In the legal literature, the water contained in the latter type is often referred to as "fossil waters".[5] It is established by hydrogeologists that a fossil aquifer is only one type of non-renewable aquifer. They describe an aquifer (whether confined or unconfined) containing water that was buried at the same time as the geologic formation in which it is trapped. A second type of non-renewable aquifer is a "connate" aquifer, which is described as:

"a confined aquifer in which the trapped water is of a different age than that of the encompassing geologic formation."[6]

It is crucial to this research that the definition of "confined aquifers" differs from that in normal hydrogeological usage. Accordingly and for the purposes of this research the term "confined aquifer" means:

---

[1]D. Lerner, A. Issar, and I. Simmers, Groundwater Recharge: A Guide to Understanding and Estimating Natural Recharge, 8 IAH (1990).

[2] G. B. Allison, A Review of Some of the Physical, Chemical and Isotopic Techniques Available for Estimating Groundwater Recharge, 222 NATO Series pp. 49-67, see also S. Foster, and S.A. Carington, The Interpretation of Tritium in the Chalk Unsaturated Zone, 46 J. Hydrol.343-364 (1980).

[3]See definition *supra* footnote 19 in Chapter 1. *See generally* C.W. Fetter, Applied Hydrogeology, (Prentice-Hall: New Jersey, 1994), H. Bouwer, Groundwater Hydrology (McGraw-Hill: New York, 1978) and M. Price Introducing Groundwater (second edition) (Chapman and Hall: London, 1996), *also see* G. Eckstein, *Hydrologic Reality: International Water Law and Transboundary Ground-Water Resources* October 1998, *found at* http://www.internationalwaterlaw.org/Articles/GlobalSouth.htm (*last visited* May 2003).

[4] See *supra* footnote 18 in Chapter 1.

[5] *Ibid.*

[6] G. Eckstein, *supra* note 3.

" an aquifer without any link to the interconnected system of waters that makes up the watercourse. A confined aquifer does not receive any perceptible replenishment from surface water or other groundwater." [7]

As for their geographical reach, aquifers' geographical scale may range from less than 10 km^2 to more than 100 km^2 or even 100,000 km^2.[8] Aquifers that lie and recharge completely within the national jurisdiction of a State are referred to as national aquifers.[9] The larger aquifers comprise the transboundary or international aquifers. Julio Barberis, an eminent legal expert, explains the various situations where an aquifer is international or transboundary (i) where a confined aquifer is intersected by an international boundary, this aquifer is not linked hydrologically with other groundwater or surface water, and consequently, only the aquifer itself can be considered a shared natural resources; (ii) where an aquifer lies entirely within the territory of one State but has interconnections and interdependence with an international river; (iii) where an aquifer is entirely situated within the territory on one state but is getting recharge in another State, and (iv) aquifers found entirely in the territory of one State but hydrogeologically connected with another aquifer in a neighbouring State.[10] The next section moves the discussion on groundwater forward. By analyzing the key features of groundwater, the need for tailor-made legal regime that governs the utilization and development of such resources becomes clear.

2.2 GROUNDWATER: KEY FEATURES

2.2.1 Background

Groundwater is like surface water in many respects; it is part of the same hydrological cycle, [11] and often ignores political boundaries.[12] However, in some

---

[7] Definition by the author.

[8] S. Foster, *Essential Concepts for Groundwater Regulators,* found in: Groundwater: Legal and Policy Perspectives, (S. Salman (ed.), World Bank Technical Paper No. 456: Washington, 1999).

[9] S. Burchi, *National Regulations for Groundwater: Options, Issues and Best Practices,* found in: Groundwater: Legal and Policy Prospects (S. Salman (ed.), World Bank Technical Paper No. 456: Washington, 1999).

[10] J. Barberis, the Development of International Transboundary Groundwater Law, 31 NRJ 167-186 (1991).

[11] The hydrologic cycle describes the continuous movement of water above, on, and below the surface of the Earth. *See* generally *supra* note 3.

[12] R. D. Hayton and A. E. Utton, *Transboundary Groundwater: The Bellagio Draft Treaty,* 29 NRJ 633-722 (1989).

other respects, groundwater is very distinct in characteristics and behaviour.[13] Most importantly, the quantity, quality and flow rates of groundwater are greatly affected by the contrasting physical and chemical environment within which it occurs.[14] Also unlike surface water systems, the losses that occur in groundwater systems through over abstraction are often irreversible;[15] while groundwaters store pollution, surface waters have some capability for self purification.[16]Additionally, the time lag between the disposal of a contaminant and the actual occurrence of pollution is generally high.[17] Aquifers, on the other hand, are considered as slow biochemical reactors that change the quality of water they store. This can be for the better, as a result of the natural attenuation of various water pollutants, or for the worse, as a result of the dissolution of toxic or troublesome trace elements from the ground. However, not all subsoil profiles and underlying strata are equally effective in contaminant attenuation. [18]

Given the foregoing, the enigma related to groundwater management and regulation is the lack of a complete understanding of the interrelationships between surface and groundwater.[19] The distinction between ground and surface

---

[13]World Bank Group on Groundwater: GW-MATE, Briefing Note No. 1, *Groundwater Resource Management: An Introduction to its Scope and Practice, found at* http://lnweb18.worldbank.org/ESSD/essdext.nsf/18DocByUnid/F5DC74EF8FF66E558525 6B9000547E5C/$FILE/BN1.pdf (*last visited* September 2003).

[14] S. Foster, *supra* note 8.

[15] J. Burke, M. Moenech, and C. Sauveplane, *Groundwater and Society: Problems in Variability and Points of Engagement,* found in: Legal and Policy Perspectives, (S. Salman (ed.), World Bank Technical Paper No. 456, 1999).

[16] L. Teclaff and E. Teclaff, *Transboundary Groundwater Pollution: Survey and Trends in Law*, 19 NRJ 629-666 (1979). Irreversibility of the losses in the context of groundwater pollution might take the form of Salinization of the resource especially in coastal aquifers due to extensive over pumping that leads to the intrusion of sea water to the replace the pumped fresh water. Additionally in intensive agricultural areas where fertilizers and pesticides are heavily used the recovery of good status of groundwater quality can be very complicated if possible at all. Other types of pollution might be caused in heavily industrialized areas where no regulations and control over waste disposal exist. In many occasions once the aquifer is seriously polluted the decision to remedy the situation is very risky and expensive to undertake. Such a situation might lead to the complete abandonment of these aquifers until they recover, a goal that requires many years and high investments to be achieved. (*See* also Fetter *supra* note 3).

[17] This fact complicates the task of pollution prediction, prevention and control in the case of groundwater. *See generally:* World Bank Groundwater Management Team, Briefing Note No. 8: Groundwater Quality Protection--defining strategy and setting priorities *found at* http://lnweb18.worldbank.org/ESSD/essdext.nsf/18DocByUnid/5C499F3B1F9D03C785256 B9000557635/$FILE/BN8.pdf (*last visited* May 2003).

[18]*Ibid.*

[19]For more background reading on the interrelationships between surface and groundwater see Groundwater Legislation in the UNECE Region. The UNECE Report on groundwater

water has only recently been subject to serious consideration by water resource engineers, managers, policy makers and regulators.[20] Efforts are increasingly oriented towards the identification of appropriate management and regulatory instruments essential to meet the special needs of optimal and sustainable development of groundwater at the technical political and regulatory levels,[21] notwithstanding the need for integrated water resources management (IWRM)[22] of surface and groundwater.[23] Finally, the hidden nature of transboundary groundwater and lack of legal frameworks, invites misunderstandings by many policy makers. Not surprisingly therefore, transboundary aquifer management is still in its infancy, since its evaluation is difficult, suffering from a lack of institutional will and finance to collect the necessary information.[24] The following sections describe the major features of groundwater that require a good and

---

legislation observes that: 'the varieties of hydrological situations involving groundwater are many and complex. They include interactions between groundwater and surface waters; between groundwater and other elements of the environment generally groundwaters are interconnected with overlying surface waters, *see* UNECE doc. UNECE/water 44, New York 1986.

[20] See generally R.D. Hayton *supra* note 12, see also S. Bogdanovic, International Law of Water Resources: Contribution of the International Law Association1954-2000, (Kluwer Law International: The Netherlands, 2000).

[21] In 1987 the United Nations Commission on Environment and Development [the Bruntland Commission] drew attention to the fact that economic development often leads to deterioration, not an improvement, in the quality of people's lives. The Commission recommended that just because it is new does not mean that it is better - or at least not for everybody. A form of sustainable development is needed that meets the needs of the present without compromising the ability of future generations to meet their own needs. There are two key issues as part of this: Development is not just about bigger profits and higher standards of living for a minority. It should be about making life better for everyone and this should not involve destroying or recklessly using up our natural resources, nor should it involve polluting the environment. See WCED, Our Common Future: the Bruntland Report, (Oxford University Press from the World Commission on Environment and Development: New York 1987).

[22] The IWRM is an approach initiated by the Global Water Partnership (GWP) that seeks to balance human, industrial, agricultural and environmental needs. In 2000, the Technical Advisory Committee of the GWP has identified IWRM as:

"a process, which promotes the coordinated development and management of water, land and related resources in order to maximise the resultant economic and social welfare in an equitable manner without compromising the sustainability of vital ecosystems".

See generally the GWP Website, *found at* www.gwpforum.org (*last visited* April 2002).

[23] A.E. Utton, the Development of International Groundwater Law, 22 NRJ 95-118 (1982).

[24] S. Puri (editor), B. Appelgren, G. Arnold, and A. Aureli, *Internationally Shared Aquifer Resources: their Significance and Sustainable Management, A Framework Document,* found at http://www.fao.org/Legal/Prs-OL/isarm.pdf (*last visited* September 2003).

comprehensive understanding prior to developing legal regimes governing its utilization, development and protection.

2.2.2 A Finite Resource

When discussing this specific feature of groundwater, quantitative as well as qualitative aspects must be considered. Considering the first aspect, it is established that amount of water that essentially reaches an aquifer at a certain point in time depends on climatologic,[25] hydrologic as well as hydrogeologic parameters.[26] Additionally, the amount that can be produced will vary under varied pumping and development.[27] A fundamental requisite of groundwater resource evaluation is to define a reliable conceptual aquifer flow model with quantitative estimates of the rates of aquifer discharge.[28] In the context of renewable groundwater aquifers, the 'safe yield' is an important concept, which, based on the ideas of many authors, could be expressed as "the amount of naturally occurring groundwater that can be withdrawn from an aquifer on a sustained basis, economically and legally, without impairing the native groundwater quality or creating an undesirable effect such as the environmental damage".[29] An evaluation of groundwater resources requires an assessment of the

---

[25] Arid and semi arid areas are characterized by low amounts of groundwater mainly due to scarce rainfall and high evaporation which implies that the portion reaching the groundwater is reduced. Groundwater is regarded as the only hope for development in many desert regions such as Libya and northern Sudan. The underground flow from the Sudan's Nubian sandstone formations into Egypt has been estimated at over seven million cubic meters annually. However, even in water abundant countries aquifers have for long been exploited, - a fact that leads to progressive deterioration in water quality. See generally Fetter, Price and Bouwer; *supra* note 3 and D. Lerner, et al *supra* note 1.

[26] A useful concept in aquifer studies is transmissivity which is

"a measure of the amount of water that can be transmitted horizontally through a unit width by the full saturated thickness of the aquifers under a certain hydraulic gradient".

Aquifer storability (storativity) is:

"the volume of water that a permeable unit will absorb or expel from storage per unit surface per unit change in head. The speed of transmission may vary from centimetres to meters a day depending on the rocks making up the aquifer".

See Fetter, *supra* note 3.

[27] *See* Fetter, *supra* note 3 at 518.

[28] *Ibid.*

[29] *Ibid* at 518. The "Safe Yield" term has been used by scientists from as early as 1915. At that time, safe yield was regarded as "the amount of water that could be pumped regularly and permanently without dangerous depletion of the storage reserve". Later other factors that need to be considered were added, such as economies of groundwater development, protection of the quality of the existing store of groundwater and the protection of the existing legal rights and potential environmental degradation. Synonyms for safe yield

groundwater recharge and processes in addition to a proper evaluation of discharge and development patterns.[30] There is an additional need to reveal the key linkages to the surface environment which include :(i) discharge to springs streams or any other freshwater systems, (ii) discharge via natural vegetation to sustain ecologically and/or economically valuable freshwaters wetlands and brackish lagoons and (iii) discharge to saline areas including coastal waters, salt lakes and playas.[31] Accordingly, a single value for the safe yield of an aquifer cannot be provided in the same sense as mean annual precipitation is calculated and provided. The safe yield evaluation should include a statement of the legal and economic constraints that were considered as well as the limiting values of environmental damage that were considered.[32] If the overall abstraction rate in a given area or aquifer system as a whole exceeds the long term average rate of replenishment, there will be a continuous decline in water-level, an overdraft of aquifer storage and consumption of aquifer reserves.[33] It is therefore evident that safe yields values are based on a number of constraints; such values require a multidisciplinary team of engineers, economists, geologists, lawyers and ecologists to prepare.

Non-renewable groundwater aquifers -including fossil water- are, if anything, more finite than renewable aquifers. When non-renewable groundwater is extracted, it will be quickly depleted, as recharge will take years to take place. [34]Accordingly, the regulation of the exploitation of these resources must be carefully planned.[35] In many arid countries, the mining of non-renewable groundwater resources could provide an opportunity and a challenge, and allow water supply sustainability within foreseeable time-frames that can be progressively modified as water related technology advances.[36]

---

appear in the literature "potential sustained yield", "permissive sustained yield" and "maximum basin yield".

[30] S. Foster, *supra* note 8.

[31] *Ibid.*

[32] S. McCaffrey, *International Groundwater Law: Evolution and Context*, found in: Groundwater: Legal and Policy Perspective, (S. Salman (ed.), World Bank Technical Paper No. 456: Washington D.C., 1999).

[33] S. Foster, *supra* note 8.

[34] UN General Assembly International Law Commission Fifty Fifth Session, *Shared Natural Resources: First Report on Outlines by the Special Rapporteur Chusei Yamada*, UN Doc. A/CN.4/533, May 2003.

[35] S. McCaffrey, *supra* note 32.

[36] Tripoli Statement to the International Conference on "Regional Aquifer Systems in Arid Zones – Managing Non-Renewable Resources", Tripoli, 20-24 of November 1999, *found at* http://www.iah.org/articles/dec1999/tripoli.pdf (*last visited* May 2003).

As for the qualitative aspect, the major concern about groundwater pollution relates to the so-called phreatic aquifers.[37] Groundwater quality deterioration mainly occurs as a result of excessive contaminant loads on the land surface and/or inadequately-controlled pumping. This deterioration is essentially an irreversible process, when considered in terms of time and money, as a result of the complex hydraulic structure and long recharge times of groundwater systems.[38] Non-renewable resources are again more vulnerable as far as pollution is concerned. When such water is contaminated, it will remain so for many years.

2.2.3 Groundwater Crosses Borders

Like rivers, aquifers cross national borders and the groundwater resources contained within them can be shared by two or more States. Very few international boundaries follow natural physical features, and water resources can cross them unhindered.[39] Groundwater has the same features and behaves in the same manner in national and international contexts; however as it crosses political boundaries it gains a new unique status. The mis-match between aquifer and socio-political boundaries can present a significant impediment to effective management.[40] Actions of one State, in connection with an international river or transboundary groundwater aquifer, may deleteriously affect the quality or quantity of water in another State. Decisions and responses concerning transboundary groundwater management and regulation require two or more States to undertake with possibly two or more legal regimes involved. The envisaged problems at this preliminary stage between national and transboundary groundwaters lie in the willingness on the sides of States to accept that the notions of groundwater regulations that apply on their domestic groundwaters can be as well adapted at the international level. To what extent state practices at the national level provide guidance to groundwater regulation in transboundary contexts will be discussed later in this chapter.

---

[37] A phreatic aquifer is defined as:
"an aquifer in which there are no confining beds between the zone of satuaration and the surface. There will be a water table in the phreatic aquifer. "uncofined aquifer" or "water table aquifer" are synonyms". *See* Fetter *supra* note 3 at 635.

[38] UNESCO: Report on the Proposal For the Establishment of an International Groundwater Resources Assessment Centre, IHP/IC-XIV/Inf.26, Paris, April 2000, *found at* http://www.iah.org/articles/aug2000/ingrace.pdf (*last visited* May 2003).

[39] S. Puri (editor), *supra* note 24.

[40] *Ibid.*

2.2.4 An Integral Component of the Environment

Awareness of the interaction of the underground environment with the surface and atmosphere is only recently becoming widespread.[41] Environment conditions in and about the aquifers in addition to the economic, political and social values are all factors that affect the efficient management of this vital resource.[42] Interest in the relation of groundwater to surface water, for example, has increased in recent years as a result of widespread concern related to water supply, contamination of groundwater, lakes, and streams, loss of wetlands due to development; and other changes in aquatic environments.[43] Where a hydrological link exists, a causal relationship can be established between surface and groundwater, affecting one section of the water system will likely affect the quantity, quality or potential economic value of a water source in another section.[44] As water tends to flow with gravity, aquifers are usually fed by surface water sources and other aquifers, and can feed other surface waters and underground aquifers.[45] That is not to say that all aquifers are interconnected to surface bodies of water.[46] It is rare, however, that an international river is not somehow linked to a groundwater source.[47]

In arid areas the effect of drought on aquifers should be determined, as well as the response of groundwater systems to the drought.[48] Additionally, changes in land use can have a severe impact on groundwater quality.[49] Poor planning, land use and wastewater management in cities in many parts of the world has polluted

---

[41] R. D. Hayton, *supra* note 12, *also* The European Water Framework Directive (WFD) of 2000 can be cited here, which demonstrates an awareness of this relationship, paragraph 20 of the preamble states that:
"The quantitative status of a body of groundwater may have an impact on the ecological quality of surface waters and terrestrial ecosystems associated with that groundwater body." The European Union WFD 2000/60/EC, published in the Official Journal of the European Communities on 22 December 2000, (entered into force on that date). The fifteen Members States of the Community are: Austria, Belgium, Denmark, Finland, France, Germany, Greece, Ireland, Italy, Luxembourg, The Netherlands, Portugal, Spain, Sweden, and the United Kingdom, *found* at http://www.europa.eu.int/eur-lex (*last visited* March 2003).

[42] R. D. Hayton, The Groundwater Legal Regime as Instrument of Policy Objectives and Management Requirements, 1 IGL 9 (1981).

[43] T.C. Winter, J.W. Harvey, O.L. Franke and W.M. Alley, *Ground Water and Surface Water A Single Resource*, USGS Circular No. 1139, Denver, 1998 *found at* http://water.usgs.gov/pubs/circ/circ1139/htdocs/title.htm (*last visited* July 2003).

[44] *Ibid.*

[45] *Ibid.*

[46] Refer to the discussion on confined aquifers containing non-renewable waters locked at great depths that are not connected to the surface water system *supra* Introduction page 7.

[47] S. Foster, *supra* note 8.

[48] *Ibid.*

[49] USGS Circular No. 1139, *supra* note 43.

town water supplies, has caused saltwater intrusion and land subsidence.[50] Urbanization for example has profound impacts on the local hydrological cycle, including major changes in groundwater recharge. Existing infiltration mechanisms are radically modified, and new ones introduced, with evidence of an overall decrease in recharge rates.[51] The next discussion employs the knowledge gained from the above analyses in order to address the main problems and challenges associated with groundwater management.

2.3 GROUNDWATER MANAGEMENT: PROBLEMS AND CHALLENGES

2.3.1 Background

The core elements in groundwater resources management are the functions and uses of groundwater aquifers, the problems and pressures (threats) and the impact of measures on the overall functioning of the water body.[52] The specification of the human uses and the ecological functioning, the identification of pressures and problems, and the determination of targets should include both quality and quantity aspects.[53] To begin with the functions and uses of groundwater it is established that it is an extremely significant source of direct water-supply for human use (urban, rural, and agricultural), and it is a source (and sink) of water for the surface environment.[54] The latter is also referred to as providing *in-situ* functions[55] and these include: (a) a buffer against periodic shortages in surface water supplies; (b) prevention or minimization of subsidence of the land surface from groundwater withdrawals; (c) protection against sea water intrusion; (d)

---

[50] S., Foster and A., Lawrence, *Hazards Induced By Groundwater Recharge Under Rapid Urbanization*, B L Morris & British Geological Survey, Wallingford UK, *found at* http://lnweb18.worldbank.org/ESSD/essdext.nsf/18ByDocName/HazardsInducesbyGroundwaterRechargeUnderRapidUrbanisation/$FILE/HazardsInducedbyGroundwaterResourcesStatus.pdf (*last visited* May 2003).

[51] *Ibid.*

[52] The UNECE Task Force on Monitoring and Assessment, under the Convention on the Protection and Use of Transboundary Watercourses and International Lakes (Helsinki 1992), State of the Art on Monitoring and Assessment of Groundwaters, Vol. 4, ISBN 9036952778, Lelystad, 1991, also *found at* http://www.iwac-riza.org/IWAC/IWACSite.nsf/7BAFC09E93A2EE9DC1256BB800476C12/$File/Task%20Force%20Vol%201.pdf (*last visited* May 2003).

[53] *Ibid.*

[54] UNESCO Report *supra* note 38, *see also* UNECE Report, *supra* note 52.

[55] The in-situ functions occur or exist as a consequence of the water remaining in place in the aquifer *See* National Research Council, *Valuing Groundwater: Economic Concepts and Approaches*, Washington D.C., 1997, (on file with author).

protection of groundwater quality by maintaining the capacity to dilute and assimilate groundwater contaminants; (e) facilitation of habitat and ecological diversity; and (f) providing discharge to support recreational activities.[56] As for the problems (threats); these are presented in the following discussion.

2.3.2 Pollution and Depletion

Given the key features of groundwater addressed in 1.2, it is established that the primary two problems relating to groundwater management are: pollution and depletion.[57] In general, aquifers are seriously threatened by depletion due to over-exploitation and water quality degradation. The declining water levels in transboundary aquifers can lead to increased costs in water pumping and utilization and may trigger conflicts between States sharing the resources.[58] The challenges posed by these two problems need to be addressed from a hydrogeologic and a legal perspectives. Firstly, a better understanding is needed on (i) the interaction between groundwater and other elements of the environment,[59] (ii) complexity of the hydrogeological regime, (iii) the vulnerability of the aquifers to pollution, (iii) the nature of replenishment or recharge, (iv) extraction and water table fluctuation, and (v) pollutants distribution.[60] From a legal perspective the issues are mainly related to (i) the identification of the scope and description of the groundwater resources[61] (ii) identification of the rules and principles that govern the allocation between uses at the national level,[62] the legal entitlements of States at the international level,[63] and

---

[56] *Ibid.*

[57] S. McCaffrey *supra* note 32, *see also* R. D. Hayton, *supra* note 42, and S. Burchi *supra* note 9.

[58] USGS Circular 1139 *supra* note 42; Excessive pumping from groundwater may lead to the lowering of the water levels in the groundwater wells. This increased depth requires more powerful pumps and technical installations which has higher costs.

[59] R.D. Hayton et al, *supra* note 12.

[60] A. E. Utton, *supra* note 23

[61] This includes the identification of the following: what are the groundwaters at issue (related groundwater, fossil waters...etc), other States that share the resource, existing legal agreements that apply, most important interests to the State and the dependence of the State on groundwater course for the State.

[62] At the national level the water code must specify the different types of water use that are considered lawful as opposed to others which the law does not recognize. This allocation can be based on many criteria depending on the State's national water policies. See D. Caponera, *the Requirement for Drafting a modern Code and Water Legislation in the ESCWA Region*, Amman, 1996, E/ESCWA/ENR/1996/WG.1/WP.1 17 November 1996.

[63] At this preliminary stage there is a need to understand that the main concern at the international level is identifying the 'Right to use Water' or the 'legal entitlements' of the States from the shared resources that must be based on the customary principle of

(iii) the implementation of these entitlements including the development of procedures, mechanisms and institutions that ensure compliance and enforcement.

2.3.3 Scarcity of Groundwater Related Data

There is an inherent lack of hydrogeological data which results in a high degree of uncertainty in the estimation of key parameters required to characterize the aquifer system.[64] In hydrogeological terms, these cross-borders resources can only be estimated through good observations and measurements of selected hydraulic parameters, analogous to the estimation process of other transboundary resources such as fisheries and wildlife, each requiring statistically sound observations.[65] While reliable data is the pillar for successful management, and good regulation is the means for achieving cooperation between States,[66] the absence of reliable information is a perpetual concern that is typical for groundwater resources.[67] The complexity of the groundwater regime and the ambiguity of its characteristics complicate the task of assessing these resources. At the national and international levels a policy decision is urgently required to identify or map the boundaries of an aquifer before beginning to accurately assess the quantity and quality of water it contains, which would require substantial investment and scientific expertise.[68] This need calls for the establishment of an information base that embraces comprehensive data on groundwater generally and on groundwater that cross political boundaries in particular.[69] Such information base requires the development of a comprehensive and credible monitoring system that can generate the type of information and data needed. Monitoring has to be supplemented by assessment systems that are capable of processing and transforming the data into reasonable and usable information for the enactment of the responsive rules and regulations.[70] The development of comparable monitoring and assessment systems in transboundary groundwater States sharing the same aquifer is essential. This will ensure the confident exchange of data and information between them and consequently fulfil the objectives of cooperation.

---

international law 'equitable and reasonable utilization'. See P.K. Wouters (ed.), International Water Law: Selected Writings of Charles Bourne (Kluwer Law International: the Netherlands, 1997).

[64] R.D. Hayton, *supra* note 12.

[65] UNECE Task Force on Monitoring and Assessment, *supra* note 52.

[66] S. Puri, *supra* note 24.

[67] These data include aquifer thickness, saturated thickness, depths, area, quantity, and quality of the area, transmissibility, permeability recharge areas and other aquifer related parameters, See Hayton, *supra* note 12.

[68] R.D. Hayton, *supra* note 12.

[69] S. Puri *supra* note 24.

[70] UNECE Task Force on Monitoring and Assessment Report, *supra* note 52.

Having understood the key features, problems and challenges associated with groundwater, and considering that groundwater behaves in the same manner in all contexts, the next discussion is illustrative for modern national regulation of groundwater. This is done with the view to identify lessons that might be adapted at the international level.

2.4 MODERN TRENDS IN NATIONAL GROUNDWATER REGULATION-RELEVANCE AT INTERNATIONAL LEVEL

2.4.1 Background

The comparative review and analysis of available national groundwater legislation illustrates the choice of mechanisms-regulatory and otherwise, or options, available to law makers in the framing of responses to challenges posed by groundwater pollution and depletion.[71] In Europe[72] the WFD[73] presents a good

[71] S. Burchi, *supra* note 9.

[72] Groundwater is a major source for drinking water around Europe. For Austria, Denmark and Portugal it represents more than 75% of the public water supply, and in Belgium (Flanders), Finland, France, Germany, Luxembourg, UK and the Netherlands between 50 and 75 % of the public water supply is abstracted from groundwater. The status of groundwater in terms of quality and quantity is therefore essential. Groundwater also plays an important role within the environment, both for aquatic and terrestrial ecosystems. *See generally* www.eeb.org/activities/water/EAF%Groundwater%issues%20%paper.pdf. (*last visited* May 2003).

[73] WFD, *supra* note 41. The WFD sets out the following basic provisions in Article 4, 7 and 11. (i) Member States shall implement the measures necessary to prevent or limit the input of pollutants into groundwater and to prevent deterioration of the status of all bodies of groundwater, (ii) Member States shall also protect, enhance and restore all bodies of groundwater with the aim of achieving good status by 2015. (iii) Member States shall implement the measures necessary to reverse any significant and sustained upward trends in the concentration of any pollutant resulting from the impact of human activity in order to progressively reduce pollution of groundwater, (iv) Member States shall ensure the necessary protection for bodies of water used for the abstraction of water intended for human consumption or intended for such future use and they shall take measures to safeguard water quality in order to reduce the level of purification treatment required for the protection of drinking water, (v) Member States have to adopt a set of basic measures that shall include a prohibition of direct discharges of pollutants into groundwater. The basic definitions for good groundwater status are set out in Annex V of the WFD. Good groundwater chemical status is considered when the chemical composition of the groundwater body is such that the concentration of pollutants do not exhibit the effects of saline or other intrusions, and do not exceed the quality standards related to Article 17. The chemical compound of the groundwater body must not be such that would result in failure

example on modern trends in legal treatment of water resources including groundwater. It is meant to be a State and Interstate[74] European Union (EU) related legal instrument, which aims at improving the protection of surface and groundwater.[75] Also in Europe the UNECE countries have adopted the Charter on Groundwater Management which gives broad support to UNECE member Governments in their common endeavours to protect ground water by providing planners and decision-makers with appropriate policy instruments.[76] The African Water Act of 1998[77] is another illustrative example for a national code that integrates surface and groundwater. Other illustrative examples on modern national regulations trends were cited in Europe, the Middle East and Africa. The primary regulatory issues encompassed within these codes are presented in the next discussion.

2.4.2 Planning, Management and Regulatory Hydrologic Unit - "Scope"

Understanding the features of groundwater namely the interaction with the surface water and the environment, is essential for the choice of planning, management and regulatory hydrogeologic unit.[78] Such a step is prerequisite for the establishment of physical and legal frameworks essential to view the entire aquifer system. After World War II, a trend developed toward administration and control of groundwater, bringing it to the mainstream of municipal water law.[79] The use of

---

to achieve the good chemical or ecological status in associated surface waters nor any significant damage to terrestrial ecosystems depending on the groundwater body.

[74] The relevance of the WFD on Interstate level will be dealt with in Chapter 4 "Transboundary Groundwater in State Practice".

[75] European Freshwater Program, *found at* www.wwfreshwater.org/initiative/wfd.html (*last visited* March 2003).

[76] Charter on Ground-Water Management as adopted by the Economic Commission for Europe at its forty-fourth session (1989) by decision E (44) United Nations, New York, 1989 E/UNECE/1197, UNECE/ENVWA/12, *found at* http://www.internationalwaterlaw.org/RegionalDocs/Groundwater_Charter.htm (*last visited* November 2003).

[77] The African Water Act of 1998, *found at* www.dundee.ac.uk/law/iwrli/Research_Document_National.html (*last visited* May 2003).

[78] L. Teclaff, *supra* note 16.

[79] *Ibid.* In 1945 the central administration Great Britain was entrusted with licensing and abstraction of underground water in areas needing water supply protection. In 1964, France subjected the capture and use of groundwater to an administrative authorization for the first time.

the "river basin"[80] and the "drainage basin"[81] concepts were advocated in many conferences and forums for planning, management and regulation.[82]

The "River Basin" Concept was originally developed for pure resource management objectives offering limited, if any, consideration to other elements of the environment.[83] It has been employed in statutes and treaties for the last 50-75 years, Teclaff confirms[84]. A "River Basin" is delimited by natural boundary called the watershed or drainage divide.[85] Accordingly, this concept as defined and employed is appropriate in the case of surface water, as aquifers' boundaries, are often very poorly known and so many transboundary aquifers remain only partly recognized.[86]

With the increasing need to include groundwater within the scope of water resources management, the "drainage basin" concept began to appear.[87] This concept was employed in the United States in the early 50's and later espoused by

---

[80] The Random House Dictionary defines 'River Basin' as:
"the area of land drained by a river and its branches, or tributaries".
This concept is an old as the earliest recorded attempts to navigate watercourse systems, control floods and irrigate cropland. From these and later developments throughout the ages, there emerged the idea of community of interest of water users which blossomed into two great systems of domestic water law and eventually into the first treaties reflecting the unity of the basin. *See* Random House Dictionary of the English Language 1661 (2nd ed. Unabridged, 1987) cited in L. Teclaff, *Evolution of the River Basin Concept in National and International Water Law,* 36 NRJ 359-391 (1996).

[81] In hydrology a "Drainage Basin" is defined as:
"The area of land that drains water, sediments and dissolved material to a common outlet at some point along the stream channel".
See European Environment Agency Glossary of Hydrological Terms *found at:* http://glossary.eea.eu.int/EEAGlossary/D/drainage_basin (*last visited* September 2003) *See* also *infra* chapter 3).

[82] See generally *infra* Chapter 3.

[83] L. Teclaff, *supra* note 80.

[84] *Ibid.*

[85] The limits of a given Basin area are the heights of land separating it from neighbouring drainage systems. *See* Teclaff *supra* note 80.

[86] S. Puri, *supra* note 24.

[87] L. Teclaff, *supra* note 80. Also an illustrative example is found in (i) the 1996 Albanian Law on Water Resources, which adopts the "drainage basin" identified to include surface and groundwater flowing into a common point (the sea). Article 1 of this Law provides:
"The water drainage basin is the area of land within which the water, after converging in a common flow through several surface and underground flows, pours into sea. The geographical borders of water drainage basins are determined in the topographic maps according to water separation dines."
found at http://www.dundee.ac.uk/law/iwlri/Research_Documents_National.php (*last visited* October 2003).

many conferences and forums.[88] The definition of the "drainage basin" goes beyond the "river basin" concept and is considered to better serve the needs of conjunctive water resources management over a wider area.[89] In its first attempt to codify and progressively develop international water law, the International Law Association (ILA)[90] advocated the "drainage basin" Concept. [91] According to Wescoat, the Helsinki Rules were resisted by many countries due to their rejection of the "drainage basin" concept.[92] This will be discussed in more details in Chapter Three.

Another hydrologic management unit that is hydrologically analogous to the "drainage basin" is used in some countries- the "catchment area".[93] The "catchment area" includes rivers, lakes, dams, groundwater, storm water and water related infrastructure.[94] This scientific understanding is not however always reflected in national codes and regulation. The African Water Act No. 36 of 1998[95] can be cited as an example where the terminology is employed. The Act

---

[88] President's Water Resources Policy Commission: A Water Policy for the American People 52 (1950). *See* also International Law Association, 2nd International Conference on Water Law and Administration, Caracas, Feb. 8-14 (1976), Conference Recommendation as Approved by the Plenary Session on Feb. 14, 1976. *Cited in* L. Teclaff *supra* note 80.

[89] UN Natural Resources/ Water Series, No. 1, The Management of International Water Resources: Institutional and Legal Aspects, Sales No. E.17.II.A.2 1975. *See also* J., Wescoat, *Beyond the River Basin: The Changing Geography of International Water Problems and International Watercourses Law,* 3 CJIEL &Policy 301,307 (1992).

[90] The ILA is a non-governmental organization founded in 1873, has a consultative status with a number of U.N. agencies. Its objectives, under its constitution include the study, elucidation and advancement of international law, public and private, the study of comparative law, the making of proposals for the solution of conflicts of law and for the unification of law, and the furthering of international understanding and goodwill. The ILA work on international water law started in 1954. *See* generally www.ila-hq.org/html/layout_about.htm (*last visited* September 2003).

[91] The Helsinki Rules adopted by the ILA in 1966 define the Drainage Basin As:

"A geographical area extending over two or more states determined by the watershed limits of the system of waters, including surface and underground water, flowing into a common terminus: they should contribute to the drainage basin principal river, stream or lake or other common terminus"

See relevant discussion *infra* Chapter Three section 3.2.2.1.

[92] J. Wescoat, *supra* note 89.

[93] In the field of hydrology the terms drainage basin, catchment area, watershed, and drainage area are all synonyms and they are defined as: "area drained by a stream or other body of water". The Colombia Encyclopaedia, Sixth Edition, (Colombia University Press: New York, 2002) found at http:// www.bartleby.com/65/ca/catchmen.html (*last visited* September 2003).

[94] *Ibid.*

[95] African Water Act, *supra* note 77 Part II.

defines the "catchment" in relation to a watercourse or watercourses or part of a watercourse as:

"the area from which any rainfall will drain into the watercourse or watercourses or part of a watercourse, through surface flow to a common point or common points."

2.4.3 Ownership Rights

Traditionally, and in accordance with basic principles of Roman law, groundwater has been regarded at law as the property of the owner of the land above (*nuda proprietas*). [96] Countries following the Napoleonic Code tradition, as well as countries following the Anglo-Saxon Common law tradition, equally subscribe to the same principle.[97] The Moslem tradition, by comparison, regards water as a public or communal commodity, "a gift of God", which cannot be owned.[98] Only wells can be owned, whereby exclusive or priority user rights in the water accrue

---

[96] D.A. Caponera and D. Alheritiere, *Principles for International Groundwater Law,* in Teclaff, L. & Utton A. (eds.) 25 IGL (1981) also found in 18 NRJ 589-618 (1978).

[97] *Ibid.* See also P. Malanczuk, Modern Introduction to International Law, (Oxford University Press: Routledge, 1997) at 6.

[98] Water Codes in Moslem Countries recognize that water is a public property to be administered by the States for the benefit of the Citizens. In Oman for example the 'Shari's principles are the governing legal framework of the State including those aspects dealing with water. In 1988 the Royal Decree No. 82 was issued declaring that 'water resources were a national wealth'. In Jordan for example Law No. 18 of 1988 (Jordan Water Law 1988) declares water as a public property. Article 25 states:

"Any water resources available within the boundaries of the Kingdom, whether they are surface or groundwater,.......are considered State owned property and shall not be used or transported except in compliance with this law".

In Yemen, water ownership, diversion, use are governed by a mixture of laws and customary practices depending on the water source land use economic activities and prevailing social and religious customs. Article 8 of the Constitution of Yemen states

"all types of natural resources and sources of energy, whether above ground, under ground, in territorial water, ...are the property of the State which assures their exploitation for public welfare".

In Lithuania and according to Article 2 of the Underground Law of 1995 No. I-1034 groundwater is owned by the State, "The underground shall be the exclusive ownership of the State. The basis of the exploitation of the underground resources shall be the right to exploitation, which in the procedure established by this and other laws can be granted to legal and natural persons of the Republic of Lithuania by the Government of the Republic of Lithuania or the public institution authorised by the Government, and to foreign legal and natural persons - by the Government of the Republic of Lithuania. Actions which directly or indirectly violate the right of State ownership of the underground shall be prohibited". *found at* http://www.litlex.lt/Litlex/Eng/Frames/Laws/Documents/369.HTM (*last visited* May 2004).

to the well-owners.[99] Furthermore, the ownership of wells entails ownership of an area around the well in which new wells cannot be dug (known as *harim*, or forbidden area).[100] However, economic development, the increasing need for water, the introduction of modern extraction methods, and the uncontrolled use of groundwater have compelled States to introduce groundwater regulations in order to replace the traditional ones.[101] The trend is in the direction of severing water

---

[99]D. Caponera, *supra* note 96.

[100] S. Bogdanovic, *supra* note 13 at 332, and D. Caponera *supra* note 96. Caponera and Alheritiere explain that the codified Moslem law is one important source of law in Moslem countries, whereby groundwater belongs to the community. The definition of water as a non-saleable publicly owned commodity applies to all water dug by unknown persons.

[101] In Yemen the proposed new law (on file with the author) contains many articles dealing with groundwater resource ownership, investigation, development, monitoring and Management. Article 3 states:

"All water resources which exist within the boundaries of the Republic, whether on the land surface or underground, are considered natural resources owned as public property, and the State shall orient and organize their exploitation so as to serve public welfare".

Article 3 of the 1996 Albanian Law on Water Resources, *supra* note 87, grants the government ownership rights to all water resources: "The state ownership is comprised of: a) All water resources of the Republic of Albania, as defined in the article 2, paragraph 1 of this law. b) All river beds and banks, torrents and other natural streams either temporary or permanent, channels, lakes, ponds, natural or artificial lagoons and reservoirs, islands: end accumulations of sand, stones and silt on river beds, lakes and reservoirs as well as geological formations of underground waters. c) All objects and hydrotechnic works accomplished by the state such as dams, irrigation, drainage and navigation systems, potable water stations and their respective channels and works. d) The land obtained from withdrawal of water or extended land toward the water. 2. The ownership right of the state as specified under paragraph 1 of this article, is hereby unchangeable and irrevocable."

In Lithuania Article 4 of the Water Law of 1997, *supra* note 101, states that:

"Water resources comprise a component part of the natural resources of the Republic of Lithuania. The use thereof shall comply with the Constitution of Republic of Lithuania, the Laws on Environmental Protection, the Law on the Earth Entrails, as well as this and other laws and legal acts". *found at* http://www.dundee.ac.uk/law/iwlri/Research_Documents_National.php (*last visited* December 2003).

Also see the 'African Water Act' *supra* note 77, which describes the National Government as the public trusteeship of nations water resources' in Chapter 1 paragraph 3:

"(1) As the public trustee of the nation's water resources the National Government, acting through the Minister, must ensure that water is protected, used, developed, conserved, managed and controlled in a sustainable and equitable manner, for the benefit of all persons and in accordance with its constitutional mandate. (2) Without limiting subsection (1), the Minister is ultimately responsible to ensure that water is allocated equitably and used beneficially in the public interest, while promoting environmental values. (3) The National Government, acting through the Minister, has the power to regulate the use, flow and control of all water in the Republic".

rights from the property law concept of ownership in the classical strict sense.[102] There is a major movement in the law of national and sub-national levels, in recent decades toward limiting and regulating groundwater withdrawals and especially mandating pollution prevention.[103] Much legislation have tended toward the limitation of exclusive private ownership in favour of a form of central administrative control over groundwater uses, thereby creating a formal separation between ownership (*nuda proprietas*) and use rights.[104]

2.4.4 Regulatory Instruments for Groundwater Protection

2.4.4.1 Monitoring and Assessment

A proper monitoring and assessment of the groundwater resources is essential for effective water resources management and regulation. Firstly, a policy decision is urgently required to identify or map the boundaries of an aquifer before beginning to accurately assess the quantity and quality of water it contains, which would require substantial investment and scientific expertise.[105] Modern national water codes include provisions on the monitoring[106] and assessment[107] of the waters, the

---

In Israel there is no private ownership of the water resources Section 1 of the 1959 Water Law (on file with the author) states that:

" The water resources of the State are public property; they are subject to the control of the State and are destined for the requirements of its inhabitants and for the development of the country".

Article 3 paragraph (1) of the Palestinian Water Law No 3/2002 (on file with the author) states:

" All water resources available in Palestine are considered public property".

[102] *Ibid.*

[103] *See* African Act No. 36 for 1998, *supra* note 77.

[104] D. Caponera *supra* note 96.

[105] See generally, Fetter and Price, *supra* note 3.

[106] "Monitoring" is:

"the process of repetitive observing, for defined purposes of one or more elements of the environment according to pre-arranged schedules in space and time and using comparable methodologies for environmental sensing and data collection. It provides information concerning the present state and past trends in environmental behaviour." See UNECE Task Force on Monitoring and Assessment *supra* note 52. *Also see* Article 38 of the Lithuania Water Law of 1997, *supra* note 101, provides:

"1. Users of water who extract water from water bodies or discharge waste water into the natural environment by using their own facilities and equipment and subscribing users of water shall supply statistical data on the use of water to the institution authorized by the Ministry for Environmental Protection according to the established procedure. 2. The Ministry for Environmental Protection in agreement with the Department of Statistics shall establish the procedure for submitting of statistical data, 3. The Ministry for Environmental

assessment of the effectiveness of measures taken to prevent, control and reduce pollution, and the establishment of a credible data base on water and effluent monitoring.[108] Other relevant aspects deal with the rules for setting up and operating monitoring programmes, which includes measurement systems and devices, analytical techniques, data processing and evaluation techniques. Monitoring includes also the development of systems for surveillance and early-warning systems to identify outbreaks or incidents of water-related diseases or significant threats of such outbreaks or incidents, including those resulting from water pollution or extreme weather.[109]

Protection shall establish the procedure for accounting for the amount of water extracted and used as well as the amount of waste water discharged into the natural environment."

[107] "Assessment" means:

"The evaluation of the hydrological, chemical and/or micro-biological state of groundwaters in relation to the background conditions, human effects, and the actual or intended uses, which may adversely affect human health or the environment." UNECE Task Force on Monitoring and Assessment, See generally *supra* note 52

[108] For example the UNECE Charter on Groundwater management, *supra* note 76includes provisions on the monitoring and control of groundwater aquifers. Article XIII provides:

"(1) Monitoring programmes for ground-water protection should be set up and applied. These programmes should include monitoring at the source of potential pollution which could pose a serious or chronic threat to an aquifer. There should be regular inspections to ensure compliance with protection requirements imposed. Attention should also be paid to the monitoring of ground-water quality changes brought about by air-borne pollution. 2. Systematic monitoring should be carried out for all aquifers found to be vulnerable to pollution and/or over-use, as well as for those whose particular importance has been recognized for public water supply, mineral water supply and industry. (3) Monitoring and control should be considered a public-service activity. Facilities should be set up for coordinating the assessment and availability of monitoring data and information on aquifers. The resulting collections of data should be related to information on ground-water quantity and quality characteristics of aquifers as well as details of their location, use, and exposure to various impacts from land uses such as agriculture, industry and urban development. Information should be readily available to those interested, (4) The data from monitoring should make it possible, *inter alia,* to revise periodically plans and forecasts of groundwater use, taking into account actual evolution of aquifers, and to determine measures necessary to ensure the sustainable use of ground-water resources in the long term. Legislative provisions and regulations should, as appropriate, allow for the revision of protection requirements imposed depending on the measures thus determined, and (5) Monitoring programmes should be periodically reviewed to ensure that they are achieving their stated aims and that the results have been used effectively."

[109] This strategy aims at collecting information on whether and where accidental spills may affect the drinking water supply, to determine public health hazards of "abandoned or illegal" land disposal sites, or to determine actual sources of groundwater quality deterioration. For early warning, special wells may be drilled, whereas for surveillance production wells can often be used.

See generally, UNECE Task Force on Monitoring and Assessment, *supra* note 52.

2.4.4.2 Permitting System

As a result of groundwater increasingly becoming public property or is held by the State for the public, only user or usufructuary type rights accrue to the owners of overlying land- or to the developer of the resources in other than the landowners. These rights are granted by the government subject to terms and conditions. [110] The permit system has now either spread over almost all parts of the world or being advocated for in many international forums.[111] Countries that rely mostly on

---

[110] In Jordan, Law No. 26 for 1968 mandates the competent authority to issue drilling permits. The issue of drilling permits requires compliance with the following conditions (1) Water use is linked with irrigation purposes and the land should be at least 50,000 m^2 (2) Prohibition of drilling wells to be used as standbys, unless for public benefit, (3) Adherence to the specified distance required between wells, and the depth of the wells as specified in the issued permit, in order to prevent water mining and quality deterioration (4) Pump tests have to be made under the supervision of the authority, (5) The well owner must submit a completion report detailing compliance with the well specifications and constraints, contract, and results of the pump test.(Available in Arabic on file with the author). Also in Jordan Resolution Number 26 of 1977 (on file with the author) concerning the 'Supervision of Groundwater Regulation' prohibits any drilling, abstraction and utilization of groundwater without a prior permit as stipulated from Article 10:

"Prior obtaining a work permit, no person may initiate or continue the drilling of ay well to abstract groundwater from it or may introduce alteration in a drilled well, or may install a mechanical device on it for pumping that may increase, or cause to be increased, the production of groundwater from that well"

In the African Water Act No. 36 for 1998 *supra* note 77, Chapter 4 states :

"In general a water use must be licensed unless it is listed in Schedule I, is an existing lawful use, is permissible under a general authorization, or if a responsible authority waives the need for a license. The Minister may limit the amount of water which a responsible authority may allocate. In making regulations the Minister may differentiate between different water resources, classes of water resources and geographical areas".

In the Lithuania Water Law of 1997, *supra* note 101, Article 25 provides the Conditions for the Use of Water Resources and Permits

"1. The use of water resources may be free (without any permits) and it may be subject to permits. 2. Article 26 of this Law establishes the procedure of free use of water resources. 3. Permits for use of the water resources shall be issued by the Government or an institution authorized by the Government according to established procedures. 4. No permits for use of the water resources shall be issued if no measures are provided for sustainable use of water, for appropriate protection of water bodies from pollution, for ensuring suitable conditions for aquatic flora and fauna or if the interests of other users of water are violated.5. In submitting information to obtain a permit for the use of water resources a water user must estimate and indicate the amount of water needed by the subscribing users as well as the amount and pollution of waste water discharged by subscribing users."

[111] On 9 September 1996 the European Commission presented a proposal for a Decision of the European Parliament and of the Council on an action program for integrated protection and management of groundwater (1). In that proposal the Commission pointed to the need

groundwater[112] have issued directives aimed at regulating groundwater extraction through well drilling permits, and the formulation of drinking and renovated waste water standards.[113] The permitting system has been expanded to include licenses, permits or authorizations to discharge wastes on or under the ground.[114] In Europe the UNECE groundwater permitting system includes exploration and abstraction permits, abstraction and recharge permits and pollution-control permits.[115] In the Middle East, for example, the oldest water legislation was established in 1925 in the Syrian Arab Republic and Lebanon governing public ownership of water and licensing.[116] In Oman, another Middle East country which depends almost entirely

---

to establish procedures for the regulation of abstraction of freshwater and for the monitoring of freshwater quality and quantity.

[112] Illustrative examples on such countries are Palestine, Israel, Jordan in the Middle East, Denmark and Lithuania in Europe Oman and Yemen in Asia.

[113] Groundwater utilization in Yemen was regulated by the issuance of drilling permits according to rules set forth in Law No. 320 of 1926. Individuals were required to submit information to the competent authority on the intended use of the well, method of drilling, material to be used. The increased groundwater development activities in 1970 compelled water authorities to issue a new decree (No. 14438) that addressed the organization of activities with respect to groundwater exploration, development, and protection. The decree emphasized the issuing of drilling permits, and that individuals must either obtain a prior permission to drill or exemption from a permit. The permit contained specific information on the well for which it was issued including location, type and method of drilling, as well as an estimate of the volume of discharged water.(text of the law found on file with the author).

[114] Part 2 paragraph 29 of the African Water Act of 1998, *supra* note 77 states: (1) A responsible authority may attach conditions to every general authorization or license – *(a)* relating to the protection of - (i) the water resource in question; (ii) the stream flow regime; and (iii) other existing and potential water users" See FAO Legislative Study 70 - Water Rights Administration - Experience, Issues and Guidelines *found at* http://www.fao.org/docrep/003/x9419e/x9419e00.htm#Contents (*last visited* June 2003).

[115] Charter on Groundwater Management, *supra note* 76.

[116] In Syria the public property regulation No. 144 of 1925 addresses water rights in some of its articles. Another law was issued in 1926 addressing the protection of public water sources including rules for regulating licenses and water concessions. See M.S. Al-Safady, *Water Legislation in the Arab World,* Arab Organization for Education Culture and Science Publication No. 002/03/1986 Tunisia. In Israel the Water Law of 1959 stipulates the need for acquiring a license for four areas i) the Production License ii) the Recharging License; iii) the Drilling License; and iv) The Construction License(Chapter 2 Articles 3 & 5). Among the responsibilities entrusted to the Palestinian Water Authority is the licensing of the exploitation of water resources as stated in Article 7 paragraph (5) of the Palestine Water Law No. 3/2002:

"Licensing the exploitation of water resources including the: construction of public and private wells, drilling, and monitoring and production wells." Within the Palestine Water Law 3/2002 Article 31 states that:" the Authority, in coordination with the other relevant parties, may consider any area that contains groundwater a protected area, if the quality or

on groundwater, there is a water deficit of about 275 mcm due to over pumping.[117] This situation was a consequence of the use of the random and unregulated use of modern machinery and equipment in drilling and pumping in 1970 onwards. In 1975 and as a response to this deteriorated situation the government issued a series of directives to regulate and develop these resources.[118] Most recently-in 1995- a new Ministerial Directive (No. 13/95) was issued. This directive concerns the regulation of groundwater wells and irrigation and it includes a whole chapter on the permitting of wells including all details that are needed to ensure control over the groundwater utilization and development.[119] The National water regulation in Lithuania in Europe whose dependence on groundwater is almost 100%[120] has developed a system of control for the use of the resources and permits.[121] The permitting system is presently not limited to water use purposes, rather to waste

---

quantity of water is in danger of pollution, or if achieving the objectives of the water policy requires such action, on condition that the Authority provides alternative water resources."

[117] S. Khameisi, *Water Legislation in Sultanate of Oman,* E/ESCWA/ENR/1996/WG.1/CP.8 (in Arabic).

[118] In 1975 Directive No. 45/75 which establishes the Water Resources Wealth Council, and in Directive No. 76/77 of 1977 on Water Resources Development and Directive No. 63/79 of 1979 concerning the Establishment of a Public Authority for Water Resources. In 1989 the Sultan Decree No, 100/89 established the Ministry of Water Resources. The responsibilities of this Ministry is to develop regulate and further explore the groundwater resources. This decree was followed by Ministerial Decision No. 2/90 of 1990 which regulates the registry of existing wells and regulate the licensing of new wells. In terms of Article 7 of Ministerial Decision No. 2/1990, groundwater exploration activities require a permit. A permit is necessary for: the construction of a new well, the modification of an existing well (by deepening or increasing the diameter of the well, cleaning or maintenance of the well), for the change in use of an existing well and for the installation of a pump in a borehole (Article 8). Permits are granted only to landowners (Article 9 of Ministerial Decision No. 2/1990). However, this provision does not apply to wells excavated for emergency purposes or in case of temporary water shortage; to wells constructed solely for exploration and monitoring purposes and to wells constructed for petroleum exploration activities. (Arabic texts available on file with the author).

[119] Chapter 3 of Directive 13/95 of 1995 includes 18 articles on the procedures for obtaining a permit and the organization concerned and the conditions upon which a permit can be granted.

[120] UNECE Task Force on Monitoring and Assessment, *supra* note 52.

[121] Lithuania Water Law of 1997 Article 27 and 28, *supra* note 101. Although the regulations still have provisions for free use of water under certain circumstances including the private owners of surface and stationary surface water bodies who can use the water freely for domestic uses and the use of dug wells waters (cisterns) and in case of disasters. (Article 26 paragraph 1-6), *found at* http://www.dundee.ac.uk/law/iwlri/Research_Documents_National.php (*last visited* December 2003).

discharges. The following discussion describes situations where activates affecting the quality and/or the quantity are either highly restricted or even prohibited.

2.4.4.3 Protection and Conservation Areas

There are situations where accelerated depletion of the resource known also as groundwater 'mining' exists and adequate management and legal responses established.[122] The declaration of special zones is invoked by the fact that control of abstraction and of recharge is rendered indispensable as a consequence of increased groundwater use.[123] In modern codes measures relevant to the control of water utilization to preserve groundwater include well spacing, regulation of pumping rate, monitoring and regulation practices, establishment of minimum flows in surface streams interconnected with transboundary aquifers and control of surface water divisions and declaration of protection zones.[124] The UNECE

---

[122] S. Burchi, *supra* note 9.

[123] For example the Water Law of 1997 in Lithuania, *supra* note 101, imposes strict regulation concerning the declaration of protection zones which is defined as:
"Protection Area around the Water Extraction Site denotes the protection area surrounding the water extraction site which is subject to certain limitations imposed on the activities of land users."
Additionally, Article 33 of this law states Protection Areas around Water Bodies and Water Extraction Sites:
"1. Protection strips of land shall be designated around surface water bodies; larger water bodies shall have a protection zone around them. A protection strip within a protection area shall be considered to be part of the protection area. 2. Sanitary protection areas and strips shall be designated around the extraction sites and individual drilled or dug wells of drinking water. 3. The procedures for the designation of protection areas and strips shall be laid down by the Law on Protected Areas and the regulations issued by the legal acts approved by the Government or by the institutions authorized by the Government. 4. No man-made obstacles shall be built on the protection strip around the banks of a surface water body to ensure free access for the maintenance of the hydrographic network."
In the 1996 Albanian Law on the Water Resources *supra* note 87 Article 41 provides that:
"The protected areas or surfaces can be determined and established by the water authorities, anytime: (i) when surface and underground sources are seriously jeopardized in terms of quality and quantity, (ii) when conflicts between water users are deepening, or (iii) when there is a threat from spread of diseases. The regime of these areas or surfaces as well as their borders will be established on basis that will change according to the case."
And Article 42 provides that :
"Upon the proposal of the basin council and approval of the National Water Agency certain areas such as basins, parts of basins, streams' rivers etc, can be announced as areas of special protection due to their nature features or ecological interest. The administration and protection regime of these; areas shall be defined through special bylaws and regulations."
In Israel the competent authority is mandated to declare "Rationing Areas" where the water resources are not sufficient for the maintenance of existing consumption water.

[124] See generally *supra* note 74; see also S. Burchi supra note 9.

Charter on Groundwater Management calls the States to declare protection zones that are considered as a preventive measure to avoid depletion and pollution of vulnerable aquifers.[125]

Measures for the abatement and control of groundwater pollution from land use activities should comprise effective and efficient methods for protecting groundwater resources from future contamination in rural and urban areas, and particularly in irrigation areas overlying unconfined aquifers, integrated groundwater vulnerability assessment techniques need to be developed for predicting which aquifers are more likely than others to become contaminated as a result of land use activities and changes at the surface.[126]

Groundwater protection zoning also has a key role in setting priorities for groundwater quality monitoring, environmental audit of industrial premises, pollution control within the agricultural advisory system, determining priorities for the clean-up of historically-contaminated land, and in public education generally.

---

[125] African Act, *supra note* 77 Article XVIII provides that
"(1) Where compatible with national legislation, ground-water protection zones should be established over and above the general protection of ground waters through relevant legal provisions, as a preventive measure protecting aquifers around present and future abstraction sites and in recharge areas where aquifers are vulnerable. Compliance with the prescribed restrictions should be strictly controlled. (2) Protection zones could be divided into different classes with differentiated restrictions on land-use and water-use graduated according to environmental considerations and the relative importance of the underlying aquifer. In this respect, particular consideration should be given to the establishment of well-head protection areas. Necessary measures should be taken to minimize the risks of accidental or diffuse pollution in protection zones and (3) Restrictions and/or prohibitions on land-use activities should include mining and industrial processes, manufacturing, intensive farming, including the application of fertilizers and pesticides on agricultural land, transportation, waste disposal and treatment as well as storage of dangerous substances."

[126] L. Teclaff *supra* note 18. For example integration within the WFD has a wide meaning including quantity quality integration and integration of the protection and sustainable management of water with other community policy such as energy, transport fishery…etc. The WFD treats water as an integral part of the ecosystem whereby measures taken to achieve the WFD objectives apply to ground and surface water within the same hydrogeological ecological, hydrological system. In Ireland Protection schemes are undertaken by the Geological Survey in partnership with, and at the request of, a Local Authority. Vulnerability and aquifer category are mapped across the whole area covered by the Local Authority. 'Source protection areas' around groundwater supplies are also mapped. "Protection responses" have been developed for each combination of vulnerability and aquifer category/source protection area. To date, protection responses have been developed for landfill, land spreading of organic wastes on a scale which requires a license, and on-site wastewater systems for single houses. Additional responses for underground storage tanks, etc., are planned for the near future. See V. Fitzsimons1, S. Gilroy, and J. Cudden, *Groundwater Protection Schemes and GIS – a Powerful aid in the Planning Process, found at* http://www.proteus.ie/pdf/GWPS-GIS.pdf (*last visited* June 2003).

All of these activities are essential components of a sustainable strategy for groundwater quality and quantity protection. Rationalization and seasonal limitations on groundwater utilization is another mean for resources protection and conservation. This measure is particularly important in arid and semi-arid areas which are subject to drought periods. In dry years, when the flow rate in aquifers is lower than estimations made for the most likely reference year, the use of the water resources is limited.[127]

2.4.4.4 Economic and Financial Regulatory Instruments

Economic measures such as fees on the use of water and the charges for waste-disposal are increasingly applied in co-ordination and have sufficient impact to constitute an effective incentive to use ground water rationally or be a disincentive to polluting aquifers. The abstraction of ground water is in some instances subject to differentiated fees in proportion to the volume abstracted, in relation to the available resource or according to the anticipated use of the abstracted ground water, while complying with legal provisions and regulations governing the applied permit system.[128] Costs attributable to pollution are to be borne by the polluter whenever the latter can be identified. Serious consideration should be given to all possible economic measures which could have an influence on preventing, mitigating and counteracting damage as well as those bearing on remedying critical situations caused by pollution or over-exploitation of aquifers. In Albania for example, the Council of Ministers will define the fees payable for the utilization of water on - various purposes. The decision of the Council of Ministers will take into account and will be constrained by some factors. These factors include the type of water and purpose of use, the season, and the type of technology used which determines the comparative cost/benefit or the consumed water.[129] There are some governments which provide financial incentives for water users to conserve and rationally use the waters. These incentives include the provision of loans, tax exemptions and other relevant incentives.[130]

---

[127] In the Lithuania Water Law, *supra* note 101, Article 34(9) provides:
"In a dry year, when the flow rate of rivers and the level of lakes and reservoirs is lower than estimations made for the most likely reference year, the use of the water resources of reservoirs and dammed lakes may be limited according to the rules for the use and maintenance of reservoirs and dammed lakes".
In Israel the Water Law 5719-1959 devotes a whole section (Article 4 items 36-43) for "rationing areas". The provisions authorize the Minster in charge to declare areas where consumptions may be rationed due to insufficiency of the water. Procedures and mechanisms of declaring rationing areas and their criteria are as well included.

[128] Charter on Groundwater Management, *supra* note 76, Article VII.

[129] The 1996 Albanian Law on Water Resources, *supra note* 87 Article 57.

[130] *Ibid,* Article 59 provides:

2.5 CONCLUSIONS

The preceding research finds that the intimate relationship between groundwater and surface water and the environment coupled with the distinct features, issues and concerns relating to groundwater complicate the task of management and protection of these resources in an integrated manner. Whilst it is essential to realize that groundwater cannot be managed in isolation from the other elements of the environment especially surface water, there is a clear need for a tailor-made legal response for groundwater that adequately responds to its special features. There are two main features relating to groundwaters that need to be apprehended; they cross borders unhindered and they are susceptible to pollution and depletion. Such resources cannot be replaced and once polluted or depleted losses are frequently irreversible.

The qualitative and quantitative characteristics of groundwater are found to be affected by a number of factors that are either a function of space or of time or the two together. The aquifer geometry (extension and thickness), the hydrogeologic properties of the aquifer formation, soil characteristics and climate are examples of natural/physical factors. Man made activities in the main recharge areas, are as well significant factors that need to be considered; these include land use, industrial and urban development activities in addition to in-situ waste disposal and/or main water leakage and overexploitation. This chapter demonstrates that prerequisite for regulating and imposing legal regimes in the case of groundwater is a proper understanding of natural and induced conditions of the aquifer systems and their surrounding environments. The "Drainage Basin" concept appears to incorporate all the natural and induced factors mentioned in the context of physically connected surface and groundwater.

The illustrative analysis of modern trends in national groundwater regulations demonstrates that there is an increasing awareness of the emerging water crises and the urgent need to preserve these resources. The analysis also finds that there are some common denominators in the relevant state practice at the national level

---

"The Government may provide financial incentives such as loans, tax exemptions for the water use for subjects engaged in studies or development of technology, processing, installations and new equipment that will reduce the water use and pollution. It may also give financial incentives to subjects supporting and promoting the forestation and other actions for protecting water resources. 2. These incentives may be extended to those who get engaged in desalinization and sewage treatment, using upgraded processes and methods, or establishing water recycling systems, or conduct research activities on this field."

Also see the 1994 Tunisian Water Law Chapter 95 which provides that

"the water administration has the authority to provide financial assistance to projects and measures aiming at reducing the water losses provided that the economic and technical feasibility of such measures are proven." (in Arabic, on file with the author).

in selected regions-those could be summarized in (i) the adoption of the "River Basin" concept, as it has been employed in national regulations since more than 70 years with the aim to achieve an integrated water resources management (ii) the permitting or licensing systems (iii) the declaration of protection zones (iv) the economic and financial instruments and incentives and (v) the development of monitoring, assessment and control systems.

In a transboundary context, decisions and responses require two or more States to undertake, with the possibly that two or more legal regimes are involved. The envisaged problems at this preliminary stage between national and transboundary groundwaters lie in the willingness of States to accept that the notions of groundwater regulations that apply on their domestic groundwaters can be as well adapted at the international levels.

The next chapter focuses on international groundwater. It introduced a case study whose political hostility and tension add a new a perspective to the complexity of regulating groundwater. This case study concerns the Mountain Aquifer shared between the Israelis and Palestinians in the Middle East.

3

The Water Conflict over the Mountain Aquifer Description and Related Politics

3.1 INTRODUCTION

Chapter One presented an overview of the occurrences, types and functions of groundwater, and addressed the unique hydrological/hydrogeological features that continue to distinguish this resource from surface water. Chapter one also included a description of the problems and challenges associated with the utilization, development and management of this resource, and the wide range of special responses needed for its regulation. It was demonstrated that essential to the efficient management and regulation of groundwater is a comprehensive understanding of the natural/physical features of such resource and a full appreciation of the interrelationships between groundwater and the environment.

Mindful of the complex issues relating to groundwater management and regulation presented in Chapter One, this chapter adds a new perspective to such complexity; as it introduces a case study of transboundary groundwater conditions coupled with a hostile political situation. This case study concerns the shared transboundary groundwater between Israel and the Palestinians in the West Bank territory of Palestine.[1] The specific objective of this chapter is to provide a general understanding of the case study and to explain the implications of politics on the general water situation. This will be done through an assessment of the shared transboundary groundwaters in the area of the case study, [2] the role international law can play, and in light of these factors it will explore in the next chapters the most appropriate approach for resolving the water conflict over the Mountain Aquifer.

3.1.1 The Water Conflict and the Political Setting

The rights and obligations of the Palestinians and Israelis concerning their international watercourses are not fully understood. The primary reason for this is Palestine's unique political status as an occupied land and the fact that since occupying the West Bank, including Jerusalem, and the Gaza Strip as a result of the Six-Day War of 1967,[3] Israel has constantly refused to relinquish control of any of the occupied territory and its resources.[4] On 22 November 1967, the UN Security Council passed Resolution 242, [5] which calls on Israel to withdraw from territories occupied in the Six-Day War of 1967 as well as a just settlement of the refugee problem. In addition, UN General Assembly Resolution 181[6] of 1947

---

[1] Within this thesis the term "Palestine" refers to the territory of the West Bank (including East Jerusalem) and the territory of the Gaza Strip (*see* Map 1).

[2] The shared aquifer system is referred to as the Mountain Aquifer which is comprised of the following aquifer basins: The Western Mountain Aquifer Basin (WMAB), North-eastern Mountain Aquifer Basin (NEMAB), and Eastern Mountain Aquifer Basin (EMAB). *See* Map 2, *also see infra* section 2.1.3.

[3] M. Mar'i, *Permanent Status Negotiations in the Palestinian - Israeli Context: A Case for International Intervention, found at* http://firms.findlaw.com/MariLaw/memo9.htm. (*last visited* June 2003).

[4] See *infra* section 3.1.4.

[5] Security Council S/RES/242, 22 November 1967, *found at* http://domino.un.org/UNISPAL.NSF/9a798adbf322aff38525617b006d88d7/59210ce6d04aef61852560c3005da209?OpenDocument&Highlight=2,242 (*last visited* May 2003).

[6] I. Brownlie, Principles of Public International Law, 5th Ed. (New York: Oxford University Press, 1999), pp.105-106. Brownlie expresses his opinion concerning the partition plan by saying

"It is doubtful if the United Nations has a 'capacity to convey title,' *inter alia* because the Organization cannot assume the role of territorial sovereign: in spite of the principle of implied powers the Organization is not a state and the General Assembly only has a power of recommendation. Thus the resolution of 1947 containing a partition plan for Palestine was probably *ultra vires*".

provides the legal basis for Israeli and Palestinian statehood and mandates full equality and human rights for all citizens of both the State of Israel and the State of Palestine.[7] None of these Resolutions were enforced, and Israel today continues to control all the water resources in the territories of the West Bank and Gaza Strip. This will be discussed in more detail in the next sections.

Given the complexities described above, there is consequently an entire spectrum of issues requiring discussion and further investigation prior attempting the assessment of the applicability of international law to transboundary groundwater shared between Israel and Palestine. These issues include: (i) the geography of the area and the general water situation and their contribution to the water conflict (ii) the Palestinian and Israeli official position from international law as a possible instrument for resolving the current conflict, (iii) the present political status of Palestine as an occupied territory, and its impact on the applicability of international law in the case of transboundary groundwater; and (iv) the relevance of the role played by the United Nations (UN)[8] in this regard.

3.1.2 Geography and General Water Situation

Palestine covers around 5719 km^2 of land plus 220 km^2 of water. It is located on the east coast of the Mediterranean Sea, west of Jordan and to the south of Lebanon.[9] Israel on the other hand, covers a total, 20,330 km^2 of land plus 440 km^2 of water reflecting armistice or cease-fire agreements with Lebanon and Syria, and more recent peace treaties with Egypt and Jordan. The international boundaries between Israel and Palestine have never been settled.[10] The Green Line refers to the boundary established in the General Armistice Agreement between Israel and Jordan on 3 April 1949.[11]

---

[7] R. Normand, *The Final Approach to Final Status,* 2000, *found at* (http://www.merip.org/pins/pin25.html (*last visited* March 2003).

[8] The principal organs of the UN are: the United Nations Security Council, the United Nations General Assembly, the United Economic and Social Council, the Trusteeship Council, the International Court of Justice and the Secretariat. For more details *see* The Encyclopaedia of International Law, Vol. 1 (Elsevier: Amsterdam, the Netherlands: 2000).

[9] *See* Map 1: Location of the West Bank.

[10] Foundation for Middle East Peace Report, *found at* http://www.fmep.org/reports/2002/sr0203.html (*last visited* July 2003).

[11] After the cessation of hostilities between the Arab countries and Israel in 1948, an Armistice agreement was signed in 1949. The agreement delineated the borders of each party and designated the "No Man's" land between them according to the location of their respective armies. This line demarcated the borders between Israel and the West Bank and Gaza Strip as recognized by the international community. It is worth mentioning here that Israel does not specify the boundaries of its state. Although the line became known later as the "Green line", its proper name is the "1949 Armistice Line". *See* Glossary by Applied

The territory of the West Bank-the focus of this book- covers around 5572 km^2 divided as: 5352 km^2 of land and 220 km^2 of water[12]. The water comprises half of the area of the Dead Sea, Huleh Lake,[13] and Tabariyya Lake, which is also known as the Sea of Galilee[14]. The West Bank is mostly composed of limestone hills having an average height of 700 to 900 m.

Palestine is characterized by limited surface and groundwater water resources. The main surface water system in the region is the Jordan River Basin which begins in three headwaters. The Hasbani River originates in Lebanon and has at least parts of its flow in Lebanon with an average flow of 138 mcm/yr.[15] The Dan and Banias (Nahal Hermon in Israel) Rivers originate in the Golan Heights and both flow into the Jordan above Tabariyya Lake having average annual flows of 245 and 121 mcm respectively.[16] It has an average annual flow of 1300 mcm.[17] The Jordan River Basin is considered under international law as an international river with waters shared by; Israel, Jordan, Syria, Lebanon and the Palestine.[18]

Research Institute and Arab Studies Society Land Research Centre, Monitoring Israeli Colonizing Activities in the West Bank and Gaza Strip, *found at* http://www.poica.org/glossary/index.htm (*last visited* June 2003), *See* also S. Willson, *History of Palestine and Green Line Israel*, May 2002, *found at* http://www.brianwillson.com/palest_hist.html (*last visited* June 2003).

[12] Valuable Agricultural Land In The West Bank Governorates, Emergency Natural Resources Protection Plan - Thematic Report 03 Directorate, *found at* http://planning.pna.net/agriculture/agriculture.htm *(last visited* June 2003).

[13] In Arabic "Lake Huleh" is referred to as "Buhayrat al Hulah", which is a near sea-level lake formed by a natural dam of basalt, North East Israel; the Jordan River exits from its southern end. In 1951 the All-Israel Plan included the draining of Huleh Lake and swamps, diversion of the northern Jordan River, and construction of a carrier to the coastal plain and Negev Desert - the first out-of-basin transfer for the River Basin.

[14] Palestine Net: Geography of Palestine, *found at* http://www.palestine-net.com/geography/ (*last visited* July 2003).

[15] T. Naff and R. Matson (ed.), Water in the Middle East: Conflict or Cooperation, (Westview Press: USA, 1984). The range of values measured for the Hasbani spring varies over a greater range than do the measurements at the other springs of the Jordan. Over the recent twenty year period, the flow of the Hasbani varied from 52-236 mcm/yr.

[16] *Ibid.*

[17] M. Abu-Taleb, *Environmental Management in Jordan*: *Problems and Recommendations*, 21 Environmental Conservation, 34-40 (1994).

[18] H.I. Shuval, Towards Resolving Conflicts Over Water Between Israel and Its Neighbours: The Israeli Palestinian Shared Use of the Mountain Aquifer as a Case Study, found in: J.A. Allan (ed.), Water, Peace and the Middle East: Negotiating Resources in the Jordan River Basin, (Tauris Academic Studies: London, 1996), see also H. Gvirtzman, Groundwater Allocation in Judea and Samaria in J. Issac and H. Shuval (eds.) Water and Peace in the Middle East (Elsevier: Amsterdam, the Netherlands, 1994).

The groundwater system in Israel and Palestine is divided hydrologically into seven basins, each of which is subdivided into aquifers.[19] The total annual renewable water from groundwater resources in both countries is 1454 mcm/yr,[20] of which at least 679 mcm/yr derived from shared Aquifers.[21]

3.1.3 THE MOUNTAIN AQUIFER

The Mountain Aquifer system –the focus of this book- comprises three sub-aquifers: the Northeastern Mountain Aquifer Basin (NEMAB)[22], the Western Mountain Aquifer Basin (WMAB)[23] and the Eastern Mountain Aquifer Basin (EMAB). [24] The former two aquifers form part of an international hydrologic system that is shared with Israel.[25] The EMAB almost lies entirely within the West

---

[19] Palestinian Water Authority Website: *Legal Framework of Permanent Status Negotiations: the Palestinian Water Rights*, found at www.pwa-pna.org/negotiations/negotiations.php (*last visited* December 2003), and official reports and fact sheets by the Palestinian Water Authority, April 16, 2000, (unpublished, on file with the author). These aquifers are the West Bank Mountain Aquifer, Eastern Mountain Aquifer Basin, the Northeastern Mountain Aquifer Basin, the Western Galeli, the Coastal, Carmel, Tiberia, and the Naqab.

[20] *Ibid.*

[21] Israeli-Palestinian Interim Agreement on the West Bank and Gaza Strip, Washington DC, 28th December 1995, *found at:* <http://www.miftah.org/Documents/documents/piia/interim.html and http://www.ariga.com/treaties/interim.html, (*last visited* June 2003).

[22] The NEAB covers the areas of Nablus and Jenin in the WB and the waters flows northwards to their major outlets in Bet Sha'an Springs in Northern Israel. The feeding and storage area of 700 km^2 lies completely in the West Bank (650 sq km). The NEMAB starts near Nablus (Schem) and flows towards the Gilboa Mountains and Jezreal and Bet Shean Valleys to the northeast. The Interim Agreement between the PLO and Israel estimated that the recharge of the NEMAB was said to yield 145 mcm/yr (*See* Map 2). Israeli Ministry of Agriculture, Hydrological Services Annual Report, 1998, *See also* Israeli-Palestinian Interim Agreement, *Ibid.*

[23] The WMAB- referred to as the Yarkon Tanninim in Israel- is located in the Western part of the West Bank. The recharge area is 1800 sq km of which 1400 sq km lie in the WB. The storage area of 2500 sq km lies in Israel. Two major natural outlets are Rash El Ein (in Israel Rosh Ha'ayn) and Tamaseeh Springs. WMAB is estimated to have a recharge of 362 mcm/yr. Israeli Ministry of Agriculture, *ibid.*

[24] The EMAB is located and recharged almost entirely in the West Bank, with the feeding and storage area spread over 2200 sq km mostly in the WB. A small part of recharge is located west of the Green Line including West Jerusalem. The Interim Agreement between Israel and the Palestinian Authority estimated that recharge of the is EMAB 172 mcm/yr. see *supra* note 18.

[25] According to the classification made by J.Barberis concerning the types of transboundary aquifers this statement is true. The two aquifers (NEMAB and WMAB) originate in one

Bank and is fed by recharge originating in the West Bank highlands.[26] According to the Barberis classification of transboundary aquifers,[27] the EMAB is part of an international system- namely the Jordan River Basin. The EMAB is an effluent to the Jordan River Basin i.e. it lies in the upstream portion of the rivers and feeds into the river. Despite its obvious international character, the Palestinians consider the EMAB as an indigenous aquifer that originates and recharges almost entirely in the West Bank.[28] The treatment of this aquifer is considered beyond the scope of this study. By and large, and from the view of this research, any solution to be proposed for the WMAB and NEMAB might equally apply to the EMAB. However, any future consideration of the whole Jordan River Basin must take into account the relationship between the EMAB utilization and development and that of the River Basin.[29]

State and extend to the other due to geographic and hydrologic factors. *See* J. Barberis, *The Development of International Transboundary Groundwater Law*, 31 NRJ 167-186 (1991). *See also* Chapter One section 1.1.1.

[26] H.I. Shuval, *supra* note 18.

[27] J. Barberis, *supra* note 25.

[28] Legal Framework of Permanent Status Negotiations, *supra* note 19.

[29] Many efforts to resolve the conflict over the waters of the Jordan River Basin the latest of which was the Johnston Plan which is basically an Allocation scheme proposed in 1953 by US special envoy to the Middle East, Ambassador Eric Johnston. The plan was the product of negotiations with representatives of Israel, Lebanon, Syria and Jordan that led in 1955 to a unified plan for all the riparians. The plan was never adopted or ratified partly because the Arab states (especially Jordan) did not need a comprehensive water development program that directly involved Israel to achieve their immediate development goals. Also the Arabs did not agree to the criteria that were used for dividing the shares among the parties, *See* PASSIA, *ibid*, T. Naff, *Water: An Emerging Issue in the Middle East?* 482 Annals of the American Academy of Political and Social Science 65-84 (1985), also *see* A. Wolf, Hydropolitics along the Jordan River Scarce Water and its Impact on the Arab-Israeli Conflict (United Nations University Press: New York, 1995) also *found at* http://www.unu.edu/unupress/unupbooks/80859e/80859E00.htm (*last visited* June 2003).

Map 1: Location of the West Bank

Jordan River Basin
Groundwater divide
Israeli National Water Carrier
Syria-Israel Cease Fire Line, 1967
Note: Quantities are in mcm (million cubic meters) per year
LEBANON
Upper Jordan River
GOLAN HEIGHTS
Eastern Wadis
(605) Total flow in U
Used by Israel (130)
Flow into Tiberias (475)
(80) Hydrological Balance
(55) Diverted to Tiberias by Israel
Haifa
Israeli National Water Carrier (420)
Tiberias
Israeli local use (90)
Salty Springs (40)
(90) Used by Jord
(20) Israel transfers to Jordan in summer
(10) Desalinated water
Green Line
Jenin
Northeastern Aquifer
Tulkarem
Mediterranean Sea
Coastal Aquifer
Nablus
Western Aquifer
WEST BANK
River Jordan
Zarqa
Tel Aviv
Lower Jordan River
(115) Remaing as Brackish Water
Aquifer
Ramallah
Jericho
Jerusalem
Coastal Aquifer
Green Line
Mountain
Eastern Aquifer
Dead Sea
Wadi Wala
Gaza
GAZA STRIP
Hebron
Wadi Mujib
Gaza Aquifer
ISRAEL

3.2 ISRAELI WATER POLICY AND PRACTICES: IMPACT ON PALESTINE'S DEVELOPMENT

3.2.1 Background

A great deal has already been written about the historical evolution , beginning in the early 1900's, of the Arab-Israeli water conflict. The overwhelming majority of authors and researchers agree that water has always been considered an important strategic element in terms of the Israeli policies, plans and regional development. [30] Many would state that the root of the Arab-Israeli water conflict can be traced back to the signing of the Sykes-Picot Agreement.[31] Under this particular document, which divided the regions of the Middle East between British and French control, no direct mention was made of water rights. [32] The Sykes-Picot Agreement would most certainly have left the watersheds in the region divided in a most convoluted manner; the Litani and Jordan headwaters just south of the Huleh region[33] would have come under French control, while the Lake Tabariyya would have been split into two, with one part coming under international control and the other under French control. The Yarmouk Valley, meanwhile, would have come under both British and French control, while the lower stem of the Jordan River would have found itself under international control on the West Bank and British control on the East Bank. Throughout the 1930's and the 1940's, the West further manipulated the affairs of the Middle East in order to control the

---

[30] H. Gleick, *Water and Conflict*, Occasional Paper Series of the Project on Environmental Change and Acute Conflict, Number 1 (1992). Gross, D. and A. Soffer, *International Groundwater and International Rules: the Case of the Middle East and the Agreement between Israel and Jordan and between Israel and the Palestinians about the Gaza Strip*, Occasional Papers on the Middle East series no. 12, (1996) also see S. Libiszewski, *The Environment of Conflict: Water Crises in the Jordan Basin Region,* ENCOP Occasional Paper No. 13. (1995) *found at* (http://www.fsk.ethz.ch/encop/13/en13.htm) (*last visited* June 2003). See also A. Wolf, A., *Hydro strategic Territory in the Jordan Basin: Water War, and the Arab Israeli Peace Negotiations,* Paper presented at a conference: Water, a Trigger for Conflict/ a Reason for Cooperation, 1996) *found at* <http://w3.pnl.gov.2080/ces/academic/midleas2.htm (*last visited* June 2003) and M. Lowi, *West Bank Water Resources and Resolution of Conflict in the Middle East*, Occasional Paper Series of the Project on Environmental Change and Acute Conflict, Number 1 (1992).

[31] The Sykes-Picot Agreement (concluded in 1916) *found at* www.geocities.com/CapitolHill/Parliament/2587/pico.html. (*last visited* May 2003). , which was signed between the British and French on 9 March 1916, divided the Levant into zones and constituted the first serious attempt to demarcate a Palestine-Lebanon border.

[32] A. Wolf, Hydropolitics Along the Jordan River: Scarce Water and Its Impact on the Arab-Israeli Conflict, (The United Nations University Press: 1995).

[33] *Ibid.*

resources of the region and then to create a Jewish homeland in an area long considered central to Arab nationhood.[34] However, despite the promise of a "national home" in the Balfour Declaration of 1917, [35] the Zionists were displeased with the land and water resources granted to the Jews by the British in 1919[36] and consequently embarked on a strategy of acquiring land as Israeli property in order to promote agricultural colonization based on Jewish labour.[37] The following sections attempt to summarize how Israel's policies and plans were centred on the region's water resources, and what impact these policies have on the status of and prospects for Palestine's development. This will be done with

---

[34] N. Ahmed, The War on Terror: Western Secret Strategy and the Struggle for Iraq (Clairview Books: UK, 2003).

[35] The "Balfour Declaration," was a letter dated 2 November, 1917 from the British Foreign Minister to Lord Rothschild head of Anglo Israeli community. The British commitment was widely published in the press, though censored by the British military authorities in Palestine. The operative language of which stated: "His Majesty's Government view with favour the establishment in Palestine of a national home for Israeli people. ...." *found at* http://www.lib.byu.edu/~rdh/wwi/1917/balfour.html (*last visited* June 2003).

[36] Palestine was ruled by Great Britain from 1917 to 1948, initially as occupied enemy territory and later as a mandate from the League of Nations. The Mandate was assigned to Britain at the San Remo Conference (1920) after World War I, which ratified the division of the Ottoman Empire's Arab provinces between France and Britain. France gained control over Syria and Lebanon; Britain acquired Iraq as well as Palestine. Although all these territories were designated Class A mandates, which meant that they would soon gain self-rule, Palestine was placed under unique provisions, because Britain had promised the Zionist movement in the Balfour Declaration (November 2, 1917) that Jews could establish a national home in the territory. *See* J. Robert, and S. Hadawi, the Palestine Diary. Vol. 1, 1914-1945. (The Palestine Research Centre: Beirut, 1970). *See* also The Palestinian American Research Centre Website, the *Encyclopaedia of the Palestinians, found at* http://www.parcenter.org/Resources/Encyclopedia/Encyc_pman.html(*last visited* June 2003).

[37] Although the British Mandate of under the auspices of the League of Nations has confirmed that the Mandate Administration safeguards the interests of the local community including those related to land property and natural resources. League of Nations *Mandate for Palestine*, Nineteenth Session of the Council Thirteenth Meeting Held at St James' Palace, London on July 24th, 1922, *found at* http://www.mfa.gov.il/mfa/go.asp?MFAH00pr0 (*last visited* June 2003), Article 11 states:

"The Administration of Palestine shall take all necessary measures to safeguard the interests of the community in connection with the development of the country, and, subject to any international obligations accepted by the Mandatory, shall have full power to provide for public ownership or control of any of the natural resources of the country ...."

And Article 12:

"The Administration may arrange with the Jewish agency mentioned in Article 4 to construct or operate, upon fair and equitable terms, any public works, services and utilities, and to develop any of the natural resources of the country, in so far as these matters are not directly undertaken by the Administration......"

reference to three major periods in time: 1948-1967, 1967 to 1992 and finally the 1992 to the present.

3.2.2 1948-1967

It is important to note when considering the Palestinian-Israeli water conflict that prior to the establishment of the State of Israel in 1948, a special UN Committee on Palestine provided evidence showing that Israelis owned only 7% of the land, whereas following the creation of the State of Israel, the percentage rose to 60%.[38] Also noteworthy is the fact that following the establishment of the State, the main objective of the water plan was to divert as much water as possible outside the Jordan River Basin into a central conduit leading through the coastal plain up to the northern Negev. Since 1955, Israel was tapping into the WMAB. The National Water Carrier, which was first operated in 1964, was consequently the outcome of many years of planning, the first stages of which were implemented in 1948. [39]

3.2.3 1967-1992

Between 1967 and September 1992,[40] the utilization of groundwater within the Occupied Palestinian Territories was governed solely by Israeli legislation and military orders. [41] Today, it relies on the three aquifers comprising the Mountain Aquifer- which combined, provide approximately 40% of Israel's water supply. [42] Of the total of some 679mcm/yr of water annually available from the Mountain Aquifer, Israel currently uses about 483 mcm/yr, while the Palestinians are limited

---

[38] S. Hadawi, (4th edition) Bitter Harvest: A Modern History of Palestine (Interlink Pub Group: New York, 1998).

[39] S. Libiszewski, *The Environment of Conflict: Water Crises in the Jordan Basin Region,* ENCOP Occasional Paper No. 13. (1995) *found at* (http://www.fsk.ethz.ch/encop/13/en13.htm) (*last visited* June 2003).

[40] This date indicates the beginning of official negotiations between Israel and the PLO see *infra* Chapters 4 and 5.

[41] Prior to 1967, Jordan issued rules and regulations governing the use, development and other water related matters. These laws remained in effect even after the 5th June 1967, whereby an Israeli proclamation stated that these could be subject to changes introduced by the military Commander. After 1967, successive Military Orders were issued to amend parts of these laws. Of the significant Military Order 92 of 7 June 1967, Order 158 (19 November 1967) Military Order 291 (19 December 1968) and many others, *found in* Israeli Military Orders in the Occupied Palestinian West Bank 1967-1992, (Jerusalem Media and Communication Centre: Jerusalem, 1995).

[42] M. Lowi, The United Nations University Program on Sustaining Global Life-support Systems Water and Power: The Politics of Scarce Resources in the Jordan River Basin, (Cambridge University Press: Cambridge, 1993).

to about 118 mcm/yr. [43] It is estimated that another 78 mcm/yr may be available after further exploitation in the Eastern Aquifer.[44] Due to the fact that the Palestinians have constantly been denied access to their share of the Jordan River waters and the massive imbalance in terms of current water-use rights, there being also a huge imbalance in terms of water consumption. The Palestinian domestic per capita consumption of 35-80 l/day being far below the standard established by the World Health Organization (WHO), i.e., a minimum of 100 l/day, and the Israeli per capita consumption, which exceeds 300 l/day is clearly excessive.[45] Israel, it should be mentioned, has additional sources of fresh water since as well as the 600 mcm/yr of water from the Jordan River[46] it is currently utilizing, it also has access to water originating from another five groundwater aquifers located within its territory.[47] As to the Palestinians' access to water, up until 1995, only a very few drilling and extraction licenses were granted to enhance the supply within Palestinian communities, which meant that the natural increase in water demand due to population growth and industrial and agricultural expansion had to be satisfied from the wells that had already existed prior to the occupation. [48]

In terms of the water services level, the existing water supply networks in almost all the West Bank and Gaza Strip are old with some dating back to more than 50 years. During the last 34 years of occupation, no planned maintenance or rehabilitation took place due to the limited financial and technical abilities of facilities of the local authorities[49] in charge of water services. Additionally, Israel also gives permission to conduct maintenance work.[50] This situation led to high loss percentages through pipe leakage, as unaccounted for water reaches a value of 50% in some of the municipalities.[51]

Finally, and most importantly, Israel during recent years has developed its technical, legal knowledge and capacity in the field of water resources

---

[43] Israeli-Palestinian Interim Agreement, *supra* note 21.

[44] *Ibid.*

[45] Legal Framework of Permanent Status Negotiations, *supra* note 19.

[46] The total average annual flow of the Jordan River is 1,300 mcm. The Palestinians have been denied access to their share - estimated at 200 mcm/yr - since 1967.

[47] These groundwater aquifers basins are as follows: the Lake Tabariyya, the Western Galilee, the Coastal, the Naqab, and the Carmel. The total amount used from these aquifers is 1663-1689 m cm/yr. *See* Israeli Ministry of Agriculture Annual Report, *supra* note 22.

[48] Palestinian Delegation Position Paper presented to the Working Group on Water to the Middle East Peace Multilateral Negotiations, First Round Vienna 13 May 1992.

[49] These are the municipalities in the big cities and village councils in the rural areas.

[50] L. Wong, The Israeli-Palestinian Water Crisis,(on file with author).

[51] Excluding the case of the Jerusalem Water Undertaking JWU in the Ramallah region which has proved to be the most efficient service provider in the WB). *See* JWU Website www.jwu.org *(last visited* April 2003).

management and regulations.[52] The Palestinians on the other hand had limited opportunities to develop similar capabilities. The range of military orders[53] and restrictions imposed on the Palestinians concerning the development of the water resources contributed to the creation of a serious asymmetry between the Parties.[54]

3.2.4 1992-Present

After long years of struggle, the Palestinians and Israelis started their formal negotiations in 1992.[55] The aim of the Israeli-Palestinian negotiations within the current Middle East peace process was, among other things, to establish a Palestinian interim self-autonomy in a period not exceeding three years would lead to a permanent settlement to be based on UN Security Council Resolutions 242 and 338.[56] The Government of the State of Israel and the Palestinian Liberation Organization (PLO), in September 1993, signed within the framework of the Middle East Peace Process, the "Declaration of Principles" (DOP). [57] The

---

[52] The management of Israel's water resources is governed by a number of laws of which the Water Law 5719-1959 is the core piece of legislation. Other laws include the Water Management Law, 5715-1955, the Water Drilling (Control) Law, 5715-1955, the Streams and Springs Authorities Law, 5725-1965, the Drainage and Flood Control Law, 5718-1957. *See* J. Alster, *Water Laws, Rules and Regulations*, unspecified date (on file with author).

[53] See *supra* note 39

[54] *See infra* Chapter 6 (Section 6.1.1).

[55] More analysis on the outcomes and content of these negotiations is found *infra* chapter 5 and .

[56] Declaration of Principles on Interim Self-Government Agreement, Ministry of Foreign Affairs, Washington DC, September 1993, *found at* http://www.ariga.com/treaties/dop.html (*last visited* June 2003) Resolution 242 is a binding Security Council resolution which has two main components:

"1. Affirms that the fulfilment of Charter principles requires the establishment of a just and lasting peace in the Middle East which should include the application of both the following principles: a. Withdrawal of Israeli armed forces from territories occupied in the recent conflict; b. Termination of all claims or states of belligerency and respect for and acknowledgement of the sovereignty, territorial integrity and political independence of every State in the area and their right to live in peace within secure and recognized boundaries free from threats or acts of force;2. Affirms further the necessity: (a) For guaranteeing freedom of navigation through international waterways in the area; (b) For achieving a just settlement of the refugee problem; c. For guaranteeing the territorial inviolability and political independence of every State in the area, through measures including the establishment of demilitarized zones Resolution 338 is a security council resolution, following the Yom Kippur war of 1973, calling on both sides to begin implementation of Resolution 242."

For more details on resolutions on the question of Palestine *visit* http://www.middleeastfacts.com/UNresolutions.html (*last visited* July 2003).

[57] *Ibid.*

DOP was the first initiative showing the willingness by both parties to put an end to the decades of confrontation. [58] and to live in peaceful coexistence. The "Gaza-Jericho" Agreement was signed between the PLO and the State of Israel regarding the autonomous rule of the Palestinian Authority PA, in Jericho and the Gaza Strip on May 4, 1994.[59]

In 1995, the Government of the State of Israel and the PLO, entered into an interim agreement on the West Bank and the Gaza Strip.[60] Both Parties showed a desire for put into effect the DOP. They also reaffirmed their recognition of mutual legitimate and political rights[61]. The water issue was among the contentious issues that were deferred to the final status negotiations.[62] Only temporary solutions were offered for the water crises in Palestine within the interim period. To date only minimal progress has been achieved to even fulfil the promised commitments. This topic will be revisited later in this book.

On 30 April 2003, the United States (US) presented the "road map" for peace in the Middle East to Israeli Prime Minister Ariel Sharon and newly confirmed Palestinian Prime Minister Mahmoud Abbas with the intention of solving the conflict between Israel and the Palestinians. [63] Together with the European Union (EU), the UN, and Russia, the US defined the "road map" destination as "a final and comprehensive settlement of the Israel-Palestinian conflict by 2005".[64] The settlement, negotiated between the Parties, would, the Parties hoped, "result in the emergence of an independent, democratic, and viable Palestinian State living side by side in peace and security with Israel and its other neighbours."[65]

During 2003, Israel commenced the construction of a 'separation fence' designed to cut the access of people in the Occupied Palestinian Territories. This

---

[58] The Preamble if the DOP, *supra* note 47, states that:
"The Government of the State of Israel and the PLO team (in the Jordanian-Palestinian delegation to the Middle East Peace Conference) (the "Palestinian Delegation"), representing the Palestinian people, agree that it is time to put an end to decades of confrontation and conflict, recognize their mutual legitimate and political rights, and strive to live in peaceful coexistence and mutual dignity and security and achieve a just, lasting and comprehensive peace settlement and historic reconciliation through the agreed political process".

[59] Israel/Palestine Liberation Organisation: Agreement on the Gaza Strip and the Jericho Area, 4 May 1994, Cairo, 33 I.L.M. 622 (1994), *also found* http://www.us-israel.org/jsource/Peace/gazajer.html (*last visited* July 2003).

[60] Israeli-Palestinian Interim Agreement on the West Bank and Gaza Strip, *supra* note 21.

[61] *Ibid*, Principle 1.

[62] These issues are: the Borders, Settlements, Security and Refugees and Jerusalem.

[63] E. Sheehan, The Map and the Fence, *found at* http://www.nybooks.com/articles/16411 (*last visited* August 2003).

[64] *Ibid.*

[65] *Ibid.*

fence, which follows the 1967 Green Line, cuts into Palestinian land by up to 6 kilometres in places - and restricts access by land owners to farming their land.[66] On completion it will completely encircle the West Bank. The West Bank and Gaza will then be unable to form a viable sovereign state.[67]

Most recently, Israelis and Palestinians informally signed the Geneva Accord[68] which calls for the establishment of the Palestinian State and ends the era of hostility and conflict. The Accord which was launched in Switzerland on the 1st of December 2003 is considered as a Framework Agreement that has to be negotiated and further elaborated in order to support the implementation of the "road map". In the context of water, the Geneva Accord does not include details on the principles governing the utilization and development of these waters.[69] This deficiency is serious as it leaves open all Israeli claims to the region's water resources, natural wealth, and airspace. The text makes several references to annexes, but these issues have, in effect, been deferred, and may become the "final status" issues of the Geneva understanding. Nevertheless, the Accord recognizes the sovereignty of the proposed Palestine State over territorial sovereignty in accordance with the UN Charter, and other rules of international law.[70]

---

[66] *See* the Jerusalemites Website *found at* www.jerusalemites.org/articles/english/nov2003/12.htm (*last visited* December 2003).

[67] According to Ian Brownlie, *supra* note 6, what qualifies a nation to be a State in international law, which rules that a State must possess the following (1) a permanent population; (2) a defined territory; (3) a government; and (4) the capacity to enter into relations with other states. Under the separation wall situation, the second conditions will not be valid. The Israeli Cabinet recently approved a route for the fence which will cut 22km into the West Bank so that the Jewish settlement of Ariel can be on the "Israeli" side. The result of detours like this is that 142,000 Jewish settlers living in the West Bank in contravention of international law, and 274,000 Palestinians will be in the area between the fence and the Green Line. The Palestinians will be confined to what Israel has decided will be a "closed zone". The report describes how some 13,545 Palestinians living in the "closed zone" next to a completed section of the wall have been told by the Israeli army that they must apply for permits to go on living in their own homes. The permits are valid for up to six months. *See ibid.*

[68] This is an unofficial peace treaty, which was drafted by a group of Israelis and Palestinians and issued in October 2003. Those involved included former Israeli Justice Minister Yossi Beilin and former Palestinian Information Minister Yasser Abed Rabbo. The Swiss government provided financial backing for the project, which also had support from former US president Jimmy Carter. http://www.al-bab.com/arab/docs/pal/geneva03.htm (*last visited* December 2003), [hereinafter referred to as the 2003 Geneva Accord].

[69] *Ibid,* Article 12 of the 2003 Geneva Accord to be completed. The text makes several references to annexes, but, from the view of the researcher, these issues should have been addressed promptly and clearly in the document.

[70] *Ibid,* Article 4 of the 2003 Geneva Accord provides:

Against this analysis of the past and present Israeli policy and practices in the field of water resources, and the impact these continue to have on the Palestinian prospects for development, the following sections provide an overall assessment of the Palestinian and Israeli Positions from the water conflict. The objective is to achieve an understanding of both the Israeli and Palestinian stances from the water conflict prior to proposing any approach for its resolution.

3.3 THE PARTIES' POSITION ON THE WATER CONFLICT

3.3.1 The Palestinian Position

Of relevance to the overall discussion with respect to the Israeli Palestinian water conflict, is the Palestinian position on international law in solving the current conflict. This analysis is made to assess the willingness of the Palestinians to accept solutions that are based on the rules of international law. In 1964, the PLO[71], the national liberation movement of the Palestinian people, was established by Arab heads of state. In 1988, the PLO accepted, in Algiers, in its Proclamation of Independence[72] all UN Resolutions since 1967, whilst renouncing violence and

---

"2. Sovereignty and Inviolability (a) The Parties recognize and respect each other's sovereignty, territorial integrity, and political independence, as well as the inviolability of each others territory, including territorial waters, and airspace. They shall respect this inviolability in accordance with this Agreement, the UN Charter, and other rules of international law."

See also *infra* Section 2.4 and *infra* Chapter 3.

[71] The PLO, the coordinating council for Palestinian organizations, founded in 1964 at the first Arab summit meeting. Composed of various political factions, the PLO is dominated by Al-Fateh, the largest group, whose leader, Yasser Arafat, has been the chairman of the PLO since 1968. Other groups in the PLO include the Popular Front for the Liberation of Palestine (PFLP) and the People's Party.

[72] PLO Proclamation of Independence, Algiers 1988. Over 114 states have already recognized the newly proclaimed state of Palestine, while only 93 maintain some form of diplomatic relations with Israel. Furthermore, on 15 December 1988 the UN General Assembly adopted Resolution 43/177, essentially recognizing the new state of Palestine and according it observer state status throughout the UN organization. The resolution was adopted by a vote of 104 in favour, with the US and Israel opposing the resolution and 44 states abstaining. On 15 December 1988 the United Nations General Assembly adopted Resolution 43/177, essentially recognizing the new state of Palestine and according it observer-state status throughout the United Nations Organization.. For the full text of the proclamation go to Independence *found at* http://www.danielpipes.org/article/185 (*last visited* May 2003).

recognizing Israel's right to exist.[73] As a confirmation to the PLO's intentions the Proclamation states that:

"The State of Palestine proclaims its commitment to the principles and purposes of the United Nations, and to the Universal Declaration of Human Rights. It proclaims its commitment as well to the principles and policies of the Non-Aligned Movement. It further announces itself to be a peace-loving State, in adherence to the principles of peaceful co-existence. It will join with all states and peoples in order to assure a permanent peace based upon justice and the respect of rights so that humanity's potential for well-being may be assured, an earnest competition for excellence be maintained, and in which confidence in the future will eliminate fear for those who are just and for whom justice is the only recourse"[74]

The Palestinians have continuously considered the notion of permanent sovereignty over natural resources as the legal background for their claims to water rights. Over the years, this position has been reflected in the writings of Palestinian experts and politicians and was reaffirmed in the recently adopted formal position from the water resources in a document entitled "Legal Framework of Permanent Status Negotiations: the Palestinian Water Rights".[75] This position builds on the spirit of international law,[76] and it confirms that the

---

[73] *See infra* section 3.3.2.

[74] PLO Proclamation of Independence, Algiers, 1988, *supra* note 72.

[75] Legal Framework of Permanent Status Negotiations, *supra* note 19. Presented in the form of demands, item 1 of the framework states: "the necessity of redemption of the permanent sovereignty over Palestinian water resources. This includes all water resources in the Palestinian territories (West Bank & Gaza Strip), i.e. the territories occupied by Israel following 1967 war, including Eastern Jerusalem. In another word, the State of Palestine shall have control over all water resources in the West Bank and Gaza Strip". The author believes that this document can be better presented in a shorter version with more focus on the substantive and procedural rules and principles of international law.

[76] *Ibid,* this framework document refers to the Hague Convention (IV) Respecting the Laws and Customs of War on Land and its annex: Regulation concerning the Laws and Customs of War on Land [hereinafter referred to as the 1907 Hague Convention], (entered into force Jan. 26, 1910), *found at* http://www1.umn.edu/humanrts/instree/1907c.htm (*last visited* July 2003) the Protocol 1 of the Geneva Convention (IV) Relative to the Protection of Civilian Persons in Time of War; done at Geneva, 12 August 1949 [hereinafter referred to as the 1949 IV Geneva Convention], the 1949 IV Geneva Convention: Final Record of the Diplomatic Conference of Geneva of 1949, Vol. I, Federal Department, Berne, at pp. 195-341; 75 United Nations Treaty Series 5-417. *Also found at*: http://www.icrc.org/ihl.nsf (*last visited* June 2003), the Resolutions adopted by the UN on "Palestine's Permanent Sovereignty Over Natural Resources" *infra* notes 108 and 109, the 1966 Helsinki Rules on the Uses of the Waters of International Rivers [hereinafter referred to as the 1966 Helsinki Rules, and finally the UN Convention on the Law of Non-Navigational Uses of International Watercourses, May 21, 1997 [hereinafter referred to as the 1997 UN

first step for the Palestinians is to regain their territorial sovereignty over their water resources.[77] The framework position considers the WMAB and NEMAB as joint aquifers with Israel, however, rejects to consider the EMAB as a transboundary aquifer.[78] The principle governing the future relation with Israel, as far as their shared water resources are concerned, is equitable and reasonable utilization.[79] There is however, a confusion between the current status of "occupation" and the future of the independent "Palestinian State".[80]

3.3.2 The Israeli Position

In order to support their claims regarding the waters currently being utilized by Israel, Israeli legal experts have usually relied on the doctrine of "Prior Use" or "Historical Rights" when dealing with regional water resources.[81] Moreover, they have persistently referred to all "existing uses" as non-negotiable and constantly raised the issue of the availability of "alternatives of comparable value" (desalination, wastewater reuse, and the importation of water from neighbouring countries) as a means to supply Palestinian needs.[82] What this means is that Israel's official position in terms of its water dispute with its Palestinian neighbours is based on its objection to sharing the available water resources in a

---

Watercourses Convention]. For the 1966 Helsinki Rules *See* International Association of the 52nd Conference xi, 447-533 (1967), *found at* http://www.internationalwaterlaw.org/IntlDocs/Helsinki_Rules.htm.(*last visited* In June 1999), and for the 1997 UN Watercourses Convention, *see* UNGA Res. 51/229, reprinted at 36 ILM 700 (1997).

[77] This has been reaffirmed in the Legal Framework of Permanent Status Negotiations on Palestinian Water Rights, *supra* note 19.

[78] *Ibid,* under the Heading "the International Law & Palestinian Rights to Water Resources", paragraph 2 states that: "The sources of ground water in the West Bank and Gaza Strip are distributed between the following aquifers: (i) Eastern Ground Aquifer, West Bank:
The aquifer is separate and did not form with any other aquifers a joint aquifer or watercourse. (ii) Western Aquifer and Northeastern Aquifer: both aquifers are related to historical Palestine and pass through what is so called the Green Line, (iii) Coastal Ground Aquifer: the aquifer is lying in the Gaza Strip. It is a separate aquifer and not connected with any other aquifers.

[79] *Ibid,* under the Heading "Conclusions" the framework position affirms: the right of the State of Palestine to obtain its equitable share from the joint watercourses, whether ground or surface, including in the first instance the Jordan River Basin. That shall be in accordance with the standards of the international law concerning the principles of equitable and reasonable utilization and participation and the obligation not to cause significant harm".

[80] *Ibid.*

[81] H. Shuval, *supra* note 18.

[82] Legal Framework of Permanent Status Negotiations, *supra* note 19.

fair and equitable manner, as has been revealed time and time again in the Israeli style of negotiating over the past seven years. In short, although Israel is prepared to discuss the need to meet some of the immediate Palestinian needs, it nevertheless does not appear to consider the water issue one as requiring a permanent solution. [83]

For proof of the official Israeli position regarding the water issue, one only has to consider the statement of the former Israeli Minister of Agriculture, who in December 1990 was quoted by the *Ma'ariv* newspaper as stating the following: "It is reality and need which created the Israeli control over the water resources; it would be impossible to give up a drop of water in the West Bank."[84] As to the former Israeli Water Commissioner Meir Ben Meir, he has repeatedly declared the following positions: "(i) The Palestinians can solve their water problem though pursuing non-conventional sources or through purchasing from Israel; (ii) Although Israel is prepared to discuss allocations and rights pertaining to the uses of water with the Palestinians, it refuses to discuss sovereignty over the available resources; (iii) International Law does not apply in the case of the Palestinians since Palestine is not recognized as an independent state".[85]

The following sections provide an overall understanding of the issues, opportunities, challenges and concerns associated with this unique political situation and their relevance to the Palestinian Israeli water conflict.

3.4 THE WATER CONFLICT AND THE ROLE OF INTERNATIONAL LAW

3.4 1 Introduction

It is beyond the scope of this research to elaborate on the spectrum of views that exists in relation to the adequacy of the existing laws of war to deal with environmental concerns.[86] By and large these views agree that the codified

---

[83] *Ibid.*

[84] Interview with the Minister of Agriculture Rafael Eitan in *Ma'ariv*, 17 December 1990.

[85] Unpublished material by PWA. Other examples on the Israeli official position in relation to the water issue include what David Ben Gorion said in 1956:

"We have a water war with the Arabs, the fate of the Israeli State depends on whether we lose or win this battle, if we lose it, we lose everything"

Cited in A. Samara, *The Political Economy of the West Bank* 1967-1987, Khamsin p. 79, London 1988. *Also* Menahem Kantour the Director General of Tahal said in 1979 was quoted in the Jerusalem Post Supplement, 4 May 1979:

"Against each dunum irrigated in Palestine there is one thirsty dunum in Israel".

[86] Four camps of views exist; Camp 1 considers the existing law relevant to environmental protection can be found in the Hague Conventions and Geneva Conventions is adequate to

customary laws[87] of war have been concerned with environmental protection in the sense of prohibiting wanton destruction of forests, orchards and the like, and in poisoning wells springs and rivers.[88] It is also agreed that in situations not covered within the scope of these laws the rules and principles of public international law apply. [89] The following discussion provides an overview on how international law deals with water resources under belligerent occupation and what impacts this has on the sovereignty of the occupied nation over these resources. Additionally, the discussion addresses the legal status of the numerous resolutions that were adopted by UN in the context of permanent sovereignty of Palestine over their natural resources.

3.4.2 The Laws of Belligerent Occupation

3.4.2.1 Background

The international community regarded the status of the Palestinian territory occupied by Israel as a result of the 1967 war as an occupied territory subject to

---

cover the worst environmental scenario; Camp 2, those who accept this but consider Protocol I to the 1949 Conventions also to represent customary law namely article 54 (5) and article 55; Camp 3 those who consider the existing law to be inadequate and in need for restatement; and Camp 4 those which considers the distinction between wartime and peacetime acts in this context is to be so vague, irrelevant and misleading that, if any new development to take place, it should be concerned with environmental destruction in all scenarios, whether wartime or peacetime. G. Plant (ed.), Environmental Protection and the Law of War, (Belhaven Press: New York, 1992).

[87] Certain "unwritten" rules generally accepted by states as legally binding; a general practice of states accepted as law. Customary rules/laws may be "codified" in a multilateral convention, However, these rules are legally binding for all states regardless of whether or not they are parties to such a "codification" treaty. *See infra* chapter 3.

[88] A. Roberts and R. Guelf (eds.) *supra* note 75.

[89] This is a reformulation of a Clause that appears in the preamble of the 1907 Hague Conventions, *supra* note 77, which provides that:

"Until a more complete code of the laws of war has been issued, the High Contracting Parties deem it expedient to declare that, in cases not included in the Regulations adopted by them, the inhabitants and the belligerents remain under the protection and the rule of the principles of the law of nations, as they result from the usages established among civilized peoples, from the laws of humanity, and the dictates of the public conscience".

Also it is a reformulation of Article 1(2) of the 1949 IV Geneva Convention, *supra* note 77, which provides:

"In cases not covered by this Protocol or by other international agreements, civilians and combatants remain under the protection and authority of the principles of international law derived from established custom, from the principles of humanity and from the dictates of public conscience".

the laws of belligerent occupation namely the provisions of both the Hague Convention of 1899 (II) [90] and the 1907 (IV)[91] and Geneva Conventions of 1949.[92]

3.4.2.2 The Hague Convention of January 1907[93]

This Convention is widely believed to codify customary international law and contain rules legally binding on all nations, including Israel.[94] The essence of these rules focuses on humanitarian concerns as they emphasize the obligations and responsibilities of an occupying power towards the local population, as well as various rights relating to state and privately owned property.[95] The Customary law of war includes clear indications on the need to avoid unjustifiable damage to the environment. [96] Violations of Articles 23 (g) of the Regulations attached to the Hague Convention prohibit the destruction by any occupying power of enemy property not acquired by military necessity, which gives rise to civil liability.[97] Additionally, the Regulations consider the occupying State as the guardians over natural and other resources, but does not entitle them to the right of ownership, to disposal or to transfer to its own state. This right of guardianship has its own limitations in terms of the amounts being used from these resources. [98]

---

[90] Hague Conventions on Land Warfare of 1899, 1 American Journal of International Law of 29 July 1899 and of UK Treaty Series No. 9 *also found at* http://www.yale.edu/lawweb/avalon/lawofwar/final99.htm (*last visited* July 2003) UK Treaty Series No. 9. This Convention was replaced by the 1907 Hague Convention *supra* note 77.

[91] The 1907 Hague Convention, *supra* note 76.

[92] The 1949 IV Geneva Convention, *supra* note 76.

[93] The 1907 Hague Convention, s*ee supra* note 76.

[94] Although Israel denies that Gaza and the West Bank are under its belligerent occupation the Israeli Supreme Court of Justice, sitting as a High Court with jurisdiction over the territories, has constantly held the Hague Regulations are applicable. According to Ian Scobbie the reasons behind that is the customary status of the Hague Regulations which form part of the law of Israel and thus bind the military administration of the Occupied Territories, *See* I. Scobbie, *H2O After Oslo: The Legal Aspects of Water in the Occupied Territories*, VIII the Palestine Yearbook of International law (1994).

[95] G. Plant (ed.), *supra* note 83.

[96] The 1907 Hague Convention, *supra* note 76, Article 23(g) forbids:
"To destroy or seize the enemy's property, unless such destruction or seizure be imperatively demanded by the necessities of war;"

[97] *Ibid.*

[98] Article 55 of the 1907 Hague Convention, *supra* note 76 states that:
"The occupying State shall be regarded only as the administrator and usufructuary of public buildings, real estate, forests, and agricultural estates belonging to the hostile State, and situated in the occupied country. It must safeguard the capital of these properties, and administer them in accordance with the rules of usufruct".

3.4.2.3 The Fourth Geneva Convention - Relative to the Protection of Civilian Persons in Time of War August 12, 1949

The Fourth Geneva Convention, which concluded in 1949, also deals with war crimes. Known formally as the "Convention on the Protection of Civilian Persons in Time of War", it required its parties to enact laws that made it illegal to commit or order others to commit "grave breaches" of the Convention, and to actively seek to bring such offenders to trial.[99] The Convention requires that an occupying State take full responsibility for meeting the needs of the civilian population under occupation. Care shall be taken in warfare to protect the natural environment against widespread long-term severe damage. [100] This protection includes a

[99] In the advisory expert meeting on the Fourth Geneva Convention, Gaza 8 And 9 August, 1998 the conclusions were as follows: The General Assembly expressed its view that the continuing Israeli violations of Geneva IV, particularly after 1993, in the Occupied Palestinian Territories threaten international peace and security, and that the Security Council has failed to exercise its primary responsibility. Resolutions adopted by General Assembly Emergency Special Sessions on collective measures are recommendations to member States. They do not diminish the responsibilities of these States under international agreements. The High Contracting Parties to Geneva IV remain responsible for implementing their obligation to respect and to ensure respect for Geneva IV in all circumstances through the diligent application of the full range of policy instruments relevant to this end. The High Contracting Parties to Geneva IV who voted in favour of resolution ES-10/5 expressed the view that they consider the convening of 'a conference on measures to enforce the Convention in the Occupied Palestinian Territory, including Jerusalem', as a collective measure to protect international peace and security against threats posed by Israel's persistent and systematic violations of Geneva IV and their damage to the ongoing peace process. By voting for this resolution the overwhelming majority of High Contracting Parties committed themselves to the convening of the conference recommended to them by the tenth Emergency Special Session. The recommendation of the tenth Special Session to the Government of Switzerland - a non-member State of the United Nations – does not release the High Contracting Parties from any of their individual obligations under Geneva IV, nor does it affect their above - mentioned commitment to convene a conference. The functions of the depository of a treaty are international in character. The depository is under an obligation to act impartially in its performance. The functions of a depository are administrative unless otherwise provided for in the treaty (Law of Treaties, Articles 76 and 77, which reflect customary law). Geneva IV does not provide for additional functions. In the event of any difference appearing between a State and a depository as to the latter's functions, the depository shall bring the question to the attention of the signatory States and the contracting States, or, where appropriate, of the competent organ of the international organization concerned.

[100] The 1949 IV Geneva Convention, *supra* note 77, Article 55 of Protocol I Additional to the Geneva Convention, 1977 states that:

"1. Care shall be taken in warfare to protect the natural environment against widespread, long-term and severe damage. This protection includes a prohibition of the use of methods or means of warfare which are intended or may be expected to cause such damage to the

prohibition of the use of methods or means of warfare, which are intended, or may be expected to cause the damage of natural environment and thereby to prejudice the health or survival of the population.[101] Wanton destruction of properties is considered a grave breach for which individual criminal responsibility can be attributed by virtue of Article 147 of the IV Geneva Convention.[102]

3.4.3 Palestine Sovereignty[103] over Transboundary Water Resources

The following discussion endeavours to provide an understanding of the competence of States in respect of their territory,[104] and what impact occupation has on the legal status of this competence. The section also includes a legal assessment of the role the UN played in confirming Palestine's permanent sovereignty over their natural resources. This assessment is important, given that Palestine is still under the effective occupation of Israel and many argue that international law does not apply in this particular case.[105] The objective is to assess the impact of belligerent occupation on Palestine's sovereignty over their water resources and to what extent this affects the application of international law in the case of transboundary groundwater.

---

natural environment and thereby to prejudice the health or survival of the population. 2. Attacks against the natural environment by way of reprisals are prohibited."

[101] *Ibid.*

[102] *Ibid,* Article 147 of the Geneva IV Convention *supra* note 77 states that:
"Grave breaches to which the preceding Article relates shall be those involving any of the following acts, if committed against persons or property protected by the present Convention: wilful killing, torture or inhuman treatment, including biological experiments, wilfully causing great suffering or serious injury to body or health, unlawful deportation or transfer or unlawful confinement of a protected person, compelling a protected person to serve in the forces of a hostile Power, or wilfully depriving a protected person of the rights of fair and regular trial prescribed in the present Convention, taking of hostages and extensive destruction and appropriation of property, not justified by military necessity and carried out unlawfully and wantonly".

[103] Sovereignty denotes the basic international legal status of a state that is not subject within its territorial jurisdiction to the governmental, executive, legislative, or judicial jurisdiction of a foreign state or to foreign law other than International Law. Sovereignty therefore implies the status of a state as a direct subject of International Law. For more details, see The Encyclopaedia of International Law, *supra* note 8, and I. Brownlie, *supra* note 6, at 105-106.

[104] The state territory and its appurtenances (airspace and territorial sea) together with the government and population within its frontiers comprise the physical manifestations of the primary type of international legal person, the State. *See* I. Brownlie; *supra* note 6 at 105-106.

[105] *See supra* section 2.3.2.

Usually, a State's competence is described in terms of sovereignty and jurisdiction.[106] The normal complement of State rights, the typical case of legal competence, is described commonly as "sovereignty": particular rights, or accumulation of rights quantitatively less than the norm, are referred to as "jurisdiction".[107] The exercise of sovereignty by a State is suspended when in the course of an *occupatio bellica* by enemy forces (Belligerent Occupation) its governmental institutions no longer operate or are subject to the orders of the occupying power. [108] Exiled governments may be capable of a continued exercise of their State's sovereignty depending on the effectiveness of such exercise in the specific circumstances.[109] Also according to the Hague Regulations, an occupant has limited authority over the occupied territory while sovereignty lies with the original people in the occupied territories:

"The authority of the State having in fact passed into the hands of the occupant, the latter shall do all in his power to restore and ensure, as far as possible, public order and safety, respecting at the same time, unless absolutely prevented, the laws in force in the territory".[110]

In the context of Palestine the UN[111] adopted many resolutions that confirm the right to self-determination as an inalienable right enjoyed by all peoples indiscriminately. This principle has been confirmed as such in Article 1(2) of the United Nations Charter[112] and in numerous other international documents.[113] The

---

[106] *Ibid.*

[107] *Ibid.*

[108] Encyclopaedia of International Law, *supra* note 8.

[109] *Ibid.*

[110] The 1907 Hague Regulation *supra* note 77, Article 34.

[111] According to the UN Charter the UN General Assembly is mandated to undertake all necessary studies and consider the general principles of cooperation in the maintenance of international peace and security and discuss any questions relating thereto and make recommendations. The Security Council is the organ responsible for taking all decisions to maintain or restore international peace and security. On the one hand the binding characters of the resolutions that are adopted by the General Assembly are highly debated and normally fail to be used as a confirmation of rules and principles that are internationally accepted as law. On the other hand, the resolutions of the General Assembly may be considered as evolution of an international custom or as expressing (and evidencing) a general principle of law. There is also a strong presumption that a law-declaring resolution such as "the right to self determination" is a correct statement of law if adopted with or without a negative vote or abstention. Resolutions with less than unanimous support are more questionable. The UN Charter entered into force on October 24 1945, there were 51 original members of the organization, the number is now 157 members, *found at* http://www.un.org/aboutun/charter/(*last visited* June 2003).

[112] *Ibid.*

[113] *Ibid,* Article 1(2) of the Charter states:

countless Resolutions passed by the UN General Assembly and the establishment of the UN "Committee on the Exercise of the Inalienable Rights of the Palestinian People" by the Assembly in 1975, are therefore clear evidences of the general recognition of this right in regard to the people of Palestine.[114]

Furthermore many General Assembly resolutions were adopted reaffirming the inalienable rights and permanent sovereignty of the Palestinian people over their natural resources, including land and water.[115] In its latest resolution of 2 December 2003,[116] the General Assembly reaffirmed the inalienable rights of the Palestinian people over their natural resources, including land and water; and called upon Israel, the occupying Power, not to exploit, to cause loss or depletion

---

"To develop friendly relations among nations based on respect for the principle of equal rights and self-determination of peoples, and to take other appropriate measures to strengthen universal peace".

[114] H. Koechler, The Palestine Problem In The Framework Of International Law Sovereignty as the Crucial Issue of a Peaceful Settlement of the Palestinian-Israeli Conflict, Madrid, 1999, found at http://www.library.cornell.edu/colldev/mideast/koechler.htm.(last visited June 2003).The relevant UN General Assembly and Security Council resolutions are: UN GA Resolution, A/RES/181(II) (A+B), 29 November 1947, Future Government of Palestine, (Termination of Mandate and Partition), found at http://www.un.org (last visited May 2003); UN GA/RES/194 (III), 11 December 1948, Progress Report of the UN Mediator (Establishment of the UN Conciliation Commission following the assassination of the UN Mediator, for which the Israeli government later paid the UN compensation) found at http://www.un.org (last visited May 2003); UN Sec Council Resolution S/RES/242 (1967), 22 November 1967, (recognizing the right to sovereignty of all states and requesting Israeli withdrawal from occupied territories) available at http://www.un.org (last visited March 2003); UN Sec Council Resolution S/RES/338 (1973), 22 October 1973, (cease fire and implementation of S/RES/242) found at http://www.un.org (last visited May 2003). The work of the United Nations on the question of Palestine is led by the Committee on the Exercise of the Inalienable Rights of the Palestinian People - UN GA Resolution A/RES/3376 (XXX), 10 November 1975, (establishment of the Committee on the Exercise of the Inalienable Rights of the Palestinian People), available at http://www.un.org (last visited June 2003). See the work of the Committee, found at http://www.un.org/Depts/dpa/qpalnew/committee.htm (last visited June 2003).

[115] United Nations General Assembly Resolutions 3175 (XXVIII) of 17 December 1973, 3336 (XXIX) of 17 December 1974, 3516 (XXX) of 15 December 1975, 31/186 of 21 December 1976, 32/161 of 19 December 1977, 34/136 of 14 December 1979, 35/110 of 5 December 1980 and 36/173 of 17 December 1981, 37/135 of 1982, 38/144 of 1983, 51/190 of 16 December 1996 and 52/207, December 18, 1997.

[116] UN Doc.A/C.2/58/L.36/Rev.1, December 2003 *found at* http://domino.un.org/unispal.nsf/0/bc75f0c2c306424e85256df70051eb58?OpenDocument *(last visited* May 2004).

of, or to endanger the natural resources in the occupied Palestinian territory, including Jerusalem, and in the occupied Syrian Golan Heights.[117] In this Resolution, the Assembly recognized the right of the Palestinian people to claim restitution[118] as a result of any exploitation, loss or depletion of, or danger to, their natural resources, and expressed the hope that the issue would be dealt with in the framework of the final status negotiations between the Palestinian and Israeli sides.[119] The Resolution also reaffirmed the need for an immediate resumption of negotiations within the Middle East peace process, on the basis of Security Council resolutions 242 (1967) of 22 November 1967, 338 (1973) of 22 October 1973 and 425 (1978) of 19 March 1978 and the principle of land for peace, and for the achievement of a final settlement on all issues.[120]

In addition to Resolutions, numerous United Nations reports were prepared by the Secretary General of the UN concerning the Israeli practices in the Palestinian Occupied Territories and its obligations under international law, in particular the rules embraced within the Hague Convention of 1907, and the Geneva Conventions of 1949.[121]

---

[117] *Ibid.*

[118] In the Chorzow Factory Case (Germany v. Poland) 1928, the Permanent Court of International Justice, on an issue involving expropriation, emphasized that reparation for an illegal act, "must as far as possible, wipe out all the consequences of the illegal act and re-establish the situation which would, in all probability, have existed if that act has not been committed." The Court emphasized the priority of restitution *(restitutio in integrum);* only if this is not possible, the obligation becomes that of paying the value of the property and compensation for such loss as result therefrom. Chorzow Factory Case, 1928, PCIJ, Ser. A, No. 17. The International Law Commission (ILC) has maintained these basic principles (reparation and restitution) in its *Draft Articles on State Responsibility* adopted in the second reading in 2000, A/CN.4/L.600, 11 August 2000 *found at* http://www.law.cam.ac.uk/rcil/ILCSR/ACN4L6~1.doc (*last visited* July 2003) Article 31 (1) states that :

"The responsible State is under an obligation to make full reparation for the injury caused by the internationally wrongful act".

Article 36 emphasizes on restitution by stating that:

"A State responsible for an internationally wrongful act is under an obligation to make restitution, that is, to re-establish the situation which existed before the wrongful act was committed, provided and to the extent that restitution: is not materially impossible; would not involve a burden out of all proportion to the benefit deriving from restitution instead of compensation".

[119] *Supra* note 109.

[120] *Ibid.*

[121] Illustrative examples: General Assembly Social and Economic Council, Documents A/40/381, June 1985, and A/55/84, June 2000.

3.5 CONCLUSIONS

The water conflict is a significant factor in the general politics of the area. It has constantly contributed to the creation of the conflict in the past, continues to amplify the present dispute, and is deemed to have bearings on any future political settlement. This Chapter demonstrates that the water conflict, which has many roots, has resulted in (i) the unequal level of development between the Palestinians and Israelis, (ii) the historic and current inequitable utilization of shared waters, and (iii) the absolute control by Israel of the shared resources. It is further concluded that what distinguishes the present water conflict is the hostility of the political situation, which appears to have favoured Israel, in addition to the lack of a common understanding and objectives on the available shared water resources. The emphasis by the Parties on their positions towards the water problem, rather than their interests in the shared groundwaters, continues to widen the gap between the Parties and delays the conclusion of any agreement in that regard.

A major conclusion from this chapter provides that although international law continues to apply under the belligerent occupancy, Israel appears not to have intended to employ these rules and principles to resolve the existing water conflict. On the issue of Palestine's permanent sovereignty over its natural resources, this research finds that while Palestine is not a "State" under international law, the sovereignty of the Palestinian people has not extinguished, it is simply not exercised. As the "belligerent occupier" under international law, Israel has rights and obligations for the natural resources under its control, but does not hold sovereignty over the Palestinian occupied territories. This finding is supported by the numerous resolutions adopted by the United Nations organs in favour of the Palestinian right to self determination and permanent sovereignty over their natural resources, which provide evidence (opinio juris) of States confirmation in that regard. The various legal instruments that were reviewed emphasize the Palestinian right to claim restitution -or if proven impossible- to reparation of the Palestinian water situation, which has to be part of the permanent status negotiations.

The Palestinians have realized that their only resort for resolving the water issue is by applying the applicable rules of international law. The Palestinian official position clearly states that they will abide to the principles of international law and the related UN Resolutions. The PLO Proclamation of Independence in 1988 presents a clear evidence of the Palestinians acceptance to rules and principles of international law which enhances their opportunities for formal recognition as a State. Israel's position on the other hand, declines to admit the co-relative rights of the Palestinians in the shared water resources and the applicability of international law in the case of the Palestinians. This position is built on the historically rooted ideology that water is a key factor to the existence of Israel, the control of which must be unilateral and absolute.

Despite the above position by Israel, and given the foregoing, this chapter recommends the Parties to consider the option of international law for solving the water conflict. This requires an investigation of the rules and principles of international that apply in the case of transboundary groundwater. The objective of such investigation is to gain a better understanding on the nature and status of development of this branch of international law, and to identify lessons that could be learnt for solving the current water conflict. This will be therefore discussed in the next two chapters.

4

Modern Development of Transboundary Groundwater Law

4.1 INTRODUCTION

Chapter Two provided an insight into groundwater's main features, behaviours and demonstrated the growing importance of this natural resource at the national and international levels. It was established that prerequisite to the effective regulation of transboundary groundwater is a good understanding of the characteristics and behaviour of such resources. Natural, social, economic and political factors were found to be highly relevant in hindering or allowing appropriate solutions as far as shared transboundary groundwaters are concerned. In light of these findings, Chapter Three looked into the water related politics associated with the conflict over the Mountain Aquifer that is shared between Israel and Palestine, and examined the impact of the water conflict on the future of the general Israeli-Palestinian conflict. According to the research undertaken, it was suggested that despite the political tension prevailing in the area of the case study, and in order to achieve the optimal and mutually beneficial utilization of the shared

Mountain Aquifer, the parties must consider the option of international law to resolve the conflict. This recommendation requires the parties to look into the feasibility of employing international law to resolve the current conflict. To that end, the parties are urged to investigate what lessons they could learn from the codification and progressive development of transboundary groundwater law and from State practice relevant to the utilization, development and protection of this resource.

This chapter and the next seek to identify the international law that applies to transboundary groundwater providing analyses of the sources of international law. State practice, which is an important source of international law, will be examined separately in Chapter Four. The purpose of this chapter is to identify the status of development of this branch of international law and its adequacy to meet the concerns and challenges related to the special case of groundwater, as described in Chapter Two. The methodology within this chapter is inspired by Article 38 of the Statute of the International Court of Justice (ICJ), which provides guidance on the sources of international law.[1] The methodology employs an analytical framework that is used to comparatively evaluate the contribution of the various sources of international law to the current development of international water resources.[2]

4.1.1 Sovereignty-Based Theories of Allocation and the Emergence of International Water Law

In their pursuit for their legal entitlements from international watercourses, States have traditionally appealed to four theories of allocation; each based on the concept of State sovereignty.[3] The first of these theories is "absolute territorial sovereignty", which provides that the jurisdiction of a nation within its own territory is necessarily exclusive and absolute and that a state may use the waters of an international river within its boundaries.[4] This theory was referred to as "Harmon Doctrine" after the U.S. Attorney General Harmon who promoted it in the United States-Mexico dispute over the Rio de

[1] Article 38 (1) of the Statute of International Court of Justice ICJ states, *supra* Introduction. The provisions of Article 59, judicial decisions and the teachings of the most highly qualified publicists of the various nations, as subsidiary means for the determination of rules of law, while Para. 2. states:
"This provision shall not prejudice the power of the Court to decide a case ex aequo et bono, if the parties agree thereto"
found at http://www.un.org/Overview/Statute/chapter2.html (*last visited* April 2003), *see also* Encyclopaedia of International Law, Volume IV, (Elsevier: the Netherlands, 2000) at 469 and 471, *see also generally* P. Malanczuk Modern Introduction to International Law, (Routledge: New York, 1997) at 39-41, *and* I. Brownlie, Principles of Public International Law (5th ed.),(Oxford University Press: New York, 1999) at 4.

[2] See *supra* Introduction section 8 "Analytical Framework".

[3] J. Lipper, *Equitable Utilization* in: The Law of International Drainage Basins 15, 20.

[4] *Ibid.*

Grande and therefore denied Mexico's right to the waters of this international river.[5] In Harmon's opinion, the diversion by the United States of the waters of the Rio Grande was justified despite the fact that Mexican farms were being ruined by this diversion.[6] His reply to the complaint made by Mexico in that regard asserted this attitude by justifying the United States actions based on the doctrine of absolute sovereignty over its national resources. This doctrine never gained acceptance and in fact was rejected eventually by the USA government itself.[7]

The second theory of allocation is the "absolute territorial integrity", which resembles the common law doctrine of riparian rights that was developed in the United States and United Kingdom and links water rights to land ownership.[8] It certainly favours downstream States who demand an undiminished flow in quality and quantity in their territory.[9] This doctrine does not allow upstream States to use the water within their territories in a manner that might affect the flow if the downstream States withhold their consent. It has been invoked in disputes between Pakistan and India, and between Canada and the United States, on the Indus and Columbia Rivers respectively.[10] However, as with the previous theory, this one is not widely accepted because it is too absolute in its application.

The third theory is the doctrine of "prior appropriation" or sometimes referred to as the "historical use", which asserts the proposition that "first in time - first in right". It tends to favour downstream States who have developed their use of the watercourse first. It is widely appealed to in arid and semi-arid countries where the demand for water greatly exceeds the supply. This doctrine is not well-received by the upstream States who find it unfair to give all the benefits of an international watercourse to those who are faster in developing the resource, like Egypt on the Nile and Israel on the Jordan River.

The fourth theory of allocation-the "limited territorial sovereignty" respects the equal and similar rights of all riparian states to the use of an international watercourse. The rights of each co-riparian state to the use of the water are limited and not absolute and therefore express reasonable entitlements. Some

[5] S. McCaffrey, the Harmon Doctrine after 100 years: Buried not Praised, 36 NRJ 540-590(1996), see also C.B. Bourne, Mediation, Conciliation and Adjudication in the Settlement of International Drainage Disputes, 9 CYIL 114-158 (1971).

[6] J. Lipper, *supra* note 3.

[7] D. J. Chenevert, Application of the Draft Articles on the Non-Navigational Uses of International Watercourses to the Water Disputes Involving the Nile River and the Jordan River, 6 Emory Int'l L. Rev. 495, 506 (1992).

[8] P.K. Wouters, Rivers of the World: the Fundamental Principles of International Watercourse Law, unpublished PhD thesis, 1997.

[9] *Ibid* at 111.

[10] S. McCaffrey, *supra* note 5.

commentators speak of another theory of "the community of interests".[11] This theory, derived from the "theory of the commons", focuses on the notion of a shared and common interest in the international watercourse which should be developed as an integrated whole, - irrespective of territorial boundaries.[12]

More recently, the foregoing "sovereignty-based" theories became less effective in resolving existing conflicts or avoiding potential ones, especially since the competition over the scarce water resources is continuously increasing. Accordingly, the need became more evident for a clear legal framework that outlines rights and obligations of States towards their shared waters. In response, the use and development of international watercourses has recently been given serious attention by international lawyers and legal experts.[13] The pace of that development has accelerated markedly during the last 50 years.[14] Additionally, numerous forums has responded to the growing concern of the international community for the development and proper management and legal frameworks governing international water resources. This has been reflected in the work of international governmental and non-governmental organisations (NGOs) and in the writing of scholars and publicists as means of focusing wide attention on the question of development and management of water resources.[15] The next discussion is an introductory guide to the assessment of what is claimed to be transboundary groundwater law. Assessing what is binding law and what is not relating to transboundary groundwater is essential for the purposes of this research. This will be investigated using a systematic approach that starts by understanding what international law is and how it works; what the sources of international law are? and how to identify the law governing transboundary groundwater using the relevant sources of international law.

4.1.2 International Law and its Sources

The term "international law" was first used by Jeremy Benthan in 1780 in his introduction to the Principles of Morals and Legislation.[16] Until the period between the

---

[11] *Ibid.*

[12] *Ibid.*

[13] See generally Smith, H.A., *The Economic Uses of International Rivers* (1931); Garretson, A.H., Hayton, R.D., & C.J. Olmstead (ed.), The Law of International Drainage Basins (1967) (GARRETSON); ; Bourne, C.B., *The Right to Utilize the Waters of International Rivers*, 3 CYIL 1965, 187-264.

[14] UN/Natural Resources/Water Series No. 1, Management of International Water Resources: Institutional and Legal Aspects (UN: New York, 1975).

[15] *Ibid.*

[16] P. Malanczuk, *supra* note 1 at 3-5.

two World Wars, writers found no difficulty in defining (public) international law which generally meant:
"The law that governs the relations between States amongst each other".[17]

After the World War II and with the emergence of actors[18] other than States on the international plane, the scope of international law became more complicated. The American Law Institute of the Foreign Relations Law of the United States has first responded to this development by restating what international law is and therefore defined it as:

"The rules and principles of general application dealing with the conduct of States and of international organizations and with their relations inter *se*, as well as with some other relations with persons, whether natural or judicial".[19]

In the context of relations between States and the assessment of their rights and duties towards each other under international law it is essential to understand what is considered a legally binding principle or rule in international law and what is not. This is referred to as the "normative content" of the law which is a necessary element to confer rights and obligations on States.[20] In order to facilitate the process of identifying what international law is, the ICJ Statute provided guidance for the assessment of its sources. These derive either from primary or subsidiary sources. [21] Primary sources are found in international treaties, international

---

[17] *Ibid.*

[18] *Ibid*, These actors include non-governmental organizations, the Holy See, transnational companies..etc. Some of these actors have acquired international legal personality or, at least, certain rights under international law, even if only granted by treaties concluded between States.

[19] Cited in P. Malanczuk *supra* note 1 at 3-5.

[20] International law governs the relations between and among States. It is traditionally defined as a system of principles and rules of general application dealing with the conduct of States and of international organisations and with their relations *inter se,* as well as with certain natural or juridical persons. States have an obligation to peacefully manage their transboundary resources, a duty reaffirmed under the UN Charter as one of the main principles of the United Nations. Rules of international law, which establish respective rights and obligations of States, can be found in written form (usually in the form of treaties or international agreements) or in unwritten form, such is the case with customary international law and general principles of international law. The most significant sources of international law are treaties and custom. It is important to appreciate that regardless of the source of the rule of international law – treaty law, customary law, general principles of law – the violation of any of these legal obligations can lead to a breach of international law, with consequences for the State in breach and the victim State. The law relating to state responsibility governs these consequences. *See* Brownlie, *supra* note 1 at 436, and the Charter of the United Nations, San Francisco, 26 June 1945, *found at:* www.icj-icj.org/icjwww/ibasicdocuments/ibasictext/ibasicunchart.htm (*last visited* June 2003).

[21] *See* Article 38 of the ICJ Statute, *supra* note 1.

customary law (State Practice), general principles of the law, and *jus cogens*,[22] while subsidiary means for the determination of the law [23] include the teachings of most highly qualified publicists[24] and judicial decisions.[25] Other possible sources of law include the acts of International Organizations,[26] soft law[27] and equity,[28]

Primary sources of international law comprise those sources that merit the greatest consideration in the creation of law.[29] However, it is critical to understand

---

[22] *See* P. Malanczuk, *supra* note 1 at 57.

[23] *Ibid*, paragraph (d).

[24] Although this factor was listed under the subsidiary means for the identification of the law some argue that highly qualified publicists can have a formative influence. Ian Brownlie argues that some publicists have a formative influence (see *supra* note 1)

[25] Article 38 of the ICJ Statute *supra* note 1, paragraph (c). In Borwonlie's view Article 59 came to rule out of a system of binding precedent. Judicial consistency in that context plays an important role in whether the court will consider precedent or not, *see* P. Malanzcuk *supra* note 1 at 51, and I. Brownlie, *supra* note 1 at 19-23).

[26] Many of these Resolutions are not necessarily conclusive evidence of customary law. However, in certain circumstances they may have some normative value (See ICJ Advisory Opinion on the Legality of the Threat or Use of Nuclear Weapons with reference to the series of General Assembly resolutions since 1961 which states that: "General Assembly Resolutions, even if they are not binding, may have some normative value. They can, in certain circumstances, provide evidence important for establishing the existence of a rule or the emergence of an opinion juris. To establish whether this is true given General Assembly resolutions it is necessary to look at its context and the conditions of its adoption; it is also necessary to see whether an *opinio juris* exists as to its normative character. Or a series of resolutions may show gradual evolution of the *opinio juris* for the establishment of a new rule." *See* Legality of Nuclear Weapons Case op. cit., at 826 paragraph 70.

[27] Soft Law are guidelines of conduct. They are neither binding nor irrelevant. Declarations, Statements Memorandums of Understanding can be cited as examples. See generally *supra* note 1.

[28] Equity can always be used by the judge to interpret or fill gaps in the law. In technical sense equity is not considered a source of international law, but it describes how equitable or just the implemented rule is (see Malanczuk, supra note 1 at 55). It is used in the sense of consideration of fairness, reasonableness (see Brownlie *supra* note 1 at 26). In the River Meuse Case between the Netherlands and Belgium, 1937, Judge Hudson recognized 'equity' as a principle of international law. He argued since Article 38(2) of the ICJ Statute allowed the application of general principles in international law, equity as a principle common to all national legal systems can be applied. See PCIJ, series A/B, no. 70, 76-77).

[29] In cases where customs and treaties provide no guidance, general principles of law can be used by the court as a guide. General principles of the law include those principles and rules that are applicable on the domestic practice in groundwater law and regulation should apply at transnational and international levels. *See* P. Malanczuk supra note 1, 63-67. *See also* M. Fitzmaurice and C. Flintermannn (eds.), L. Erades, *Interaction Between International Law and Municipal Law: A Comparative Case Study, 1993,* and Encyclopaedia of International Law, *supra* note 1, at 469 and 471, see also P. Malanczuk *supra* note 1 at 39-41, and I. Brownlie, *supra* note 1.

that when considering "custom" as a primary source of law, the first problem which arises is that of ascertaining the basis of the binding force of custom. [30] It is generally held that custom is composed of two elements: the first is factual (*usus*) and derives from a uniform and continuous repetition of the same behaviour, while the second is psychological (*opinio juris*) and lies in the conviction that such behaviour is legally necessary. The combination of two provides an evidence for the existence of a specific custom.[31] Subsidiary sources of the law supplement the efforts to determinate what the law is. [32]

4.1.3 What is Transboundary Groundwater Law

When investigating the primary sources of the international law governing transboundary groundwater, and considering the absence of a "multilateral law-making treaty", the analysis of the work of international organizations and experts groups in the field of codification and progressive development comprises a first step. This covers the modern development made by the International Law Institute (IDI),[33] the International Law Association (ILA),[34] the Bellagio Draft Treaty[35] and the International Law Commission (ILC)[36]. The outcome of all these organizations and groups can be classified under the

---

[30] See Encyclopaedia of International Law, *supra* note 1.

[31] *Ibid.*

[32] *Ibid,* at 469 and 471.

[33] The IDI (Institut De Droit International) was established in Ghent, Belgium in 1873 to develop and implement international law as a codified science. See generally www.britannica.com (*last visited* September 2003). The extensive work of the institute culminated in the adoption of the 1961 Salzburg Resolution. The Special Rapporteur for the drafting of the Resolution was Professor Juraj Andrassy. Salzburg Resolution on the Utilization of Non-maritime International Waters [Hereinafter referred to as the Salzburg Resolution], 11 September 1961, 49 Annuaire de l'Institute de Droit International, Tome II, 381 (1961) Salzburg Resolution, 49-II *AIDI* 1961, 381. The Resolution was passed by a vote of 50 to 0, with 1 abstention, see 49-II *AIDI* 1961, 162-192.

[34] The "Association for the Reform and Codification of the Law of Nations" (later renamed the International Law Association (ILA) was founded in Brussels in October 1873. Its objects include: "the study, elucidation and advancement of international law, public and private, the study of comparative law, the making of proposals for the solution of the conflicts of law and for the unification of law, and the furthering of international understanding and goodwill". This task is accomplished through committee work and resolutions adopted at the ILA's plenary session meetings. The ILA has a consultative status with a number of U.N. agencies. *See* generally www.ila-hq.org/html/layout_about.htm (*last visited* September 2003).

[35] R.D Hayton and A. E. Utton, *Transboundary Groundwaters: the Bellagio Draft Treaty,* 29NRJ (1989) [hereinafter referred to as the Bellagio Treaty].

[36] The International Law Commission [hereinafter called the ILC] was established by the General Assembly in 1947 to promote the progressive development of international law and its codification. More information *found at* http://www.un.org/law/ilc/introfra.htm (*last*

teachings of highly qualified publicists. Non-governmental institutions, it has to be mentioned, have no formal character and therefore their efforts on the codification of the law is often referred to as "restatement"; that is the unofficial codification.[37] The ILC, on the other hand, as an organ of the UN, has a more authoritative character. The ILC has a mandate to codify and progressively develop the law.[38] Consequently, the work of the ILC is reviewed with the view that this institution has a more formative influence than the other non-governmental organisations. It is proposed within this book to analyze the work of these institutions and groups as evidence of the law - giving more weight to the work of the ILC due to the explained authoritative institutional character.[39]

The next sections present the historical overview of the development of transboundary groundwater law and the main actors that played a role in its development.

---

visited January 2003). The Statute of the ILC identifies the two terms in Chapter II Article 15:
"progressive development of international law is used for convenience as meaning the preparation of draft conventions on subjects which have not yet been regulated by international law or in regard to which the law has not yet been sufficiently developed in the practice of States. Similarly, the expression "codification of international law" is used for convenience as meaning the more precise formulation and systematization of rules of international law in fields where there already has been extensive State practice, precedent and doctrine".
found at http://www.un.org/law/ilc/texts/statufra.htm (*last visited* January 2003).

[37] R.D. Hayton, *The Formation Of The Customary Rules Of International Drainage Basin Law*, in Garretson, Hayton and Olmstead, The Law of International Drainage Basins, (834 Oceana Publications; New York, 1967). The writing of qualified publicists only constitutes an evidence of law. In some subjects eminent writers had a formative influence on the progressive development of the Law. Brownlie emphasizes that National courts lean on secondary sources including the writings of publicists for their judgements, International courts on the other hand tend to use these opinions without much citations. He further mentions the work of the International Law Institute IDI and the ILC as having authoritative influence on the content of the law. This point is still the subject of heated debate among legal writers see Brownlie, *supra* note 1 at 25.

[38] *Ibid.*

[39] The Statute of the ILC identifies the two terms in Chapter II Article 15:
"progressive development of international law is used for convenience as meaning the preparation of draft conventions on subjects which have not yet been regulated by international law or in regard to which the law has not yet been sufficiently developed in the practice of States".
Similarly, the expression "codification of international law" is used for convenience as:
"meaning the more precise formulation and systematization of rules of international law in fields where there already has been extensive State practice, precedent and doctrine" ,*found at* http://www.un.org/law/ilc/texts/statufra.htm (*last visited* June 2003)

4.2 HISTORICAL EVOLUTION OF TRANSBOUNDARY GROUNDWATER LAW

4.2.1 Implicit Consideration of Transboundary Groundwater

Many writers and legal experts agree that transboundary groundwater law is considered far less developed than the law dealing with international surface waters.[40] The delay in its development has several reasons. Firstly, the insufficient knowledge of groundwater and the complexity of the hydrogeological regime pose great challenges regarding the assessment of the availability of the resources that cross international boundaries. Secondly, States have serious concern over their territorial sovereignty and they normally hesitate to accept limited sovereignty over shared groundwater especially when water rights are linked to land ownership. The latter fact contributes to the lack of comprehensive agreements concerning the transboundary groundwater resources.[41] Thirdly, in situations of acute shortages as surface water is threatened by pollution and depletion States are becoming increasingly precocious and heavily dependent on groundwater. This fact increases States' resistance and unwillingness to negotiate solutions for their shared groundwaters.[42]

Prior to 1966 groundwater was dealt with as secondary concern of surface water study, and in this respect, there is little evidence of international aquifers being treated as a separate area of study.[43] In the early to late 1950's, institutions such as the IDI and the ILA began to consider the law regarding non-navigational uses of international watercourses. Many declarations and principles have been adopted for the benefit of international water resources optimisation, management and regulation.[44] In 1961, the IDI adopted the Salzburg Resolution regarding the utilization of non-maritime international waters without directly mentioning groundwater.[45] The Resolution is presented in the form of recommendations confirming some general rules of international law.

---

[40] S. McCaffrey, *International Groundwater Law: Evolution and Context*, found in: Groundwater: Legal and Policy Perspective, (S. Salman (ed.), World Bank Technical Paper No. 456: Washington D.C., 1999).

[41] *Ibid.*

[42] *Ibid.*

[43] R.D. Hayton, *The Law of International Aquifers,* 22 NRJ 71-92 (1982).

[44] *Ibid.*

[45] Salzburg Resolutions, *supra* note 33.

4.2.2 Explicit Consideration of Transboundary Groundwater

The year 1966 presents a remarkable shift in the manner groundwater was dealt with on the international plane. As an output of their extensive work, the ILA [46] adopted a set of rules concerning the Uses of the Waters of International Rivers known as the Helsinki Rules which include connected groundwater within their scope.[47] In the context of the work of the ILC, the instruments that were reviewed are: the 1991[48] and 1994[49] sets of

---

[46] The entire ILA record on international water resources has been collected and republished by the Finnish Branch of the ILA in one volume, including commentaries and beginning with the Helsinki Rules. See E. Manner and V. Metsalampi, the *work of International Law Association on the Law of International Water Resources* 257-97. For the work on the Law of International Groundwater see *supra* note 28. The ILA at its Forty-Sixth Conference held in Edinburgh in 1954 brought up the topic of "River Law". Clyde Eagleton invited the ILA to "undertake a study concerning the development and uses of waters in international rivers". The Association endorsed this idea by a resolution adopted at the same Conference and appointed C. Eagleton President of the Committee created to study the topic. Participation in the ILA's meetings included leading experts in the field, representatives of, and legal counsel for, most of the disputing and other watercourse States. This combination provided for lively debate, if not outright confrontation. Nonetheless, the Committee succeeded in producing a series of reports which, together, led to the text presented at Helsinki. See C. Bourne, *The ILA Contribution to International Water Resources Law,* 36 NRJ 155-216(1996). *See also S.* Bogdanovic, International Law of Water Resources: Contribution of the International Law Association1954-2000 (Kluwer Law International : Dordrecht, the Netherlands, 2001).

[47] International Association of the 52nd Conference xi, 447-533 (1967), *Found at* http://www.internationalwaterlaw.org/IntlDocs/Helsinki_Rules.htm.(*last visited* In June 1999 the Water Resources Committee at Campione d'Italia prepared a revised set of articles as a consolidation for the Helsinki Rules under the chairmanship of Professor Charles Bourne. These articles comprise the rules on international water resources as adopted by the International Law Association between 1966 and 1999. The Water Resources Committee is now considering revision of these rules to reflect developments that have occurred since the original rules were adopted. The major revisions were based on emerging treaty practice in the maintenance of adequate stream flow as an obligatory requirement consistent with the requirement of equitable and reasonable utilization.

[48] Report of ILC on the Work of its Forty third Session, U.N. Doc. A/46/10 (1991), reprinted in 2(2) YILC 66-70(1991) [hereinafter referred to as the 1991 Draft]. Part I embraced the introduction, which reflected the scope of the articles, use of terms, and watercourse agreements. Part II incorporated the general principles namely equitable and reasonable utilization and participation, the obligation not to cause appreciable harm, the general obligation to cooperate, regular exchange of data and information and the relationship between uses. Part III dealt with planned measures encompassing the obligations for information, notification and consultation as far as planned measures are concerned. Part IV handled the issues of protection and preservation of the ecosystems, prevention and control of pollution. Part V harmful conditions and emergency situations and finally Part VI included some miscellaneous provisions on management, regulation,

draft articles, which treat connected transboundary groundwater equally as surface water and the "Resolution on Confined Transboundary Groundwaters" of 1994,[50] which provides that the principles applying to international watercourses might be as well applied to transboundary confined groundwater. The commentary to 1994 draft indicates the extent to which the different articles reflect pre-existing customary law and agreed views of States. [51] The legal articles of the ILC were the outcome of many actors including: (i) the special Rapporteurs and their main products; (ii) States contribution[52] in the form of replies[53] and comments[54] to the ILC's work; and (iii) the Sixth Committee (Legal) of the UNGA.[55]

---

installations, international watercourses and installations in times of armed conflicts, data and information vital to national defence or security and non-discrimination.

[49] Report of the ILC to the Secretary General of the UN on the Law of Non-Navigational Uses of International Watercourses (1994), GAOR, 49th session, Supp 10 UN Doc A/49/10, 179, YILC 1994, vII (2). [Hereinafter referred to as the 1994 Draft] *also found at* http://www.un.org/law/ilc/texts/pagemode=bookmarks *(last visited* July 2003), in the Commentary on the 1994 Draft the Commission has referred to international court cases such as the Corfu Channel, the Lake Lannoux and the Trail Smelter Cases, Other conferences and declarations were important references in their work. The valuable work of some remarkable private institutions even not explicitly referred to can be traced throughout the draft articles. These institutions include the ILA namely the 1966 Helsinki Rules *supra* note 47 and the Salzburg Resolution *supra* note 11.)

[50] *Ibid.*

[51] To that end the ILC appointed highly respected scholars as Special Rapporteurs. The Five Special Rapporteurs that were appointed delivered 13 reports. Richard Kearny (1974-1976, 1 report); Stephen M. Schwebel (1977-1981, 3 reports); Jens Evensen (1982-1984, 2 report); Stephen McCaffrey (1985-1991, 3 reports) and Robert Rosenstock (1992-1994, 2 reports).

[52] Robert Hayton explains that every declaration made by a foreign office, every decision of a municipal tribunal, every vote taken in the legislature is, of course " state conduct", some of which may be cognizable at the international level. He further explains that the real behaviour of a state in the actual conduct of its international relations is distinguished from what they say they did, say they are doing or say they will do. He adds that there is not necessarily a great gap between the actions a nations says it pursues or will pursue as a matter of policy position and what it in fact does concretely vis-à-vis other states. But there can be and often it has been consequential difference. Only if its is the customary rule in the international community not to take into account a nation's real deeds, and only to look to the official representations about state conduct, can state performance be ignored in the study of formation of rules of international law, which all agree are the outgrowth of state practice. See R. D. Hayton, *supra* note 37.

[53] This questionnaire was prepared by the Commission on the topic of non-navigational uses of international watercourses and was submitted to governments for their comments. Particular emphasis was given to the Scope and matter of priority for regulation. ILC, *Replies of Governments to the Commission's Questionnaire*, (U.N. Doc. A/CN.4/294 and Add.1 of 1 April 1976, U.N. Doc. A/CN/314 of 23 June 1978, U.N. Doc. A/CN.4/324 of 13

The work of the ILC has in 1997 finally culminated in the adoption of the "Convention on the Law on Non-Navigational Uses of International Watercourses".[56] Although the 1997 UN Watercourses Convention has been developed with the view to present this type of "law-making treaty", it is unfortunately not yet in force.[57]

4.2.3 Special Consideration of Transboundary Groundwater

After extensive work for 20 years, the ILA adopted the Seoul Rules in 1986[58] which provide guidelines for the application of the 1966 Helsinki Rules for international groundwater aquifers.[59] These Rules play a complementary role to Helsinki Rules and augment their scope by including transnational groundwater not necessarily related to the

---

July 1979, U.N. Doc. A/CN.4/329 and Add. 1 of 10 March and 3 July 1980, U.N. Doc. A/CN.4/352 and Add. 1 of 18 February and 28 June 1982)

[54] Comments and Observations Received from States on the Draft Articles on the Law of Non-Navigational Uses of International Watercourses of 1991 (Doc. A/CN.4/447 0f 3 March 1993, Doc. A/CN.4/447/Add.1of 15 April 1993, Doc. A/CN.4/447/Add.2 of 18 May 1993, Doc. A/CN.4/447/Add.3 of 14 June 1993) and Comments and Observations Received from States on the Draft Articles on the Law of Non-Navigational Uses of International Watercourses and Resolution on Confined Transboundary Groundwater of 1994 (Doc. A/51/275 of 6 August 1996, Doc. A/51/275/Add.1 of 12 September 1996, Doc. A/51/257/Corr.1 of 4 October 1996, Doc. A/51/275/Add.2 of 14 October 1996, Doc. A/51/275/Add.3 of 14 October 1996).

[55] After the adoption of the draft articles in 1994 the UN General Assembly decided that this text should be considered by the Sixth (Legal) Committee of the UN, convened as a Working Group (WG) of the Whole, with a view to finalizing it in the form of a multilateral treaty. The first session was held in New York for a period of 3 weeks from 7-25 October 1996 (GA Res. 49/52) and the second session held also in New York between "24 March to 4 April 1997" (GA Res. 51/206). *See also* UNGA Res. 49/52, U.N. Doc. A/RES/49/52

[56] UN Watercourses Convention on the Law of Non-navigational Uses of International Watercourses, May 21, 1997, UNGA Res. 51/229, reprinted at 36 ILM 700 (1997) [Hereinafter 1997 UN Watercourses Convention]. A majority of States voted in favour (103) which indicates that the rules embodied in the Convention were acceptable; only three States voted against (Burundi, China and Turkey) and 26 absented: Andorra, Argentina, Azerbaijan, Belgium, Bolivia, Bulgaria, Columbia, Cuba, Ecuador, Egypt, Ethiopia, France, Ghana, Guatemala, India, Mali, Monaco. Mongolia, Pakistan, Panama, Paraguay, Peru, Rwanda, Spain, Tanzania, and Uzbekistan. Article 36 of the Convention stipulates that it would require 35 instruments of ratification, acceptance, approval or accession so that the Convention enters into force.

[57] *Ibid.*

[58] Seoul Rules on International Groundwater, 1986 (Adopted by the ILA at the 62nd Conference Held at Seoul in 1986).

[59] S. Bogdanovic, International Law of Water Resources, (Kluwer Law International, 2001).

surface waters of the drainage basin.[60] Despite the preceding efforts in addressing the subject of transboundary groundwater, an expert group confirmed that there was still a gap in comprehensive treatment of the issues concerned.[61] The so called "Bellagio Draft Treaty" of 1989 is the product of scholarly efforts that attempted to respond to the lack of legal and institutional arrangements, and the weakness of the few existing institutions dealing with international aquifers.[62] The product is considered a recommendation for a framework treaty that can be modified to respond to the particular requirements of the shared groundwater resources. It embraces the essential procedural mechanisms for implementing the equitable and reasonable utilization, development and protection of groundwaters in the border region.[63]

The "Resolution on Confined Transboundary Aquifers", which was adopted by the ILC in conjunction with the 1994 Draft Articles, includes confined aquifers within the scope of its application. This Resolution was not adopted as part of the final text of the UN Watercourses Convention. In May 2003, the ILC returned to the question of transboundary confined aquifers. Chusei Yamada, the special Rapporteur, in his first report on "Shared Natural Resources" emphasized that in order to formulate rules regulating confined transboundary groundwaters, an inventory of these resources worldwide and a breakdown of the different regional characteristics of the resources was needed. National, regional and international organizations are currently studying and assessing such major aquifer systems was necessary.[64] He further emphasized the need for a multitude of studies to be conducted which include;

(a) Socio-economic importance: groundwater is becoming increasingly important for all populations, but particularly for the populations of the developing world. The development aspects of groundwater are being extensively studied by the World Bank Groundwater Management Advisory Team;

(b) The practice of States with respect to use and management;

(c) Contamination: causes and activities which adversely affect the resources as well as its prevention and remedial measures;

(d) Cases of conflicts;

---

[60] *Ibid.*

[61] The Bellagio Draft Treaty, *supra* note 35.

[62] *Ibid.*

[63] *Ibid*, Art. II (1) states:

"The Parties recognize their common interest and responsibility ensuring *the* reasonable and equitable development and management of groundwaters in the border region for the well being of their peoples".

[64] UN General Assembly: ILC Fifty Fifth Session, *Shared Natural Resources: First Report on Outlines*, First Report by the Special Rapporteur Chusei Yamada, UN Doc. A/CN.4/533, May 2003.

(e) Legal aspects: existing domestic legislation and international agreements for management of the resources; and

(f) Bibliography of materials of direct relevance to the work of the ILC.[65]

The next sections examine the relevance of the codification and progressive development undertaken by the various institutions mentioned in the above discussion

4.3 SCOPE AND USE OF TERMS

4.3.1 Introduction

The "scope and use of terms" is critical a element in determining "what" waters are included and "which" areas are to be regulated in any legal arrangement. In the context of this research the key issues that need to be investigated are (i) whether or not including transboundary groundwater within the scope of an international watercourse agreement is considered generally an accepted practice by States, and (ii) whether or not such waters include connected as well as confined transboundary groundwaters. The answer to these two main questions will be sought in the work of the ILC, IDI, ILA and the Bellagio Draft Treaty.

4.3.2 The work of the ILC

4.3.2.1 Background

The debate on the "scope and use of terms" within the ILC was one of the most contentious issues during the process of drafting. In 1980 the ILC after several years of deliberation, reported several tentatively approved articles to the GA adopting a "system approach" and a working hypothesis or Note of Understanding as to what was meant by the term "international watercourse system". In that Note of Understanding "groundwater" was for the first time expressly listed as one of the hydrographic components of any watercourse system.[66] Prior to the adoption of the 1991 Draft Articles[67] on the first reading, the fourth special Rapporteur Stephen McCaffrey recommended that a decision must be made concerning the scope of the term "international watercourse".[68] The basic rules at that time were clear and the task was to

---

[65] *Ibid.*

[66] See Text and narrative report in II YILC 110-136 (1980), UN Doc. A/35/10. At that stage the term 'international watercourse' was not endorsed and it was again deferred until a later stage on the topic.

[67] The 1991 Draft *supra* note 48.

[68] *Ibid.*

determine the scope of their application.[69] The ILC investigated the appropriateness of using the "Drainage Basin" or the "International Watercourse" as terms to define scope. After lengthy discussions on this matter, the ILC concluded that the former concept was problematic and could result in resistance from States to its acceptance.[70] As McCaffrey explains:

"the reason behind ILC's decision to reject the "Drainage Basin" is States' resistance to accept it. In the view of certain governments and Commission members this concept was considered to include land territory as well as watercourses."[71]

In 1991, the members of the ILC agreed to define a watercourse as a "system" of water. The essence of this definition in their view was the interdependence of its different parts, which made the system a unitary whole.[72] On the other hand, given the multitude of terms used to refer to an international water resource, the ILC established that the "watercourse" or "river system" are not novel terms- affirming that these expressions have long been used in international agreements to refer to a river, its tributaries and related canals.[73]

---

[69] Seventh Report on the Law of Non-navigational Uses of International Watercourse, by S. McCaffrey, special Rapporteur, UN Document A/CN.4/436, March 1991. The report included two alternatives on the use of terms: Alternative 1: "a watercourse system: is a system of waters composed of hydrographic components including rivers, lakes, groundwater, and canals, constituting by virtue of their physical relationship a unitary whole. An international watercourse system: is a watercourse system parts of which are situated in different States. A watercourse System State: is a State in whose territory part of an international watercourse system is situated". Alternative 2:
"a watercourse is a system of waters composed of hydrographic components including rivers, lakes, groundwater, and canals, constituting by virtue of their physical relationship a unitary whole. An international watercourse is a watercourse system parts of which are situated in different States. A watercourse State is a State in whose territory part of an international watercourse system is situated".

[70] Many States disclosed their position concerning the extension of the distinctive status of the frontier water to the whole drainage basin or the full length of the current as an unjustified encroachment for their sovereignty. Among the countries that accepted the 'Drainage Basin' concept are: USA, Argentina, Ecuador, Finland, Indonesia, Netherlands, Greece, and Luxemburg, Niger. On the other hand many countries rejected this concept like France, Spain, Colombia, Brazil, Germany, Libyan Arab Jamahiriya, Bangladesh. (See Documents A/CN.4/294 1976 and A/CN.4/447).

[71] Seventh Report by Stephen McCaffrey, *supra* note 69. Although McCaffrey further explains that the decision was taken notwithstanding the fact that, it is almost impossible to exclude totally actions on land from the scope of the draft (except to the extent that they would have no effect, through an international watercourse, upon another watercourse State.

[72] Comments and Observations by Member States, *see generally supra* note 54.

[73] 1991 Draft, *supra* note 48.

4.3.2.2 States' Replies to the ILC Questionnaire

The States' replies to the ILC questionnaire are useful to look at.[74] Only 32 Member States[75] responded to the request by the ILC and most of these replies were scanty and inadequate. Countries like Finland the Netherlands,[76] Spain,[77] Poland,[78] and Hungary [79] submitted comprehensive replies covering most of the pressing issues relevant to the topic of non-navigational uses of international watercourses. In most of the replies the concept of "hydrological unity" was emphasized implying the need to include groundwater within the watercourse term. In the opinion of the Netherlands, for instance, the need to include brackish water in the definition of watercourses and not to limit them to fresh water resources was raised. The Netherlands drew the attention to the significance of groundwater not only as to how the use of such waters might affect the quantity and quality of the watercourse, but also to the situation where groundwater shows characteristics different from the surface water and above all when it is not even connected with it.[80] As for the use of the "drainage basin" concept, State opinions were divided. Some states were in favour of enlarging the geographical scope of the articles,[81] while others stressed a narrower approach in the definition of scope and use of terms. Poland for example recommended the inclusion of frontier subsoil waters and those subsoil waters which successively cut across the territories of two or more States.[82] Spain was very broad in its reply and confirmed that the term "basin" should not only encompass surface water but also underground and atmospheric water, as well as frozen resources, thus covering the concept of international water resources system. Spain has also made its reservation regarding the use of the "drainage basin" term, characterizing it as being vague. It confirmed that the traditional term "watercourses" was more appropriate for the purposes of the study.[83] Spain finally recommended that the ILC should as well prepare draft articles relating to the utilization of confined groundwater in cases such waters cross boundaries.[84] Greece, on the other hand, welcomed the use of the

[74] ILC Questionnaire, *supra* note 53.

[75] As of 15 June 1976, 32 Member States have submitted replies to the questionnaire formulated by the Commission in 1974: Argentina, Austria, Bangladesh, Barbados, Brazil, Canada, Colombia, Ecuador, Finland, France, Federal Republic of Germany, Greece, Hungry, Indonesia, Libyan Arab Jamahiriya, Luxemburg, Netherlands, Nicaragua, Niger, Pakistan, Philippines, Poland, Portugal, Spain, Sudan, Swaziland, Sweden, Syrian Arab Republic, United States of America, Venezuela, Yemen and Yugoslavia.

[76] U.N. Doc. A/CN.4/294 and Add.1 of 1 April 1976, Reply dated 12 April 1976, p.157.

[77] *Ibid*, Reply dated 22 Sep. 1975 p. 159-160.

[78] *Ibid*, Reply dated August 27 1975, p. 104.

[79] *Ibid*, Reply dated 14 July 1975.

[80] *Ibid*, Reply dated 12 April 1976.

[81] Nordic Countries are Iceland, Finland, Denmark, Norway

[82] *Supra* note 54, Reply dated 27 August 1975.

[83] *Ibid*, Reply dated 22 Sep. 1975.

[84] Spanish reply to the questionnaire, 22 Sep. 1975 Document A/CN.4/294.

"drainage basin" concept, particularly because this new concept embraced the tributaries of the international river and groundwater.[85] Among the countries that accepted the "drainage basin" concept were: USA, Argentina, Ecuador, Finland, Indonesia, Netherlands, Greece, Luxemburg and Niger. On the other hand, many countries like Spain, Colombia, Brazil, Germany, Libya and Bangladesh rejected this concept.[86]

4.3.2.3 The 1991 Draft Articles

The definition of a "watercourse"[87] in the 1991 Draft covers only connected/related groundwaters reflecting the physical unity approach and the need to regulate these resources in one set of regulations.[88] The provisional text of the 1991 Draft Articles included the notion of "common terminus" concept which limits the scope of the Articles to connected surface and groundwater ultimately flowing into the same terminus.[89] The 1991 draft Articles apply to uses of international watercourses and of their waters for the purposes other than navigation and to conservation measures related to the uses of those watercourses and their waters.[90] Navigation is not within the scope of the articles except in cases where other uses affect or are affected by navigation.[91]

---

[85] UN Document A/CN.4/329, *Replies of Governments to Questionnaire*, Reply of Greece dated 8 June 1979.

[86] *Ibid*, Hungary for example explained that:
"there is no geographical term that can be applied to the description of all of the legal relations relating to the water or drainage basin which are on the territory of more than one State. The terms "International Rivers" or "drainage basins" or "frontier waters" are all geographically exactly determined and well explainable. The question is what term is suitable to the regulation of certain relations".

[87] *Ibid.*

[88] However Charles Bourne confirms that physical unity cannot explain why and which legal norms and principles of international law lead to the transformation of the physical unity into legal unity. In the Lake Lannoux Arbitration Tribunal stated that:
"the physical unity of the river is not a basis for recognizing the existence if a legal unity between States."
See Lac Lannoux Award, 12 RIAA 1957 (Award, 16 November 1957).

[89] 1991 Draft Articles *supra* note 48, Article 2 states that:
"A watercourse was defined to mean a system of surface and underground waters constituting by virtue of their physical relationship a unitary whole and flowing into a common terminus". See discussion in chapter 2 on the 'common terminus' concept.

[90] 1991 Draft, *supra* note 48, Article 1 states that:
"the present articles apply to uses of international watercourses and of their waters for purposes other than navigation and to measures of conservation related to the uses of those watercourses and their waters".

[91] *Ibid*, paragraph 2 of Article 1. From 1976 to 1991 attention within the Commission was devoted to the formulation of the general principles applicable to legal aspects of the uses of these resources. It wasn't until early 1991 that the Commission considered the need for deciding upon the 'international watercourse' term. *See* S. McCaffrey *supra* note 70.

In the consideration of non-renewable transboundary confined groundwater, Robert Rosenstock, the Special Rapporteur who succeeded McCaffrey[92] in 1992, initiated a new discussion on the issue of groundwater in 1993. His first report[93] was in favour of including confined groundwaters within the scope of the new articles. He was convinced that the principles and norms applicable to surface water and related groundwaters were equally applicable to confined groundwaters. In his view, the changes required in the Draft to achieve this wider scope were relatively few and uncomplicated and he prepared such changes as required to the Draft Articles. Some of the members agreed with Rosenstock's proposal to include confined groundwaters in the scope, while many did not see how "confined" groundwaters could be envisaged as part of a system of water that constituted a unitary whole. In their view, little was known about the confined groundwaters, and they agreed that the law relating to confined groundwater was more analogous to that governing the exploitation of mineral natural resources, especially oil and gas.[94]

4.3.2.4 States Comments and Observations to the 1991 Draft Articles

In addition to the States' replies to the questionnaire, a review of the comments and observations of States to the 1991 of the Law of non-navigational uses also reveals important insights.[95] In response to the 1991 Draft Articles, some States disclosed their stance. The United Kingdom, [96] Germany[97] and Greece[98] commended the fact that connected groundwaters of a system have been included within the scope of the Draft. To the United Kingdom, for example, where 40% of the abstracted water for irrigation comes from groundwater, it was important to support such approach. According to its opinion, this application reflected a scientific and geographic reality that the ILC should

---

[92] Within the context of the 1991 Draft McCaffrey raised some questions to be considered by the Commission: (i) whether groundwater should be included within the definition of the watercourse; and if so (ii) whether the draft articles should apply both to groundwater related to surface water ("free" groundwater) and to groundwater confined to surface water ("confined" groundwater), or (iii) whether they should apply only to "free" groundwater. Seventh Report by Stephen McCaffrey, *supra* note 69.

[93] First Report of the Special Rapporteur, Robert Rosenstock (45th session of the Commission, 1993), U.N. Doc. A/CN.4/451 Commission report A/48/10.

[94] See YILC, 1993, vol. I: Summary Records of the 2309th, 2111th-2316th and 2322nd meetings; and YILC, 1994, vol. I, Summary Records of the 2334th-2339th, 2353rd-2354th and 2356th meetings.

[95] 1994 Draft Articles, *supra* note 49.

[96] U.N. Doc. A/CN.4/447. Comments and Observations by the United Kingdom of Britain and Northern Ireland dated 15 January 1993.

[97] *Ibid*, Comments and observations by the Federal Republic of Germany dated 11 Jan. 1993.

[98] *Ibid*, Comments and Observations by Greece dated 3 Feb. 1993.

be commended for.[99] Germany in its comments to the 1991 Draft Articles affirmed the importance of including groundwater, except for "confined aquifers" within their scope. According to Germany, the need to do that so is apparent when one bears in mind the fact that connected groundwaters are part of the hydrological cycle that feed the watercourse. Germany proceeded by stating that the exclusion of "confined aquifers" was also wise, as they have no physical relation to surface water and thus does not form part of a whole in need of protection. Contrary to the foregoing positions, the Nordic countries emphasized that it did not follow that aquifers without any physical relationship with surface waters- i.e. transboundary confined groundwater aquifers- must be left outside the scope of the Draft.[100] If this was done then these countries proposed that the question of "confined aquifers" be considered for separate Draft Rules. Spain also recommended preparing Draft Articles relating to the utilization of confined groundwater in cases where such waters cross borders.[101]

In assessing the impact of States on the final definition of the ILC on the scope of the 1991 draft articles, Portugal[102] appears to have influenced the modification of the definition of the "watercourse system". Its comments to the 1991 draft articles included a suggestion for the addition of the word "normally" before the phrase "flowing into a common terminus". This was explained to minimize the flaws of the expression "international watercourse".[103] This suggestion has been reflected in the changes made to the 1991 Draft as will be addressed in the next discussion.

4.3.2.5 The 1994 Draft Articles

The scope and the definitions of terms of the 1994 draft articles as compared with the 1991 draft were changed. The word "normally" was added before the phrase "flowing into a common terminus". [104] This addition is quite controversial and implies that the requirement of common terminus can be overlooked in certain circumstances.[105] This

---

[99] *Ibid.*

[100] *Ibid*, Comments and Observations by the Nordic Countries dated 19 Dec. 1993 (these countries are Iceland, Finland, Denmark, Norway and Sweden).

[101] *Ibid*, Comments and Observations by Spain dated 27 Jan. 1993.

[102] Portugal is one of the countries in Europe whose supply is almost completely dependent on groundwater. Portugal has many aquifers shared with Spain.

[103] U.N. Doc.A/CN.4/447.

[104] 1994 Draft, *supra* note 49, Article2 paragraph 2 states that:

"a watercourse means a system of waters and groundwaters constituting by virtue of their physical relationship a unitary whole and normally flow into a common terminus".

[105] G. Eckstein, *Hydrologic Reality: International Water Law and Transboundary Ground-Water Resources* October 1998, *found at* http://www.internationalwaterlaw.org/Articles/GlobalSouth.htm. (*last visited* June 2003). Eckstein argues that the common terminus requirement does not have strong foundations especially in the case of transboundary groundwater Due to certain geological features; groundwater may

change was a compromise between two extreme opinions: one favouring the common terminus qualification in order to narrow the scope of the Draft, and, the other proposing the deletion of the entire phrase "flowing into a common terminus", to maximize the amount of waters that are part of an international watercourse system.[106] In effect, the scope of application of the 1994 draft articles is the same as those of the 1991 draft with some additions related to measures aimed at solving watercourse problems, a matter that was addressed in the 1994 Draft. [107] The draft Resolution on Transboundary Confined Groundwater recognized such water "a natural resource of vital importance for sustaining life, health and the integrity of the ecosystem". Thus, it was important to continue efforts to elaborate rules pertaining to them. The Resolution recommended that the principles contained in the 1994 Draft Articles could be applied equally to transboundary confined groundwater. The Resolution further recommends that States be guided by these principles and proposes that they enter into agreements regarding those shared confined transboundary groundwater.[108]

4.3.2.6 States Comments and Observations to the 1994 Draft Articles

Within the States' responses to the 1994 Draft Articles, many government representatives supported the adoption of the "watercourses" term making it clear that the use of all components of this system should be covered by the same legal regime.[109] A wide majority of the States indicated the need to include connected groundwater in the scope of the definition and commended the fact that confined groundwater was not included.[110] Furthermore, most of the countries emphasized their disagreement regarding inclusion of the transboundary confined groundwaters within the scope of the articles.[111] Some States even suggested that groundwater should be wholly deleted from the scope of the articles.[112] Hungary, on the contrary, was happy with the fact that groundwater was

---

occasionally flow in a direction different from, or have an outlet different from surface waters of the same area. The fact that such related water resources flow toward disparate termini does not negate the possibility of transboundary consequences.

106 The Seventh Report of McCaffrey, *supra* note 69.

107 These measures include: flood control, erosion sedimentation and salt intrusion in addition to measures taken to deal with degradation of water quality and quantity. *See* the 1994 Draft, *supra* note 49, commentary to article 1.

108 2 YILC 135, (1997).

109 *Ibid.*

110 *Ibid.*

111 Comments and Observations of States, *supra* note 54, in Guatemala's Comments and Observations dated 28 June 1996 they stated that

"it is the opinion of Guatemala that groundwater forming part of the Drainage Basin is that which contributed to the main river, stream or lake or other terminus".

112 *Ibid*, Colombia in its Comments and Observations dated 10 July 1996 has suggested to delete groundwater from the definition of the Watercourse claiming that this matter is under

included and referred to its overriding interest in a well-founded regime of an equitable and reasonable utilization of international watercourses for navigational and non-navigational uses.[113] Ethiopia, on the other hand, rejected the definition of "watercourse", which embraces all components of the hydrographic basin. They considered this as an unjustified encroachment on the State Sovereignty.[114]

4.3.2.7 The 1997 Watercourses Convention

Within the two sessions of the Sixth Committee (Legal) the use of terms was not one of the issues that were highly contested. However, in disclosing their statements for not adopting the Convention some States confirmed that part of their objection was related to the use of terms and scope, namely the definitions in article 2(a) and (b).[115] Pakistan, for example, objected to the use of the term "groundwaters" in the definition, which, in its view, was impossible to measure in precise terms as surface water. [116] It further emphasized that different laws should be applied in the case of groundwater.[117] Rwanda as well had objections to Article 2 for the same reasons. In its opinion the inclusion of underground waters within this article transforms the Convention into a flawed agreement.[118] Moreover, in the context of scope and the applicability of the Convention, most States encouraged the idea of a "Framework Convention". Accordingly in the final text of the 1997 Watercourses Convention the ILC adopts the "watercourse" term defined in the same manner as in the 1994 Draft Articles. This term was collectively viewed to respond to the needs of optimal and sustainable utilization of international watercourse, especially employed in combination with the "system" term. The Resolution on Confined Transboundary Groundwater was not adopted in conjunction with the UN Watercourses Convention which excludes such waters from the regulation by this instrument. It is not clear why this Resolution was dropped out but one logical explanation for not doing so is States' resistance to adopt it.

---

the exclusive jurisdiction of the State which exercises sovereignty over the subsoil and is therefore not subject to international regulation. Sudan also disclosed its comments which provide that :

"the inclusion of "groundwaters" in the definition of an international watercourse given in article 2(b) would enable any State in which such groundwaters are located to claim that it is an international watercourse States and hence to claim a share in its waters. It is therefore that this text might create problems of a new kind, and we accordingly propose that the Draft Articles should deal only with surface water."

[113] *Ibid*, Comments and Observations by Hungary dated 25 June 1996.

[114] *Ibid*, Comments and Observations by Ethiopia dated 28 June 1996.

[115] Turkey, Pakistan, See statements of these States in Press Release GA/9248 of 17 May 1997.

[116] *Ibid.*

[117] *Ibid.*

[118] *Ibid,* Statement by Rwanda.

4.3.3 The work of the IDI and ILA

The Salzburg Resolution[119] applies to the "utilization of waters which form part of a watercourse or "hydrographic basin" concept, which extend over the territory of two or more States."[120] Transboundary connected groundwater is implicitly included in the "hydrographic basin" but does not respond to the special requirements of these resources. Accordingly, the Resolution is only considered as an adequate foundation for the formulation of more specialized rules at the international level concerning groundwater. The 1966 Helsinki Rules adopt the "drainage basin" approach which is similar to the "hydrographic basin" approach under the IDI Salzburg Resolution. The Helsinki Rules, however, explicitly include transboundary groundwater as part of the entire "drainage basin"[121] which was defined as follows:

"A geographical area extending over two or more states determined by the watershed limits of the system of waters, including surface and underground water, flowing into a common terminus: they should contribute to the drainage basin principal river, stream or lake or other common terminus".[122]

The above definition implicitly excludes the non-renewable resources in confined aquifers.[123] The requirements of "common terminus" and the physical interaction between surface and groundwater were restrictive factors in the drainage basin term. The notion of "common terminus" was meant, in part, to provide a geographic limitation whereby two different watercourses connected by

---

[119] 1961 Salzburg Resolution, *supra* note 33.

[120] *Ibid,* Article 1.

[121] Historically, the concern regarding use of an international river was almost completely for navigation, and there was little necessity for dealing with any portion of an international drainage basin other than the navigable channel of the stream. With the relatively recent multi-use development of international rivers, the concern is no longer limited to the navigable portion of the international river, but rather encompasses all waters included in the entire system comprising the international drainage basin. The drainage basin is an indivisible hydrologic unit which requires comprehensive consideration in order to effect maximum utilization and development of any portion of its waters when it is recognized that a state, although not riparian to the principal stream of the basin, may nevertheless supply substantial quantities of water to that stream; such a state thus is in a position to interfere with the supply of water through action with respect to the water flowing within its own territory. Therefore, in order to accommodate potential or existing conflicts in instances of multi-use development and to provide the optimum rational development of a common resource for the benefit of each State in whose territory a portion of the system lies, the drainage basin approach has become a necessity. See ILA, 1966 Helsinki Rules, Commentary, in (779 Garretson, Hayton and Olmstead; Oceana Publications, NY: USA, 1967).

[122]The 1966 Helsinki Rules, *supra* note 47, Art II.

[123]S. McCaffrey, *supra* note 40.

a canal could not be regarded as a single watercourse for the purposes of the Rules.[124] As a complementary to the Helsinki Rules, the Seoul Rules[125] consider aquifers to be international whether or not these aquifers are interrelated with surface water system flowing into a common terminus:

"The waters of an aquifer that is intersected by the boundary between two or more states are international groundwaters and such an aquifer with its waters forms an international basin or part thereof". [126]

The Seoul Rules stipulate that States are considered "Basin States" within the meaning of the 1966 Helsinki Rules whether or not the aquifer and its waters form with the surface waters part of a hydraulic system flowing into a common terminus.[127] The term "aquifer", as was employed in the Seoul Rules, encompasses all underground water bearing strata capable of yielding water on a practicable basis, whether or not these waters are connected with the surface water system. On the other hand, the Seoul Rules spell out relevant and important factual interrelationships under the rubric of "Hydraulic Independence": "1) An aquifer that contributes water to, or receives water from, surface waters of an international basin constitutes part of that international basin for the purpose of the Helsinki Rules. 2) An Aquifer intersected by the boundary between two or more State that does not contribute water to or receive water from, surface waters of an international drainage basin constitutes an international drainage basin for the purpose of the Helsinki Rules. 3) Basin States, in exercising their rights and performing their duties under international law, shall take into account any interdependence of the groundwater and other waters, including any interconnections between aquifers and any leaching into aquifers caused by activities under their jurisdiction."[128]

4.3.4 The Bellagio Draft Treaty

The Bellagio Draft Treaty applies to international aquifers in critical areas on the basis of mutual agreement rather than continuing to be subject of unilateral decisions. Issues like aquifer contamination, depletion, drought and transboundary transfers, as well as withdrawals and recharge issues are addressed by the draft treaty.[129] The overriding goal

[124]G. Eckstein, *supra* note 105.

[125] Seoul Rules *supra* note 58.

[126] 1961 Salzburg Resolution, *supra* note 33, Article 3:

"If states are in disagreement over the scope of their rights of utilization, settlement, will take place on the basis of equity, taking particular account to their respective needs as well as of other pertinent circumstances."

[127] Seoul Rules *supra* note 58 Article 1.

[128] *Ibid*, Art. 2.

[129] *Ibid*.

of the draft treaty is to achieve an optimum utilization of shared groundwaters. It provides procedures for the avoidance or resolution of differences over shared groundwater.[130] The draft treaty defines transboundary groundwater in the same manner as the 1986 Seoul Rules,[131] including both connected and confined aquifers within its scope. The draft treaty has also introduced the term "underground environment", which does not only encompass the quality and quantity of the waters, but also, the geologic structure of the water body and the aquifer itself affecting these properties.[132]

4.3.5 Summary

The definition of scope and the uses of terms within a specific legal instrument are determining factors as to whether or not transboundary groundwater is covered and *ipso facto* the manner in which it is treated. The scope of the ILC 1991 and 1994 draft articles and the adopted 1997 UN Watercourses Convention includes connected groundwater. There is evidence that term "watercourse system" was favoured by the ILC members and the representative of States in the comments and observations to the Draft. The "drainage basin" concept was avoided by many States, as creating problems in relation to States' territorial sovereignty.

The analysis of the work in the field by the IDI, ILA and the expert group on the Bellagio Draft Treaty reveals some crucial differences between the various groups' approach to scope and use of terms. The IDI Salzburg Resolution and the ILA Helsinki Rules employed quite similar concepts, which consider groundwater as an integral component of the environment, subject to the same principles and rules as the other physical components of the hydrological cycle. In effect, these instruments did not include connected transboundary groundwaters within their scope and further, did not treat groundwater as a special component of the hydrological cycle. The 1986 ILA Seoul Rules, which are complimentary to the Helsinki Rules, were developed with a view that groundwater is special in its characteristics and behaviour and thus requires special treatment for management and protection. Moreover, the Seoul Rules include confined groundwaters within

---

[130] *Ibid.*

[131] *Ibid*, Article I(1) identifies "Aquifer" as:

"a subsurface water bearing geologic formation from which significant quantities of water may be extracted".

Article 1 (20) identifies a transboundary aquifer as: "an aquifer intersected by a common boundary". The commentary to this Article explains that The term "aquifer" as here employed comprehends all underground water-bearing strata capable of yielding water on a practicable basis, whether these are in other instruments or contexts called by another name such as "groundwater reservoir", "groundwater catchment area", etc. including the waters in fissured or fractured rock formations and the structures containing deep, so called "fossil waters".

[132] *Ibid*, Comment (3) to Art. II.

their scope. Both connected confined transboundary groundwaters are covered within the scope of the Bellagio Draft Treaty.

4.4 SUBSTANTIVE RULES

4.4.1 Introduction

The legal importance of substantive rules was described in the Introduction. In brief these are the rules that determine the rights and duties of States towards each other with respect to transboundary groundwaters. The key issues to be investigated within this chapter are: (i) what are these rules and to what extent they present rules of customary international law and (ii) how these rules have evolved in the work of the ILC, IDI, ILA and the Bellagio Draft Treaty.

4.4.2 The Work of the ILC

4.4.2 1Background

The study of the evolution of the substantive rules governing the utilization and development of international watercourses within the work of the ILC has great significance in understanding the legal authority and normative content of these rules. In the context of this research it is crucial to determine whether such rules are equally applicable to transboundary groundwater or not. In 1982, the second Special Rapporteur Schwebel proposed two fundamental articles for the ILC's consideration.[133] One of these articles - namely- Article 6, demonstrated the fact that no State may be denied its equitable participation in the utilization of the waters of the international watercourse system of which it is a system State.[134] For him this right is fundamental and has primacy over the obligation not to cause appreciable harm.[135] Jens Evensen, the third Special Rapporteur supported the inclusion of equitable and reasonable utilization principle as an obligatory rule of customary international law.[136] However, Schwebel's proposed articles raised a crucial question as to which of the two rules have primacy over the other?

---

[133] Third Report of the Special Rapporteur, Stephen Schwebel, U.N. Doc. A/CN.4/348 (1982) and Corr.1 reprinted in 2(1) YILC, 1982.

[134] *Ibid.*

[135] *Ibid*, Article (6) Paragraph 1 states:

"The right of a system State to use the water resources of an international watercourse system is limited to the duty no to cause "appreciable" harm to the interests of another system State, except as may be allowable under the determination for equitable participation for the international watercourse system involved".

[136] First Report of the Special Rapporteur, Jens Evensen (35th session of the Commission, 1983), U.N. Doc. A/CN.4/367 (1983) reprinted in 2 YILC 1983 and the Second Report of

The Special Rapporteurs adopted the position that there is a fundamental requirement on States in general international law not to undertake activities within their territories that might cause "harm" beyond their jurisdiction.[137] Whether any harm is prohibited or whether some can be tolerated will be discussed below. According to Schwebel, "appreciable harm" might be permitted if it is essential for fulfilling the requirements of implementing equitable participation in the utilization of the international watercourse. Schwebel was in favour of qualifying the term harm by the word "appreciable", which in his view was more than "perceptible" but less than "substantial".[138] For him, harm was only one important factor in judging whether the utilization is equitable or not: he actually never considered obligation not to cause harm as a basic rule. Schwebel strongly proposed that the equitable and reasonable utilization principle is the guiding principle that determines the right of a system State in utilizing the waters of an international watercourse.[139]

Evensen endorsed qualifying harm with the term "appreciable" but emphasized that no other qualification of the no-appreciable harm rule should be accepted. Thus, Evensen introduced a new wording of article 9, implying that the no appreciable harm rule is the guiding rule.[140]

Stephen McCaffrey, who became the Special Rapporteur in 1985, endorsed both Schwebel's and Evensen's propositions concerning equitable and reasonable utilization principle, but disagreed with the latter, regarding the relationship between this principle and the obligation not to cause harm above a certain threshold. McCaffrey had to deal with the two opposing positions and was asked to reconcile them. McCaffrey stated that the principle of "no harm" itself is embodied within the principle of equitable and reasonable utilization.[141] A

---

the same Rapporteur (26th session of the Commission, 1984) U.N. Doc. A/CN.4/381 and Corr. 1 and 2 reprinted in 2(1) YILC 1984.

[137] Schwebel Report, *supra* note 133.

[138] The Law of the Non-Navigational Uses of International Watercourses, U.N. Doc. A/CN.4/381 (1984).

[139] Schwebel Report, *supra* note 133.

[140] Evensen's Report, *supra* note 136, Evensen's new article 9 (replacing article 7 in the 1991 and 1994 drafts) reads as follows:

"A watercourse States shall refrain form and prevent within its jurisdiction uses or activities with regard to an international watercourse that may cause appreciable harm to the rights and interests of other watercourse States, unless otherwise provided for in a watercourse agreement or other arrangement".

[141] S. McCaffrey, *supra* note 40. The three cases that are normally cited with regard to the no harm principle are : of cases most often cited in support of the international obligation known as the no-harm rule includes the Trail Smelter arbitration (United States v. Canada), Arbitral Tribunal, Montreal 16 April 1938 and 11 March 1941; 3 UN RIAA 1905 (1947); 33 AJIL 182 (1939), also *found at* http://www.jura.uni-muenchen.de/

comprehensive survey of all available general practice of States revealed to him the strong legal foundation of the rule of equitable and reasonable utilization and consequently he confirmed its predominance over the no harm rule.[142] He further elaborated that a minor rewording of Evensen's article (9) would still make equitable and reasonable utilization principle the dominant principle. In his view, the core issue lies in understanding what "injury" means and when it is permitted and why.[143] For McCaffrey, harm was allowable so long as it was consistent with the equitable and reasonable utilization. According to him, what is not permitted is injury in its legal sense, but not in its factual sense, i.e. when a watercourse State deprives another State from the enjoyment of its equitable and reasonable utilization of the international watercourse system.[144]

As far as pollution is concerned and in relation to the no significant harm principle, it was confirmed that watercourse States either individually or jointly shall prevent, reduce or control pollution of an international watercourse that may cause appreciable harm to other watercourse States or to their environment. This appreciable harm included harm to human health or safety, to the use of the waters for any beneficial purpose or to the living resources of the watercourse.[145] In his fourth report to the ILC on the matter, McCaffrey took the position that "appreciable harm" to other watercourse States or to the ecology must be prohibited.[146] He believed that the draft article 21 prohibiting pollution,

---

einrichtungean/ls/simma/tel/case8.htm (*last visited* May 2003) and the Corfu Channel case, ICJ Reports 1949, 4-169 (Judgement, merits, 9 April 1949), and the Lac Lannoux arbitral award, 12 RIAA 1957, 281-317 (Award, 16 November 1957); 24 ILR .

[142] Second Report of the Special Rapporteur, Stephen McCaffrey, U.N. Doc. A/CN.4/399 (1986). Caflish *supra* note 9 argues that the 'No harm Rule' is a derogation from the general principles of international law namely neighbourliness and State responsibility. The precise origins of the no-harm rule have been attributed to such diverse foundations as the Roman Law *maxim sic utere tuo ut alienum laedas* which means that as a legal duty to exercise one's rights in such a way as not to cause harm to the interests of other persons.

[143] McCaffrey *ibid*, in his explanation on the distinction between harm and injury McCaffrey explained in his second report:

"Crucial to the understanding of the latter statement, and indeed of the doctrine of equitable utilization in general, is an appreciation of the meaning of the term injury in the context: the term is used in its legal, as opposed to its factual sense. Thus an allocation of the uses and benefits of the waters of an international watercourse between two or more States may entail a certain degree of harm, and still be 'equitable'.... Thus, where there is a conflict among the water needs the States making beneficial use of those waters, that conflict is to be resolved on the basis of equity, taking all relevant factors into account".

[144] *Ibid.*

[145] *Ibid*, Article 21 of Part IV.

[146] Fourth Report on the Law of Non-Navigational Uses of International Watercourses, UN GAOR ILC 40th Session, at 14, UN Doc. A/CN.4/412/Add2.

automatically contravenes draft article 5 which contains the requirement of equitable utilization.[147]

4.4.2.2 *The 1991 Draft Articles*

A. Equitable and Reasonable Utilization

The provisional text of the 1991 Draft came as a natural outcome of the foregoing debate of the Special Rapporteurs and their understanding of the legal foundation for the substantive rules governing the utilization and development of international watercourses,. Article 5 of the 1991 ILC Draft contained the ILC's expression of the principle of equitable utilization. It stated that:

"1. Watercourse States shall in their respective territories utilize an international watercourse in an equitable and reasonable manner. In particular, an international watercourse shall be used and developed by watercourse States with a view to attaining optimum utilization thereof and benefits therefrom consistent with adequate protection of the watercourse.2. Watercourse States shall participate in the use, development and protection of an international watercourse in an equitable and reasonable manner. Such participation includes both the right to utilize the watercourse and the duty to co-operate in the protection and development thereof, as provided in the present articles".

Article 6 sets forth the guidelines for the implementation of Article 5. It includes a non-exhaustive list of the factors that need to be taken into account when determining what is equitable and what is reasonable.[148]

---

[147] *Ibid.* McCaffrey's actual statement on this was as follows:
" In light of the need to protect the environment in order to permit sustainable development and to preserve the earth for future generations...water uses that cause appreciable pollution harm in other watercourse States and the environment could well be regarded as being per se inequitable and unreasonable....In the view of the Special Rapporteur, the Commission should...adopt a rule of 'no appreciable pollution harm' that is not qualified by the principle of equitable and reasonable utilization. This position is without any prejudice to any decision the Commission may take with regard to whether there should be an equitable use exception to the general rule of "no appreciable harm" contained in draft article 9(7 now)."

[148] Article 6 reads:
"Utilization of an international watercourse in an equitable and reasonable manner within the meaning of article 5 requires taking into account all relevant factors and circumstances including: (a) geographic, hydrographic, climatic, ecological and other factors of a natural character. (b) the social and economic needs of the watercourse States concerned; (c) the effects of the use or uses of the watercourse in one watercourse State on other watercourse States; (d) existing and potential uses of the watercourse; (e) conservation, protection, development and economy of use of the water resources and the costs of measures taken in that effect; (f) the availability of alternatives, of corresponding value, to a particular planned or existing use.2. In the application of article 5 or paragraph 1 of this article, watercourse

B. No Harm Rule

The ILC approved Evensen's approach to understanding the relationship between equitable and reasonable utilization and no harm. Accordingly, the harm factor within the 1991 Draft was contained in Article 7 that does not allow a certain magnitude of harm. The provision in question reads:

"Watercourse States shall utilize an international watercourse in such a way as not to cause appreciable harm to other watercourse States".

Article 7 prohibits uses that might cause appreciable harms thus giving primacy of the no-appreciable-harm rule over the equitable and reasonable utilization principle. This primacy was further emphasized by the ILC's provision on Article 21(2) on pollution.[149]

4.4.2.3 States Comments and Observations to the 1991 Draft Articles

Some States resisted the adoption of the term "appreciable" as not appropriate to describe the harm that occurs due to activities by other States. The UK, for example considered the term to be "far from clear" and recommended that "significant" should replace "appreciable".[150] Germany, also considered that the term "appreciable" has the disadvantage of having a double meaning and proposed the use of "significant" instead.[151] By and large, States supported the principle of equitable and reasonable utilization and participation as laid down in article 5 of the 1991 Draft Articles. They agreed that this, in their opinion, ensures the goal of optimum utilization and that the notion of equity which is a legal standard imposed by customary international law.

States concerned shall, when the need arises enter into consultations in a spirit of cooperation."

[149] Article 21(2) states that:

"Watercourse States shall, individually or jointly, prevent reduce and control pollution of an international watercourse that may cause appreciable harm to other watercourse States or to their environment, including harm to human health or safety, to the use of the waters for any beneficial purpose or to the living resources of the watercourse. Watercourse States shall take steps to harmonize their policies in this connection"

[150] UN Doc. A/CN.4/447, *supra* note 97 The UK elaborated in this context stating that:

"the commentary accompanying draft article 7 states that "appreciable" embodies a factual standard which requires harm to be substantiated by objective evidence; it is not "insignificant" or "barely detectable" not it is necessarily "serious". The UK is in agreement with the meaning attributed to "appreciable" in the commentary but is not sure whether or not "appreciable" convey this meaning"

Ibid, Germany commented that the term "appreciable" may mean "detectable" or "significant" in connection with risk or harm. As this double meaning involves considerable differences in substance we suggest that the term "significant" be used, especially as this also corresponds to the ILC's interpretation of "appreciable".

Equitable participation complements the said principle, which ensures cooperation between States in the protection and development of the watercourse.[152]

Some States favoured using the word "significant" or "perceptible", instead of "appreciable", which gives a greater threshold to harm and introduces a due diligence test. Some States insisted on the necessity that equitable utilization has primacy over the no appreciable harm.[153]

4.4.2.4 The 1994 Draft Articles

In light of the above discussions within the ILC and the comments received from the Member States the ILC was faced with difficulty of deciding which of the rules should have priority; the equitable and reasonable utilization principle or the no appreciable harm rule. The ILC's Drafting Committee's task became more difficult when the issue of pollution was raised in the context of the equitable utilization vs. the no harm principle discussion.[154] The ILC revised the original text of the 1991 by raising the threshold of harm from "appreciable" to "significant" and adding the "due diligence" requirement, which basically shifted the responsibility of States from an obligation of result to an obligation of behaviour.[155] The 1994 draft articles gave primacy to the rule of no harm over the principle of equitable and reasonable utilization. Article 6 remained unchanged as in the 1991 Draft. The Commentary to that provision notes that it "is a specific application of the general obligation contained in Article 7 not to cause appreciable harm to other watercourse States".[156]

---

[152] *Ibid*, 1993, Germany in her comments on this specific issue commended the fact that article 5 of the 1991 Draft Articles contains both a right to utilise the water and a duty to limit other States in their right to equitable utilization. She also welcomed the important aid to interpretation in deciding what constitutes "equitable and reasonable" use. Also Greece noted that articles 5 and 6 are the keystones if the draft articles, for in reality it determine the conduct of watercourse States with respect to the utilization of a watercourse in their respective territories. It follows on by clarifying that equity is a legal standard imposed by customary international law. Equity in the point of view of Greece a verifiable concept, meeting the specific criteria which are listed in article 6.

[153] *Ibid.*

[154] 1994 Draft Articles, *supra* note 49, Commentary.

[155] Greece for example in its comment to the new version of article 7 emphasized that the changes introduced were built on an uncalled criterion. Which allows a State to cause significant harm to other States of the same watercourse provided that it confines itself within the limits of what article 7 calls due diligence. Consequently everything becomes imprecise and subjective. Accordingly Greece asked the reinstatement of the older version of article 7. For exact statement see U.N. Doc A/51/275/Add.2.

[156] Commentary to 21(2), para. 3

4.4.2.5 States Comments and Observations to the 1994 Draft Articles

In the 1994 Draft Articles the question was raised concerning the relationship between equitable and reasonable utilization on one hand and the obligation not to cause harm. States noted that the first principle must not be subordinated to the prohibition of causing appreciable harm. In the view of Nordic Countries, for example, in cases of use not involving pollution, the obligation not to cause appreciable harm should be subject to the principle of equitable utilization.[157] The UK and Northern Ireland rejected this argument and commended the ILC for providing that the obligation of equitable use in draft article 5 is subordinate to the obligation not to cause significant harm.[158] Other States, like Turkey, demanded more balance in the formulation of article 5 that favours them as an upstream State and gives primacy to the rule of equitable utilization.[159] In their view this balance could be obtained by adding a clause that restricts uses (particularly new uses) of the lower riparian States.[160] What matters, Turkey added, is the conformity with article 5 and 6 in order to ensure that equality of rights has been achieved for the concerned States. Ethiopia, Spain, Finland, and Switzerland supported the primacy of the equitable utilization over the no harm. Spain affirmed that the equitable utilization principle has the force of a customary norm of general scope. Hungary, on the other hand rejected the amendments in the 1994 draft articles 5 and 7 which in their opinion gave primacy to equitable utilization rule.[161]

---

[157] *Ibid.* In their comments to Articles 5 and 7 of the 1994 Draft the UK stated:
"…International watercourse States may utilize an international watercourse in an equitable and reasonable manner, but this obligation of equitable and reasonable use in draft article 5 is subordinate to the obligation not to cause appreciable harm "

[158] *Ibid,* Turkey stated in its comments:
"Although Article 5 of the draft contains some positive elements for the upper riparian State, it needs to be more balanced. For the sake of a well-balanced article it will be appropriate to expand this paragraph to include a clause which restrict the use (and especially new uses) of the lower riparian States. If this balance cannot be obtained, the subject of participation should be excluded from article 5…."

[159] *Ibid*, Ethiopia, Spain, Finland, Turkey and Switzerland supported the primacy of the equitable utilization over the no harm. Spain affirmed that the equitable utilization principle has the force of a customary norm of general scope. Hungary, on the other hand rejected the amendments in the 1994 Draft art. 5 and 7 which in their opinion gave primacy to equitable utilization rule. Turkey believed that the rule of equitable utilization should not be limited by the second criterion. What matters is the conformity with article 5 and 6 in order to ensure that equality of rights has been achieved for the concerned States.

[160] *Ibid.*

[161] UN Doc. A/51/275.

4.4.2.6 The 1997 UN Watercourses Convention

Article 5 and paragraphs 1 and 2 of Article 6 remained unchanged in the final text of the Convention. However, a new paragraph 3 was added to the latter:

"The weight to be given to each factor is to be determined by its importance in comparison with that of other relevant factors. In determining what is reasonable and equitable use, all relevant factors are to be considered together and a conclusion reached on the basis of the whole".

One of the issues that were seriously debated within the Sixth Committee is the relationship between articles 5, 6 and 7 (on equitable utilization, the factors relevant to equitable utilization and the duty not to cause harm respectively).[162] A new paragraph to Article 7 was introduced:

"Where significant harm nevertheless is caused to another watercourse State, the State, whose use causes such harm shall, in the absence of agreement to such use, take all appropriate measures, having due regard for the provisions of articles 5 and 6, in consultation with the affected State, to eliminate or mitigate such harm and, where appropriate, to discuss the question of compensation."

The new articles as revised by the Sixth Committee gave primacy to the equitable utilization rule over the no significant harm rule. The new articles allowed for significant harm on the basis of fulfilling the obligation of equitable and reasonable utilization as the governing principle. The above provision Articles 5, 6 and 7 were voted for as a package and were adopted by 38 States for, 4 contrary and 22 abstentions.[163]

4.4.3 The Work Of The IDI, ILA and The Bellagio Draft Treaty

Under the IDI Salzburg Resolution, the basic general principle governing the neighbourly relations in general and the use of shared waters in particular is the obligation not to cause "unlawful harm" to others.[164] In its approach, the Resolution applies the "limited sovereignty theory", whereby the right of use of any State from the

---

[162] 11 States submitted their comments on articles 5, 15 on article 6 and 18 on article 7 in addition to the WG chairman's proposal on each article.

[163] China, France, Tanzania and Turkey voted against. Turkey for example voted against the Convention as a whole and in its statement to the GA on the 21 May 1997 it justified this voting by objecting to the fact that equitable and reasonable utilization was given the primacy over the obligation not to cause significant harm. Israel on the other hand stated that her objection to the convention stems partly from the lack of clarity regarding the balance between the two principles. Neither principle in the opinion of (Israel) should be subservient to the other. The balance between them should be based on the specific case. (Detailed statements found in GA/9248, 1997). Bolivia abstained in the voting also namely because it had reservations regarding articles 5, 6 and 7.

[164] 1961 Salzburg Resolution, *supra* note 32, paragraph 3 and 4 of the Preamble.

waters flowing across or bordering its territory is limited by the right of use by the other States concerned with the same river or "hydrographic basin".[165] In their utilization of the waters States are obliged not to seriously affect the possibility of utilization of the same waters by other States.[166] In the case when a serious damage or loss occur States causing the harm are obliged to assure that the affected State/s enjoy the advantages of their utilization rights on the basis of: equity, their respective needs and other pertinent circumstances[167] and they are entitled to adequate compensation.[168] The principle of equitable and reasonable utilization is a governing rule of the 1966 Helsinki Rules and all of the work by the ILA. The factors determining equitable and reasonable sharing are listed under Article V, whereby harm is only one factor.[169] Under the 1966 Helsinki

---

[165] *Ibid,* Article 2:
"Every State has the right to utilize waters which traverse or border its territory, subject to the limits imposed by international law and, in particular, those resulting from the provisions which follow."

[166] *Ibid*, Art. 4
"No State can undertake works or utilizations of a waters of a watercourse or hydrographic basin which seriously affect the possibility of utilization of the same waters by other States except on condition of assuring them the enjoyment of the advantages to which they are entitled under article 3, as well as adequate compensation for any loss or damage".

[167] *Ibid* Article 3:
"If the States are in disagreement over the scope of their rights of utilization, settlement will take place on the basis of equity, taking particular account of their respective needs as well as of other pertinent circumstances."

[168] The requirement of compensation is however not an established principle in international law. As Charles Bourne argues the matter of compensation is not obligatory, *see* P.K.Wouters (ed.), Selected Writing in International Water Law: Charles Bourne, (Kluwer Law International: the Netherlands, 1997) at 82.

[169] The 1966 Helsinki Rules, *supra* note 47 Chapter 2 Article V states that:
" I. What is a reasonable and equitable share within the meaning of article IV to be determined in the light of all the relevant factors in each particular case. II. Relevant factors which are to be considered include, but are not limited to: 1. The geography of the basin, including in particular the extent of the drainage area in the territory of each basin State; 2.The hydrology of the basin, including in particular the contribution of water by each basin State;3.The climate affecting the basin; 4.The past utilization of the waters of the basin, including in particular existing utilization; 5.The economic and social needs of each basin State; 6.The population dependent on the waters of the basin in each basin State; 7.The comparative costs of alternative means of satisfying the economic and social needs of each basin State; 8.The availability of other resources; 9.The avoidance of unnecessary waste in the utilization of waters of the basin; 10.The practicability of compensation to one or more of the co-basin States as a means of adjusting conflicts among uses; and 11.The degree to which the needs of a basin State may be satisfied, without causing substantial injury to a co-basin State. III. The weight to be given to each factor is to be determined by its importance in comparison with that of other relevant factors. In determining what is reasonable and

Rules, pollution is viewed to constrain the beneficial and equitable utilization of such waters and therefore there is an obligation to abate pollution.[170] As for the Seoul Rules it is clearly stated that the same principles that govern utilization in the 1966 Helsinki Rules equally apply to transboundary groundwater.[171] The principle of equitable and reasonable utilization and the no harm rule govern in the Bellagio Draft Treaty.[172]

4.4.4 Summary

The previous discussion presented the various approaches by the different organizations for identifying the substantive rules governing the utilization, development and protection of transboundary groundwater. By and large, there is a consensus among the different organizations' approaches, which provides that the right for using the waters of international watercourse derive from the fundamental right of States to equitably use waters that traverse or border their territory. The extent to which this right is limited by the obligation not to cause a certain degree of harm varies from one approach to another.

The 1991 and 1994 draft articles, which apply equally to surface water and connected transboundary groundwater, considered equitable utilization a fundamental principle of customary international law. Both Drafts, however, gave primacy of the no harm rule over the principle of equitable and reasonable utilization. This was corrected in the final text of the 1997 UN Watercourses Convention, whereby the equitable and reasonable utilization principle became predominant, though including a degree of factor of harm. The foregoing does not apply to confined transboundary groundwater owing to the exclusion of such waters from the scope of the final text of the 1997 UN Watercourses Convention.

The 1961 IDI Salzburg Resolution allows for harm to occur if such harm result from the requirements of ensuring equity of rights of States to water utilization. The 1966 Helsinki Rules consider harm as one factor in defining equitable and reasonable utilization. Due to its complementary nature, the Seoul Rules apply the same substantive rules as the 1966 Helsinki Rules to transboundary confined

---

equitable share, all relevant factors are to be considered together and a conclusion reached on the basis of the whole".

[170] The 1966 Helsinki Rules, *supra* note 47, Chapter 3 Art. IX and X.

[171] The 1986 Seoul Rules, *supra* note 58 at 342.

[172] The Bellagio Draft Treaty, *supra* note 35, Preamble, paragraph 5

"Seeking to provide for the utilization, protection and control of those groundwaters on an equitable basis and to that end, for the creation and maintenance of an adequate database"

and Article VI (1):

"the Parties undertake cooperatively to protect and to improve, insofar as practicable, the quality of transboundary aquifers and their waters in conjunction with their programs for surface water quality control, and to avoid appreciable harm in or to the territories of the Parties."

groundwaters. The Bellagio Draft Treaty considers equitable and reasonable utilization as the primary principle governing the utilization, development and protection of both connected and confined transboundary groundwater.

4.5 PROCEDURAL RULES

4.5.1 Introduction

Procedural rules are essential in any watercourse agreement. They provide the means through which the substantive rules are implemented. The distinction between the "substantive" and "procedural" obligations is made mostly for analytical purposes to better understand the treaty structure and requirements. This does not mean that "procedural" obligations are less binding than "substantive" obligations. Procedural rules encompass a range of obligations: from a general duty to cooperate to obligations concerning data and information exchange, prior notification and consultations.

4.5.2 The Work of the ILC

4.5.2.1 Background

The ILC efforts to identify the substantive rules and principles pertinent to the use and development of international watercourses had to be complemented by a set of proper procedural steps to achieve the objectives of these rules. The legal origins of such procedures and what States accept as obligatory procedural rules were not fully understood by the members of the ILC. To this end Schwebel in his third report included a survey of the agreements, decisions of international courts and tribunals, declarations and resolutions adopted by intergovernmental organizations relating to the principle of cooperation. [173] In the latter survey it was illustrated that cooperation between

[173] An important legal instrument that emphasized the importance of cooperation in relation to the utilization of international watercourses is the Charter of Economic Rights and Duties of States. The Charter obliges States sharing a natural resource to cooperate among themselves by establishing a system of information exchange and prior consultation in order to optimize the use of these resources without causing damage to the interests of other States. *See* GA Res. 3281(xxix), UN GAOR, 29th Sess., Supp. No. 31 (1974) found at: http://www.dal.ca/~wwwlaw/kindred.intllaw/EcRtsandDuties.htm (last visited April 2003) Article 3 reads as follows:

"In the exploitation of natural resources shared by two or more countries, each State must co-operate on the basis of a system of information and prior consultations in order to achieve optimum use of such resources without causing damage to the legitimate interest of others'. The United States Mexico on co-operation of 1983 is another example for the importance of cooperation for the protection, improvement and conservation of the environment and the problems, which affect it."

Watercourse States on the utilization of international watercourses is an essential requirement.[174] Cooperation encompasses an obligation to exchange information, prior notification and the need for consultation. Opinions are divided when discussions arise about prior notification. Most of the debates among legal experts are centred on when and what measures states need to notify other States about.[175] Some scholars favour the rule requiring notice of all utilizations of the waters of an international watercourse.[176] While the majority opinion supports the narrower obligation, which stipulate that notification is only need of utilizations that might cause serious injury to other States.[177]

4.5.2.2 The 1991 Draft Article

The procedural rules within 1991 Draft were embodied in Articles 9 to 19. These procedures are a natural consequence of the obligation to cooperate that is embodied in Article 8 on the duty to cooperate.[178] The eleven articles are considered as a road map for States for the implementation of equitable and reasonable utilization of their shared watercourses including connected transboundary groundwater.

A. Duty to Cooperate

Cooperation between watercourse States with regard to their utilization of an international watercourse including connected transboundary groundwater was agreed to be an important basis for the attainment and maintenance of an equitable allocation of the uses and benefits of the watercourse and for smooth functioning

---

See Agreement on Co-Operation for the Protection and Improvement of the Environment in the Border Area Signed at La Paz, Baja California, on 14 August 1983, XXII (5) ILM 1983 also found at http://www.Usembassy-Mexico.Gov/Elapaz.Html. Article 1 reads: 'The United States of America and the United Mexican States, hereinafter referred to as the Parties, agree to cooperate in the field of environmental protection in the border area on the basis of equality, reciprocity and mutual benefit. The objectives of the present Agreement are to establish the basis for cooperation between the Parties for the protection, improvement and conservation of the environment and the problems which affect it, as well as to agree on necessary measures to prevent and control pollution in the border area, and to provide the framework for development of a system of notification for emergency situations. Such objectives shall be pursued without prejudice to the cooperation which the Parties may agree to undertake outside the border area'. Agreement between the Federal Republic of Nigeria and the Republic of Niger concerning the equitable sharing in the development, conservation and use of their common water resources there is a duty to cooperate'.

[174] Schwebel's Third Report, *supra* note 133, at 43-58.

[175] P.K Wouters, *supra* note 8.

[176] H.A. Smith, *supra* note 12.

[177] *Ibid.*

[178] *Ibid*, Article 8 reads:

"Watercourse States shall cooperate on the basis of sovereign equality, territorial integrity and mutual benefit in order to attain optimal utilization and adequate protection of an international watercourse"

of the procedural rules.[179] Cooperation as stipulated from Article 8 primarily stems from the States' sovereign equality, territorial integrity and mutual benefit.[180] According to the Article 8 formulation, cooperation in itself lacks the "normative content" as it relies on general legal principles including sovereign equality and territorial integrity that have a normative content in international law. Accordingly, failure of States to cooperate does not confer rights and obligations on them unless such violation is considered a violation to these general principles of international law.

B. Duty to Exchange Information

The ILC members agreed that as a logical requirement for the attainment of cooperation there is the duty on watercourse States to exchange information whether regular data on the conditions of the watercourse or specific information concerning planned measures on the conditions of an international watercourse.[181] While Paragraph 1 of Article 9 obliges States to exchange "reasonably available" data on the condition of the international watercourse, paragraph 2 of the same article deals with the case when data are not reasonably available. In the latter situation States are obliged to employ their best efforts to comply with the request by other States with the possibility to condition their compliance upon payment by the requesting State of the reasonable costs of collecting and where appropriate, processing such data or information. Paragraph 3 of the same article sets the duty on States to employ their best efforts to collect and where appropriate, to process data and information in a manner, which facilitates its utilization by other watercourse States to which it is communicated.

C. Duty to Prior Notification

Articles 12-16 of the 1991 Draft deal with the terms and conditions relating to prior notification on planned measures. The 1991 Draft gave the notified State a period of six months to study and evaluate the effects of planned measures.[182] The 1991 Draft introduce three situations where the former notification procedures are slightly different. The first situation was embodied where a planned measure is of the utmost urgency in order to protect public health, public safety or other equally important interests. The Article provides that the implementing States must transmit to the other watercourse States a formal declaration of the urgency of the measures together with relevant data and information, thereafter the normal requirements for consultation and negotiation apply. The second exception was found in Article 25 which deals with emergency situations and calls upon States to immediately notify other interested States and take all practical steps to prevent,

---

[179] *Ibid.*

[180] The 1991 Draft, *supra* note 48 Article 8. .

[181] 1991 Draft Articles, *supra* note 48, Article 11 of Part III.

[182] *Ibid,* Article 14.

mitigate and eliminate the harmful effects of such emergency. The third exception is when disclosing data and information might threaten a watercourse national defence or security.[183] In such case the governing rule when implementing a planned measure is the equitable and reasonable utilization.

D. Consultation and Negotiations

The 1991 Draft emphasized the obligation to enter into negotiations and consultations when a watercourse State strongly believes that certain planned measure to which this State has not received a notice for might cause "adverse effects" upon it.[184] If within the six months the notified State objects to the planned measures, both States are to enter into consultations and negotiations.[185] According to paragraph 2 of Article 17 there is no clear evidence that entering into consultation and negotiations is a customary duty. The principle as presented in Article 17 lacks the "normative content",[186] as it had to be supported by the general principle of "good faith". The Notifying States are further obliged to refrain from implementing any planned measure for another period not exceeding six months from the start date if consultation if so requested by the notified State/s.[187]

4.5.2.3 States Comments and Observations to the 1991 Draft Articles

In general, there was a unanimous acceptance by States of the obligation to cooperate. However, some States demanded the broadening of the applicable principles giving the importance of the principle of cooperation that should underline the relationship between watercourse States.[188] The core procedural rules that provoked the bulk of the discussion were the provisions concerning the planned measures. Some States believed it was inappropriate to include such detailed procedures within a framework type instrument

---

[183] *Ibid,* Article 31

[184] *Ibid,* Article 18.

[185] *Ibid,* Articles 15 and 16.

[186] *Ibid,* Article 17 paragraph 2 reads:

" The consultations and negotiations shall be conducted on the basis that each State must in good faith pay reasonable regard the rights and legitimate interests of other States".

[187] *Ibid,* Article 17 paragraph 3 reads:

"During the course of the consultations and negotiations, the notifying State shall, if so requested by the notified States at the time it makes the communication, refrain from implementing or permitting the implementation of the planned measures for a period not exceeding six months".

[188] Hungary, Portugal affirmed the need to include the principle of good faith-neighbourliness in order to attain optimal utilization and adequate protection of the international watercourse. . A/CN.4/447 Of 3 March 1993, Doc. A/CN.4/447/Add.1of 15 April 1993, Doc. A/CN.4/447/Add.2 of 18 May 1993, Doc. A/CN.4/447/Add.3 of 14 June 1993

and suggested that exchange of information and notification are essential only in the cases of water utilizations liable to cause significant harm.[189] Others insisted that notification, consultation and information exchange should be required in all cases regardless whether or not the proposed measures exceed the threshold of possible harm. [190] The latter opinion was based on the worry that States would be given the authority to decide what amounts to significant adverse effects. Suggestions were also made concerning the cessation of work during the notification period and until the period of reply to notification has expired.[191]

4.5.2.4 The 1994 Draft Articles

The procedural rules within the 1994 Draft found in Articles 9-19 did not vary much from that of the 1991 Draft. According to paragraph 2 of the commentary on Article 12 these procedures are triggered by the criterion that measures planned by a watercourse State may have a "significant adverse effect" upon other watercourse States. The threshold established by this standard is intended to be lower than the "significant harm" under article 7. On the regular exchange of data and information, the only change compared to the 1991 Draft is the use of the word "readily available" instead of "reasonably available". The terms "readily" as used in paragraph 1 and 2 of Article 9 are terms of art having a meaning corresponding roughly to the expression "in light of all the relevant circumstances" or to the word "feasible" rather than "rationally" or "logically".[192] Another change that was introduced to the 1994 Draft concerns the period for a reply to notification which became more flexible in the 1994 Draft. In addition to the first six months period, the 1994 Draft gave more flexibility for the notified State to request an extension in the period of reply, which was agreed not to exceed another six months. [193] Another change in the procedures as compared to the 1991 Draft is the case of absence of reply to notification (article 16). A new paragraph 2 was introduced to this article, which stipulates that no compensation would be eligible to a State which fails to reply within the said period.[194]

---

[189] Turkey, Colombia, Ethiopia, *ibid.*
[190] Hungary's comments, *ibid.*
[191] Columbia's comments, *ibid.*
[192] *Ibid*, Commentary to article 9.
[193] 1994 Draft Articles, *supra* note 49 article 13 Para (b).
[194] paragraph 2 of Article 16 reads:
"Any claim to compensation by a notified State which has failed to reply maybe offset by the costs incurred by the notifying State for action undertaken after the expiration of time for a reply which would not have undertaken if the notified State had objected within the period applicable pursuant to article 13."

4.5.2.5 States' Comments and Observations to the 1994 Draft Articles

As the changes to the procedural rules within the 1994 were marginal States had similar concerns and worries as to the 1991 Draft. The main comments polarized around the criterion that triggers the procedures concerning planned measures: is it the significant harm or the equitable and reasonable utilization principle and whether the "adverse effect" notion included in article 12 serves as an appropriate standard for the activation of such procedures. In the opinion of some States, including Switzerland, such procedures must be activated by the latter principle.[195] Turkey, Sudan,[196] on the other hand, considered it essential to undertake such procedures only in situations liable to cause significant harm.[197] Other States, such as Hungary believed that notification should be extended to all planned measures affecting international watercourses irrespective of the threshold of possible harm.[198] Another major issue that was commented on related to the requirement by a notifying State to refrain from implementing the planned measures during the period of consultation and negotiations for a period not exceeding six months. Some States, including Hungary, Portugal and Niger suggested that States must refrain from implementing planned measures during the whole period of consultation and negotiation.[199] Switzerland further suggested that failure to enter into negotiations and consultation by a State must have legal consequences on that State. Other States, like Turkey, proposed to reduce the whole part III into a minimum and further suggested to include only general principles regarding planned measures.[200]

4.5.2.6 The 1997 UN Watercourses Convention

Paragraph 1 of Article 8 on the duty to cooperate remained almost the same as in the 1994 Draft. The phrase "good faith" was included as a basis for cooperation in addition to sovereign equality, territorial integrity and mutual benefit. A new paragraph 2 was also introduced to Article 8. Accordingly the new Article 8 reads as follows:

"1.Watercourses States shall cooperate on the basis of sovereign equality, territorial integrity, mutual benefit and good faith in order to attain optimal utilization and adequate protection of an international watercourse. 2. In determining the manner of such cooperation, watercourse States may consider the establishment of joint mechanism or commissions, as deemed necessary by them, to facilitate cooperation on relevant measures and procedures in light of experience gained through cooperation in existing joint mechanisms and commissions in various regions".

---

[195] UN Document A/51/275/Add.1 reply dated 2 April 1996.
[196] UN Document A/51/275/Add.3 reply dated 14 October 1996
[197] *Ibid,* Turkey's reply dated 5 July 1996.
[198] Hungary's reply *supra* note 192, reply dated 25 June 1996.
[199] *Ibid.* Portugal's reply dated 26 June 1996, Niger's reply dated 21 September 1996..
[200] *Ibid,* Turkey's reply dated 5 July 1996.

The first and second parts of Article 8 emphasise that the obligation to establish cooperation mechanisms is an aspiration rather than something requiring immediate effect and that States are not obliged to establish joint mechanisms and commissions, but rather *may* consider such actions to be taken. The procedural rules as embodied in the 1997 UN Watercourses Conventions did not change when compared to the 1994 Draft.

4.5.3 The work of the IDI and the ILA

The 1961 Salzburg Resolution emphasises the need for consultation, joint planning and the application of the principle of reciprocity among the States sharing water resources.[201] The Resolution recommends States not to undertake utilizations or works without a previous notice to interested State/s.[202] The Resolution does not refer to a specific period for the reply but mentions that it must be within a reasonable time.[203]

The Procedural Rules in the1966 Helsinki Rules are referred to as "preventive measures". These provisions encompass the exchange of relevant and reasonably available information concerning the "hydrographic basin", its use and the activities with respect to its waters. The Rules also encourage state to notify other co-basin states concerning any proposed construction or installation which would affect the regime of the basin and which might give rise to conflict. [204] The same procedural rules that apply within the 1966 Helsinki Rules equally apply in the Seoul complementary Rules. Under the Seoul Rules it is provided that in order to realise the objective of the Rules,[205] Basin States are urged to consult and exchange relevant available information and data. Basin States are also recommended to cooperate for the purposes of collecting and analysing additional needed information and data pertinent to the international groundwaters or their aquifers.[206] Consultation and exchange of relevant information are essential

201 Salzburg Resolution, *supra* note 33, Preamble.

202 *Ibid*, Article 6.

203 The 1966 Helsinki Rules *supra* note 47, Arts. 5, 6 and 7.

204 *Ibid*. Chapter 6 Art. XXIX.

205 The objective of the Seoul Rules *supra* note 58 as stipulated from Article 1 is: "to prevent or abate the pollution of international groundwaters with special emphasis on the long-term pollution'.

206 *Ibid,* Art. 3 (2-3): "2. Basin States shall consult and exchange relevant available information and data at the request of any of them. (a) for the purpose of preserving the groundwaters of the basin from degradation and protecting from impairment of geologic structure of the aquifer, including recharge areas; (b) for the purpose of considering joint or parallel quality standards and environmental protection measures applicable to international groundwaters and their aquifers. 3. Basin States shall cooperate at the request pf any of them, for the purpose of collecting and analyzing additional needed information and data pertinent to the international groundwaters and their aquifers."

procedures in the context of pollution prevention or abatement of international groundwater.[207]

4.5.4 The Bellagio Draft Treaty

The Bellagio Draft Treaty is based on a comprehensive integrated approach to defining what rules of procedures are needed to ensure successful management of a shared water resource. These are tailor-made to transboundary groundwater and therefore constitute an important relevant guidance for this thesis.

A. The Duty to Exchange Information

The Draft Treaty, similar to the ILA Rules, urges States to establish and maintain reliable data and information concerning transboundary aquifers and their waters in order to use and protect these waters in a rational and informal manner.[208] Article V is devoted to the "Establishment and Management of the Database". It provides that the Commission is charged with the creation and maintenance of a comprehensive and unified database pertaining to transboundary groundwaters. Greater knowledge about the quality and quantity of transboundary groundwater is essential for the improvement of the management of such resources. Regular and systematic collection of hydrometeorological, hydrological and hydrogeological data needs to be promoted and be accompanied by a system for processing quantitative and qualitative information for various types of water bodies.[209]

B. Monitoring and Assessment

In order to fulfil the objectives of data and information exchange between watercourse States the Draft Treaty calls on State to establish observation networks and strengthen existing systems and facilities for measurements and recording fluctuations in groundwater quality and level; organize the collection of all existing data on ground water (borehole logs, geological structure, and hydrogeological characteristics, etc.); systematically index such data, and attempt a quantitative assessment so as to determine the present status of and gaps in knowledge; increase the search for, and determination of, the variables of aquifers.

B. Establishment of Joint Commissions

At the outset of the treaty negotiations the Basin States are encouraged to establish a "Joint Commission" responsible for the implementation of an agreement that

---

[207] *Ibid.*

[134] The Bellagio Draft Treaty, *supra* note 34, Art II Para 2:

"the Parties have entered into this Agreement in order to attain the optimum utilization and conservation of transboundary groundwaters and to protect the underground environment. It is also the purpose of the Parties to develop and maintain reliable data and information concerning the transboundary aquifers and their waters in order to use and protect in a rational and informed manner".

[209] *Ibid,* commentary to Art. V.

includes responsibility for all transboundary groundwaters related matters.[210] The joint Commission is mandated to undertake a variety of tasks and functions to fulfil the requirement of the Treaty.[211]

C. The Duty of Prior Notification

Among the tasks entrusted to this Commission is groundwater quality and quantity protection. This is ensured through the prompt notification on any actual or planned activity that may cause "appreciable harm" to the transboundary groundwaters or recharge areas.[212]

D. Declaration of Transboundary Groundwater Conservation Areas[213]

Also as part of the quality and quantity conservation plan the Joint Commission shall determine the desirability of declaring any area within the border region containing transboundary groundwaters to be a Transboundary Groundwater Conservation Area. These areas are determined based on selected criteria that the Commission has to follow.[214] The Joint Commission is also encouraged to prepare

---

[210] *Ibid,* Article III paragraphs 1-8. The Parties may already have a joint international institution whose powers and functions may readily be expanded to deal with the added responsibilities of transnational groundwater. If no such institution is in place, or the Parties choose to create a separate commission or agency, the name to be given the new entity is, similarly, to be inserted in Paragraph 1. In the case of the United States and Mexico, it is likely that the Parties would choose to designate their existing International Boundary and Water Commission, with which the Governments have a long and satisfactory experience. *See Ibid* commentary to Article III.

[211] *Ibid.*

[212] *Ibid*, Art. VI paras 1-3:1.

"The Parties undertake cooperatively to protect and to improve, insofar as practicable, the quality of transboundary aquifers and their waters in conjunction with their programs for surface water quality control, and to avoid *appreciable harm* in or to the territories of the Parties. 2. The Governments shall promptly inform the Commission of any actual or planned, significantly polluting discharge into transboundary groundwaters or recharge areas, or of other activity with the potential for significant leaching into transboundary groundwaters."

[213] *Ibid* Article III VII (1-7) states

"The Commission is authorized to declare Transboundary Groundwater Conservation Areas, Drought Alerts, Drought Emergencies and Public Health Emergencies and to promulgate the corresponding plans and Depletion Plans, in accordance with the provisions of this Agreement. The Commission shall have jurisdiction over such additional matters concerning the border region as are from time to time referred to it by the Governments jointly".

[214] *Ibid*, Art. VII paras 5 and 6 provide guidance for States on when and why an area should be declared as 'Conservation Area':

"In making its determination, the Commission shall consider whether: a) groundwater withdrawals exceed or are to exceed recharge even so as to endanger yield or water quality or are likely to diminish water, the quantity or quality of interrelated surface waters; b) recharge

a Comprehensive Management Plan for each declared Transboundary Groundwater Conservation Area to ensure the rational development, use, protection and control of the waters in the Transboundary Groundwater Conservation Area.[215] The Master Plan is to include a description of what measures are needed (i) to prevent, eliminate or mitigate degradation of transboundary groundwater quality and quantity;[216] (ii) to allocate the uses of groundwaters and interrelated surface waters taking into account the other allocations previously made applicable within the transboundary groundwater conservation area; [217] and (iii) to limit pumping, set criteria for well placement and number of wells, decisions on retirement of existing wells, imposition of extraction fees, planned depletion regimes[218] or reservations of groundwaters for future use.

---

has been or may become impaired; c) the use of the included aquifer(s) as an important source of drinking source of water has been or may become impaired water; d) the aquifer(s) have been or may become contaminated and; e) recurring or persistent drought conditions necessitate management of all or some water supplies in the particular area. 6. In making its determination, the Commission shall take into account the impact of implementation of the declaration under consideration on the sources and uses of water previously allocated by agreement between the Parties or under the Drought Management Plan".

[215] *Ibid*, Art. VIII. This article includes a list of possible elements to be included in the Plan. Allocation of uses under the Plan shall take into account some factors that were listed under Art. VIII Para (3). This list is based on two sources: the 1966 Helsinki Rules factors *supra* note 47 and Schwebel's Third Report *supra* note 124 to the ILC on the determination of equitable use. The factors include (a) hydrogeology and meteorology; (b) existing and planned uses; (c) environmental sensitivity; (d) quality control requirements; (e) socio-economic implications (including dependency); (f) water conservation practices (including efficiency of water use); (g) artificial recharge potential; and (h) comparative costs and implications of alternative sources of supply. The weight to be given to each factor is to be determined by its importance in comparison with that of the other relevant factors.

[216] *Ibid,* Article VIII Para. 2 stipulates that these measures include the classification, assessment and monitoring of transboundary groundwaters according to use and establishing criteria for safe storage of all possible sources that may cause pollution to aquifers, monitoring of water quality, providing the requirement for establishing protection zones.

[217] *Ibid,* Art. VIII Para. 3, In making any allocation of water uses within a comprehensive management plan, the Commission shall consider all relevant factors such as: (a) hydrogeology and meteorology, (b) existing and planned uses, (c) environmental sensitivity, (d) quality control requirements, (e) socio-economic implications (including dependency), (f) water conservation practices (including efficiency of water uses), (g) artificial recharge potential, and (h) comparative costs and implications of alternative sources of supply. This list resembles in a way the list of factors determining equitable and reasonable utilization in Article 5 of the UN Watercourses Convention, *supra* note 51.

[218] Article X paragraph 1 reads:

4.5.5 Summary

By and large, the legal instruments and documents discussed above have common denominators as far as procedural rules are concerned. These procedures are built on establishing adequate mechanisms of cooperation with special emphasis on the generation of reliable data and information base at the national level and the maintenance of adequate information exchange at the international level. The basic procedural rules that are considered obligatory in the case of connected transboundary groundwater are the duty to exchange information and prior notification. Based on the preceding research, cooperation in itself does not have a normative content. It is a duty that derives from the general principles of sovereign equality, territorial integrity, mutual benefit and good faith of States. States are further encouraged based on their "good faith" to consult and negotiate.

The Bellagio Draft Treaty, which specifically deals with transboundary groundwater, contains more detailed mechanisms and procedures that provide a better guidance in the case of transboundary groundwater. These include in addition to those mentioned earlier the requirement to establish Joint Commissions, the declaration of conservation areas, the integrated master planning, the duty for monitoring and assessment and conjunctive planning with surface water and, finally, procedures on planned depletion.

4.6 INSTITUTIONAL AND DISPUTE SETTLEMENT MECHANISMS

4.6.1 Introduction

Institutions are the vehicle for cooperation,[219] they can as well be the platform where dispute resolution can take place. It is a relatively common feature of legal arrangements established to manage the transboundary watercourses. The functions and authorities of the institutions established for the purpose of implementing the obligations under the legal arrangement may range from pure technical cooperation to full authority over water management and protection plans and activities.[220] The territorial competence may include: the whole drainage basin, including surface waters, ground water, or both, more

---

"The Commission, after evaluation of all relevant considerations, may prepare and, with the consent of the Governments, may approve a plan for the depletion of an aquifer over a calculated period. The plan may apportion the uses and specify the rates and means of extraction of the transboundary groundwaters, and may authorize advances, exchanges and transboundary transfers of water consistent with the objectives of the depletion plan".

[219] UN Natural Resources Series No., *supra* note 13. See also relevant discussion *supra* Introduction section 8.

[220] *Ibid.*

than one drainage basin (multi-basin), part of a drainage basin (sub-basin), an area otherwise defined and clearly delimited or all or part of boundary waters. The institutional mechanism may have a form of a permanent joint body[221], a joint commission convened on a regular basis,[222] or regular meetings of national representatives.

In international law the need for mediation in relation to international watercourses arise mainly when watercourse States are in disagreement and cannot themselves solve their disputes using negotiations. In the words of article 33 of the Charter of the United Nations,[223] they must "Seek a solution by mediation, conciliation arbitration, judicial settlement, resort to regional agencies or arrangements or other peaceful means of their own choice".

In the context of this research, the key issues are to understand whether or not there is a customary duty to establish mechanisms for institutional cooperation and for dispute settlement. It is essential to understand how the various codification and progressive development attempts addressed these issues and what guidance can be derived from that in the case of transboundary groundwater.

---

[221] Illustrative legal arrangements establishing permanent committees (a) The Agreement For The Establishment Of The Organization For Management And Development Of The Kagera River Basin, concluded At Rusomo, Rwanda 24 August, Signatories: Tanzania, Rwanda, Burundi (Entered into force 5th February 1978), and the TFDD treaty collection at http://ocid.nacse.org/qml/research/tfdd/toTFDDdocs/200ENG.pdf (English) United Nations Treaty Series No. 16695, Vol. 1089, pp 172-177., (b) the Convention Relating to the Status of the River Gambia Signed at Kaolack, on 30 June 1978, Parties: The Republic of the Gambia, The People's, The Republic of Senegal (Guinea became a party to the Convention by virtue of resolution 2/CEG/CG of 6 June 1981 of the Conference of Heads of State and Government of the Gambia River Basin Development Organization).

[222] Illustrative examples on such mechanisms are found in: (a) Agreement Between The Government Of The Federal People's Republic Of Yugoslavia And The Government Of The People's Republic Of Albania Concerning Water Economy Questions, Together With The Statute Of The Yugoslav-Albanian Water Economic Commission And With The Protocol Concerning Fishing In Frontier Lakes And Rivers. Parties: Albania, Yugoslavia 12 May 1956Came Into Force on 6 August 1957.*fouund at* http://ocid.nacse.org/qml/research/tfdd/toTFDDdocs/97ENG.htm (*last visited* October 2003). (b) Republic Of Syria and the Hashemite Kingdom Of Jordan Concerning The Utilization Of The Yarmuk Waters. Signed At Damascus, On 4 June 1953. Came into force on 8 July 1953, *found at* http://ocid.nacse.org/qml/research/tfdd/toTFDDdocs/82ENG.htm (*last visited* October 2003).

[223] Charter of the United Nations, Articles 1 and 2 which was confirmed in the preamble of the Convention.3 supra note 4. found at www.icj-icj.org/icjwww/ibasicdocuments/ibasictext/ibasicunchart.htm (last *visited* June 2003).

4.6.2 The Work of the ILC Work

4.6.2.1 The 1991 and 1994 Draft Articles

The 1991 and 1994 articles recommended States to establish joint management mechanisms for the attainment of consultation objectives.[224] It has been left to the discretion of States to decide on the type of arrangement they want to pursue in terms of these mechanism. Management in the context of the Article 24 of the 1991 and 1994 Articles refers to the "(i) planning of sustainable development of an international watercourse and providing for the implementation of plans adopted and (ii) otherwise promoting rational and optimal utilization, protection and control of the watercourse".[225] Dispute settlement mechanisms were not addressed with the 1991 Articles because of lack of time. The 1994 Draft Articles introduced in article 33 a set of procedures that are obligatory in the cases of disputes. These provisions are only applicable in the absence of agreement between disputants on the type of procedures to be followed in situations of conflict.[226] The adoption of the provisions on negotiations is merely codifying a basic rule of international law.[227] The Draft Articles give the flexibility to any of the disputants to request the establishment of a Fact-Finding Commission. This type of Joint Commissions are gaining wide acceptance by States and the UNGA has demonstrated its acceptance of such mechanisms by adopting a Declaration on Fact-Finding in the Field of Maintenance of International Peace and Security.[228] The UNGA had the desire to encourage States to bear in mind the role that competent organs of the UN can play in ascertaining the facts in relation to disputes or situations. [229] In the context of the said Declaration fact-finding meant any activity designed to obtain detailed knowledge of the relevant facts of any dispute or situation, which the competent UN organs need in order to exercise effectively their functions in relation to the maintenance of international peace and security.

---

[224] Article 24 Para (1) of the 1994 Draft Articles, *supra* note 49 read:
"Watercourse States shall, at the request of any of them, enter into consultations concerning the management of an international watercourse, which may include the establishment of a joint management mechanism".

[225] *Ibid,* Para (2) and (3).

[226] First paragraph of Article 33 of the 1994 Draft, *supra* note 49 reads: "In the absence of agreement between the watercourse States concerned, any watercourse dispute concerning a question of fact or the interception or application of the present article shall be settled. …etc."

[227] The ICJ in the North Sea Continental Shelf case between the Federal Republic of Germany and Denmark ruled that: "the parties under dispute are under an obligation so to conduct themselves that the negotiations are meaningful, which will not be the case when either of them insists upon its position without contemplating any modification of it".

[228] A/RES/46/59, 9 December 1991, *found at* http://www.un.org/documents/ga/res/46/a46r059.htm (*last visited* June 2003)

[229] *Ibid*, annex Para 4.

4.6.2.2 Comments and Observations by States to the 1994 Draft Articles

While some States (e.g. Hungary, Finland, Venezuela Sudan) have commended the fact that the 1994 draft articles included provisions concerning the settlement of dispute between parties, which was not the case with the previous draft other States (e.g. Turkey, Israel) considered it inappropriate to foresee compulsory rules as regards the settlement of dispute and to leave that to the discretion of the States concerned.[230] The final contested issue was the new provision on dispute settlement and the discussion focused on the extent that a framework convention should include binding provisions in that regard. Turkey for example expressed its opinion that it was not appropriate for a framework convention to foresee any compulsory rules regarding the settlement of disputes, a matter that should be left to the discretion of States.[231] Israel also had problems with the articles on dispute settlement noting that States must settle their disputes peacefully.[232] Other States like Sudan for example proposed that the submission of a dispute to arbitration or judicial settlement should be made obligatory since making arbitration or judicial settlement subject to the agreement of States concerned may cause damage to the wronged States because of the refusal of another States or States to refer the dispute to arbitration of judicial settlement.[233]

4.6.2.3 The 1997 UN Watercourses Convention

One of the issues at the centre of debate within the Working Group in the Sixth Committee was the new provision on dispute settlement.[234] States like China, India, Israel and Rwanda rejected the compulsion element in the said provision and recommended that a framework convention should not include binding provisions in that regard.[235]

4.6.3 The IDI, ILA and the Bellagio Draft Treaty

The IDI Salzburg Resolution recommended that an appropriate international institution should be established with a mandate to plan the development of the common resource, facilitate economic development and prevent and settle disputes.[236] In case of a conflict,

---

[230] The 1994 Draft Articles *supra* note 49.
[231] U.N. Doc. A/51/275/Add.3.
[232] *Ibid.*
[233] *Ibid.*
[234] Article 33 of the 1994 Draft Articles.
[235] Press Release, GA/9248 21 May 1997.
[236] Salzburg Resolution, *supra* note 33 Art. 9:
"It is recommended that States interested in particular hydrographic basins investigate the desirability of creating common organs for establishing plans of utilization designed to

especially disagreement on the rights of States to use the waters, equity shall be the basis for conflict settlement. [237] If States fail to reach any agreement after negotiating with good faith, the Resolution provides that the case shall be submitted to judicial settlement or arbitration. [238] During negotiations and on the basis of good faith, States are recommended to refrain from undertaking any action such actions without reaching a consensus with the other State/s.[239] Failure of the previous efforts the notifying State can undertake their works ensuring their compliance with the equity requirement.[240] According to the ILA Helsinki Rules States are under the obligation to settle international disputes as their legal rights or other interests by peaceful means consistent with the Charter of the UN. [241] These provisions stipulate that in situations of conflicts States are recommended to follow a similar stepwise approach as in the IDI Salzburg Resolution. Upon failure of negotiations the dispute is to be referred to a joint agency to formulate plans or recommendations for the benefit of interested States. The next step is to seek good offices, or jointly request the mediation and conciliation..[242] The final resort, having failed to resolve the dispute, is to submit the dispute to an ad hoc arbitral tribunal, to a permanent arbitral tribunal or to the ICJ.[243]

According to the Bellagio Draft Treaty, it is recommended to make use of already existing joint bodies where transboundary groundwater element can be integrated within the work of such bodies.[244] In the absence of any form of institutions the newly established "Joint Commission" shall have authority over all matters related to transboundary groundwater.[245] The Joint Mechanism is considered the vehicle through which the implementation of the Agreement shall be made possible. The Joint Commission is not empowered with the authority for

---

facilitate their economic development as well as to prevent and settle disputes which might arise."

[237] *Ibid*, Article 3 states that:

"If states are in disagreement over the scope of their rights of utilization, settlement, will take place on the basis of equity, taking particular account to their respective needs as well as of other pertinent circumstances. "

Equity here is referred to as a general concept to implement the law when defining the scope of water utilization rights of co-basin States taking particular account to their respective needs as well as of other pertinent circumstances.

[238] Salzburg Resolution, *supra* note 11, Art. 6-8.

[239] Article 7: "During the negotiations, every State must, in conformity with the principle of good faith, refrain from undertaking works or utilization which is object to dispute or from taking any other measures which might aggravate the dispute or render agreement more difficult."

[240] *Ibid.*

[241] The 1966 Helsinki Rules *supra* note 47 Chapter 6 Articles XXVI-XXX.

[242] *Ibid*, Articles XXX, XXXI, XXXII.

[243] *Ibid*, Article XXXIV paras a-d.

[244] The Bellagio Draft Treaty, *supra* note 34, Commentary to Article III Para 1, 2 and 3.

[245] *Ibid.*

enforcement and oversight of water quality and quantity measure and related land use within the territory of each Party. The enforcement is the responsibility of the Parties themselves. In this context the role of the Joint Commission is restricted on monitoring and evaluation of these measures as reported by each Party.[246] In cases of differences between the Parties it is stipulated that the Joint Commission shall pursue to resolve these disputes amicably through good faith consultation.[247] Failure of the Joint Commission to bring the disputants into agreement within 12 months allows any of the Governments to invoke Article XV of the Draft Treaty. The content of Article XV is primarily the proposed mechanism for appointing a Commission of inquiry for the purpose of verification of the facts of the situation.[248] Upon failure to implement the requirements of Article XV I the parties are to enter into direct negotiations, mediation, conciliation and finally arbitration to the International Court of Justice or any other means of peaceful settlement.[249]

---

[246] *Ibid,* Article IV
"1. The enforcement of water quality and quantity measures and related land use controls within the territory of each Party shall be the responsibility of that Party or of its political subdivisions as appropriate. 2. The Commission shall biennially conduct a review of the water quality and quantity control measures taken within each Party' s territory affecting the border region and shall issue a Report containing its assessment of the adequacy and effectiveness of programs for the protection and improvement of the transboundary aquifers and their waters and withdrawal and land use controls including with respect to any Transboundary Groundwater Conservation Areas, Depletion Plans, Drought Emergency Plans and Health Emergencies…."

[247] Montevideo Convention on the Rights and Duties of States (Signed at Montevideo, 26 December 1933, entered into Force, 26 December 1934), Article 10 'The primary interest of states is the conservation of peace. Differences of any nature which arise between them should be settled by recognized pacific methods' *found at* http://www.taiwandocuments.org/montevideo01.htm (*last visited* June 2003) The Charter of the United Nation Article 33 states that
"1. The parties to any dispute, the continuance of which is likely to endanger the maintenance of international peace and security, shall, first of all, seek a solution by negotiation, enquiry, mediation, conciliation, arbitration, judicial settlement, resort to regional agencies or arrangements, or other peaceful means of their own choice. 2. The Security Council shall, when it deems necessary, call upon the parties to settle their dispute by such means"
found at http://www.taiwandocuments.org/charter.htm#chapte11. (*last visited* May 2003).

[248] *Ibid,* Art. XV paras 1-5.

[249] *Ibid* , Art XVI

4.6.4 Summary

The 1994 Draft Articles and the 1997 UN Watercourses Convention encourage the establishment of proper institutional mechanisms to facilitate the implementation of the legal arrangements. The final text of the UN Convention did not include details on types or organizational structures of such institutions. As for the dispute settlement mechanisms, these were introduced as compulsory provisions thus presenting a highly controversial progressive development by the ILC. The IDI Salzburg Resolution offers minimum guidance on the type and structure of institutions that should be responsible for the implementation of the Agreements, however the recommendation is clear and calls for establishing a joint institution for the implementation of the Agreement. The Helsinki Rules do not include a section on institutional mechanisms although the role of joint agency was discussed under the prevention and settlement of disputes. Such an institution is envisaged to play a role in situations of conflict. The Bellagio Draft Treaty on the other hand deals with this issue extensively, whereby the joint institution is considered as the backbone of cooperation.

4.7 OTHER SUBSIDIARY SOURCES FOR THE DETERMINATION OF THE LAW

4.7.1 Introduction

As mentioned in the introduction to this chapter, subsidiary means for the determination of the law in the field of transboundary groundwater law include the teachings of most highly qualified publicists, judicial decisions and the outcomes of international conferences and meetings.[250] Of these sources, the teachings of publicists have been mostly covered in the previous discussion of this chapter. The next discussion focuses on judicial decisions and all previous and current efforts to adopting non-binding declarations, principles resolutions that promote equity and fairness in the use of an international groundwater.

4.7.2 Judicial Decision

Federal courts in interstate water disputes employed the principle of equitable apportionment, which is considered as the predecessor to the equitable utilization principle.[251] In the Donauversinkung case[252] involving the German federal States of Prussia against Baden the Staatsgerichtshof had to determine whether the manipulation

[250] These declarations are regarded as soft law – not binding. Soft law is law, but its vague, unclear language doesn't establish responsibilities and duties.

[251] L. Caflish, preamble to P.K Wouters *supra* note 9.

[252] Wurttemberg and Prussia v. Baden (18 June 1927), 4 ADPILC, 1927-1928, 128-133

of the Danube river, so as to augment (in Baden) or to decrease (in Wurttemberg) the effects if the river's natural seepage, was legally permitted.[253] The Court held that neither side could actively increase or decrease the resources of the Danube. The Court found that "the exercise of sovereign rights by every State in regard to international rivers traversing its territory is limited by the duty not to injure interests of the other members of the international community. ....The interests of the States in question must be weighed in an equitable manner against one another. The court ruled that one must consider not only the absolute injury caused to the neighbouring State, but also the relation of the advantage gained by one to the injury caused by the other".[254]

The Trail Smelter Arbitration[255] offers a good example on how tribunals refer to the "no harm" principle by ruling that when there is a serious harm, obligations of States arise.[256] The Convention that was signed by the US and Canada called for the creation of a Tribunal to determine the answers to four questions: "1. Had damage been done to Washington State by the smelter since January 1, 1932? 2. If the smelter was found to have done damage, should it be made to refrain from doing so in the future? 3. Should the smelter operate under any restrictions? 4. Should any compensation be paid in light of the answers".[257] Whilst the Tribunal's decision provided that States have the obligation not to cause harm to others, this decision was balanced by Article 4 of the *Compromis,* which states:

"The Tribunal considered that it should endeavour to adjust the conflicting interests by some just solution which would allow the continuance of the operation of the Trail Smelter but under such restrictions and limitations as would, as far as foreseeable, prevent damage in the United States, and as would enable

---

[253] P.K. Wouters *supra* note 9.

[254] *Ibid,* at 130.

[255] Trail Smelter arbitration (United States v. Canada), Arbitral Tribunal, Montreal 16 April 1938 and 11 March 1941; 3 UN RIAA 1905 (1947); 33 AJIL 182 (1939), also *found at* http://www.jura.uni-muenchen.de/einrichtungean/ls/simma/tel/case8.htm (*last visited* May 2003).

[256] Trail Smelter Arbitration, *Ibid,* the Arbitral Tribunal was asked to determine the extent of Canada's obligations for damage caused in the State of Washington by deleterious fumes emitted by a smelter company in British Columbia. the second decision of the Tribunal was: "In the absence of international cases on the subject, there are certain decisions of the Supreme Court of the United States dealing with both air pollution and water pollution which may legitimately be taken as a guide in this field of international law where no contrary rule prevails in international law and no reason for rejecting such precedents can be induced from the limitation of sovereignty inherent in the Constitution of the United States. The Tribunal finds that under the principles of international law, as well as the law of the United States, no state has the right to use or permit the use of its territory in such a manner as to cause injury by fumes in or to the territory of another or the properties or persons therein, when the case is of serious consequence and the injury is established by clear and convincing evidence".

[257] Trail Smelter Arbitration, *supra* note 255.

indemnity to be obtained, if in spite of such restrictions and limitations, damage should occur in the future in the United States".[258]

Maybe the most remarkable court decision in the field of transboundary water resources is the one taken by the International Court of Justice in the Gabcikovo-Nagymaros case in 1997.[259] The decision was based on an analysis of obligations and duties of Hungary and Slovakia under the Treaty of 16 September 1977 concerning the construction and operation of the Gabcikovo-Nagymaros System of Locks,[260] and in light of the recently adopted 1997 United Nation Watercourses Convention.[261] The Court specifically referred to the modern development of international law, which has strengthened the principle of equitable and reasonable sharing for non-navigational uses of international watercourses as well, -as evidenced by the adoption of the 1997 United Nations Watercourses Convention. The court considered that Slovakia, by unilaterally assuming control of a shared

---

[258] *Ibid.*

[259] Treaty between the Hungarian People's Republic and the Czechoslovak People's Republic on 16 September 1977 Concerning the Construction and Operation of the Gabcíkovo-Nagymaros System of Locks.

[260] Treaty between the Hungarian People's Republic and the Czechoslovak People's Republic on 16 September 1977 Concerning the Construction and Operation of the Gabcíkovo-Nagymaros System of Locks. The Treaty essentially aimed at the production of hydroelectricity, the improvement of navigation on the relevant section of the Danube and the protection of the areas along the banks against flooding. At the same time, by the terms of the Treaty, the contracting parties undertook to ensure that the quality of water in the Danube was not impaired as a result of the Project, and that compliance with the obligations for the protection of nature arising in connection with the construction and operation of the System of Locks would be observed, *found at:* http://www.icj-j.org/icjwww/idocket/ihs/ihsjudgement/ihs_ijudgment_970925_frame.htm (*last visited* June 2003). The Hungarian Government suspended the works at Nagymaros on 13 May 1989 and finally abandoned it on 27 October 1989. Some of Hungary's justifications for abandoning the project and terminating the treaties were related to groundwater. It was considered that as a result of a reduction of the flow in the old bed of the Danube by diversion of water into the bypass canal the groundwater level would have fallen in Hungary. According to Hungary, the principal ecological dangers, which would have been caused by this system, were as follows: At Gabcíkovo/Dunakiliti, under the original Project, as specified in the Joint Contractual Plan, the residual discharge into the old bed of the Danube was limited to 50 m^3/s, in addition to the water provided to the system of side-arms. That volume could be increased to 200 m^3/s during the growing season. Additional discharges, and in particular a number of artificial floods, could also be effected, at an unspecified rate. In these circumstances, the *groundwater* level would have fallen in most of the Szigetköz. Furthermore, the groundwater would then no longer have been supplied by the Danube — which, on the contrary, would have acted as a drain — but by the reservoir of stagnant water at Dunakiliti and the side arms, which would have become silted up. In the long term, the quality of water would have been seriously impaired.

[261] The UN Watercourses Convention *supra* note 33.

resource, and thereby depriving Hungary of its right to an equitable and reasonable share of the natural resources of the Danube- with the continuing effects of the diversion of these waters on the ecology of the riparian area of the Szigetkoz - failed to respect the proportionality which is required by international law.

4.7.3 Summary

The influence of judicial decisions or "case law" on the modern development of international water law cannot be overlooked. [262] Although very few examples that involve transboundary groundwater can be cited, nonetheless, considering the physical unity of the watercourse system, these decisions are apt to give guidance on the relevant rules and principles governing the utilization and development of international water resources including transboundary groundwater.

4.7.4 International, Declarations and Resolutions

The following sections provide an analysis of the relevant international declarations, meetings and resolutions following an ascending chronological order. The analysis employs the same analytical framework used above for each instrument reviewed.

In 1972, the UN adopted the Stockholm Declaration, [263] which encompasses principles that inspire, and guide the people of the world in the preservation and enhancement of the human environment. The 1972 Stockholm Declaration provides an implicit reference to renewable and non-renewable groundwater resources in the context of preserving and safeguarding these resources from depletion and pollution. [264] The Declaration derives its principles from the

---

[262] Case law is not regarded as a means capable of creating international legal rules. They might contribute to the process of forming a custom, but custom itself is a primary source of law. Case law simply sets out either to make known or note the existence of legal rules. For more details see Encyclopaedia of International Law, *supra* note 1 at 474.

[263] Declaration of the UN Conference on the Human Environment(Stockholm Declaration), June 16 1972 *found at* http://www.unep.org/Documents/Default.asp?DocumentID=97&ArticleID=1503 *(last visited* May 2003). [hereinafter referred to as the 1972 Stockholm Declaration]. The natural resources of the earth as stipulated from principle 2 of the Declaration include the air, water, land flora and fauna and especially the representative samples of natural ecosystems.

[264] *Ibid*, Principles 3:
"the capacity of the earth to produce vital renewable resources must be maintained and, wherever practicable, restored or improved."
Principle 5 also provides:
"the non-renewable resources of the earth must be employed in such a was as to guard against the danger of their future exhaustion and to ensure that benefits from such employment are shared by all mankind."

fundamental right of people to freedom, equality and adequate conditions of life and the responsibility to safeguard the environment.[265] There is a further reference to the sovereign rights of States in their exploitation of their own resources based on the principles of international law and those embodied within the Charter of the UN. This is a clear reference to the equitable and reasonable utilization rule that includes a degree of factor of harm.[266] There is an obligation to protect all natural resources for the benefit of present and future generations.[267] Pollution prevention is considered as a goal that States should endeavour to achieve recognizing the role of water resources in social and economic development.[268] The procedural rules embodied within the 1972 Stockholm Declaration emphasize the role of science and technology. It is confirmed that such role is essential for the identification, avoidance and control of environmental risks.[269] The Declaration recommends States to follow an ecosystem-integrated coordinated approach stressing the need for rational planning to reconcile conflicts between needs of development and the need to protect the environment.[270]States are encouraged to deal with international matters concerning the protection and improvement of the environment in a cooperative spirit and on equal footing and to conclude cooperative arrangements essential for serving their mutual interests.[271] There is also a need to establish the appropriate institutions to undertake the planning, management and protection of the resources.[272]

UNEP has been active also in the development of non-binding instruments in the field of the environment. Those non-binding instruments adopted by the

[265] *Ibid*, Principle 1 and 2.

[266] *Ibid*, Principle 21.

[267] *Ibid*, Principle 2:

"The Natural resources of the earth, including air, water, land, flora and fauna and especially representative samples of natural ecosystems, must be safeguarded for the benefit of present and future generations through careful planning or management, as appropriate".

Principle 13:

"States should adopt an *integrated and coordinated approach* to their development planning so as to ensure that development is compatible with the need to improve environment for the benefit of their population".

[268] *Ibid*, Principles 8, 9 and 10.

[269] *Ibid* , Principle 18, 19 and 20.

[270] *Ibid* , Principles 13, 14 and 15.

[271] *Ibid*, Principle 24:

"International matters concerning the protection and improvement of the environment should be handled in a cooperative spirit by all countries, big and small, on equal footing. Cooperation through multilateral arrangements or other appropriate means is essential to effectively control, prevent, reduce and eliminate adverse environmental effects resulting from activities conducted in all spheres, in such a way that due account is taken of the sovereignty and interests of States."

[272] *Ibid*, Principle 17.

Governing Council of UNEP or intergovernmental meetings convened by UNEP include the Principles of Conduct in the Field of the Environment for the Guidance of States in the Conservation and Harmonious Utilization of Natural Resources Shared by Two of More States.[273] Such resources implicitly include groundwater as a natural resource which is an integral part of the environment.[274] The Principles stressed the importance of applying the rule of equitable and reasonable utilization in controlling, preventing, reducing or eliminating adverse environmental effects which may result from the utilization of such resources.[275] The principles also encourage States to conclude bilateral agreements to secure the harmonious utilization of shared natural resources. UNEP's Principles also point out the importance of establishing institutional structures, such as joint commissions.[276] The principles encompass procedural rules that are recommended for the fulfilment of the objectives of equitable and reasonable utilization. These include the duty to exchange information and consultation,[277] prior notification[278] to be undertaken on the basis of the principle of good faith and in the spirit of good neighbourliness. The principles further introduce mechanisms for dispute settlement on the basis of UN Charter and Declaration of Principles of

---

[273] Governing Council Decision 6/14, Principles of Conduct in the field of the Environment for the Guidance of States in the Conservation and Harmonious Utilization of Natural Resources Shared by Two or More States, Nairobi, 19 May 1978 *found* in FAO, Sources of International Water Law, Rome, 1998.

[274] Interpretation by the author.

[275] *Ibid*, Principle 1: "It is necessary for States to co-operate in the field of the environment concerning the conservation and harmonious utilization of natural resources shared by two or more States. Accordingly, it is necessary that consistent with the concept of equitable utilization of shared natural resources, States co-operate with a view to controlling, preventing, reducing or eliminating adverse environmental effects which may result from the utilization of such resources. Such co-operation is to take place on an equal footing and taking into account the sovereignty, rights and interests of the States concerned".

[276] *Ibid*, Principle 2: "In order to ensure effective international co-operation in the field of the environment concerning the conservation and harmonious utilization of natural resources shared by two or more States, States sharing such natural resources should endeavour to conclude bilateral or multilateral agreements between or among themselves in order to secure specific regulation of their conduct in this respect, applying as necessary the present principles in a legally binding manner, or should endeavour to enter into other arrangements, as appropriate, for this purpose. In entering into such agreements or arrangements, States should consider the establishment of institutional structures, such as joint international commissions, for consultations on environmental problems relating to the protection and use of shared natural resources".

[277] *Ibid*, Principle 5.

[278] *Ibid*, Principle 6(a-d).

International Law concerning Friendly Relations and Co-operation among States.[279]

In 1981, the participant in the UN interregional Meeting of International River Organizations held, in Dakar, emphasized the need to recognize the interrelationships between surface waters and groundwater as well as the interrelation with the underground environment and the atmosphere.[280] This meeting confirmed that States have not yet developed a good understanding of the significance of these interrelationships and urged them to entrust their international river and lake organizations with the task to initiate studies and to call for hydrogeologic data. Additionally, the meeting encouraged governments to look into the possibilities of conjunctively use surface and groundwater resources with a view to develop conservation and protection measures of the underground environment.[281]

In 1987, the UNECE adopted the "Principles Regarding the Cooperation in the Field of Transboundary Waters".[282] The UNECE Principles demonstrated awareness of the interrelation between surface and groundwater and declared that through cooperation effective prevention and control of pollution is possible.[283] The UNECE went on to declare that transboundary waters "do not lend themselves to purely national approaches because natural phenomena and human activities… may make themselves felt across borders and require cooperation".

In 1988, a group of responsible government officials and interregional experts elaborated on the principles that were discussed in Dakar in 1981. At an interregional meeting held in Addis Ababa many issues that relate to

---

[279] *Ibid,* Principle 11: "The relevant provisions of the Charter of the United Nations and of the Declaration of Principles of International Law concerning Friendly Relations and Co-operation among States in accordance with the Charter of the United Nations apply to the settlement of environmental disputes arising out of the conservation or utilization of shared natural resources. 2. In case negotiations or other non-binding means have failed to settle a dispute within a reasonable time, it is necessary for States to submit the dispute to an appropriate settlement procedure which is mutually agreed by them, preferably in advance. The procedure should be speedy, effective and binding. 3. It is necessary for the States parties to such a dispute to refrain from any action which may aggravate the situation with respect to the environment to the extent of creating an obstacle to the amicable settlement of the dispute".

[280] Experiences in the Development and Management of International River and Lake Basins, UN publication, 10 Natural Resources/Water Series (1983).

[281] *Ibid* Conclusion No. 6, at 14.

[282] Resolution I (42), UN doc. E/ece/(42)/L.19, 30 April 1987.
"that prevention and control of transboundary pollution in rivers and lakes...and in related groundwater aquifers, as well as prevention and control of floods are important and urgent tasks whose effective accomplishment can only be ensured by enhanced cooperation among riparian countries."

[283] *Ibid*, at 13.

transboundary groundwater were evoked.[284] The objectives of these deliberations meeting was achieving the optimum cooperation between or among basin States. It was recommended that the "drainage basin" and the "system approach" concepts are keys to achieving cooperation and agreements. The meeting declared the need to apply the general principles of international law that to transboundary water resources. These rules include the equitable utilization principle, which encompasses the factor of degree of harm, the duty to exchange available relevant information and data, the duty to notify and to consult reciprocally with co-basin States that may be adversely affected by a project or program planned by one or more States and the duty to consult concerning the institutionalization of cooperation or collaboration for basin development upon request of any other basin State.[285]

In 1992, the Dublin Statement on Water and Sustainable Development[286] was adopted. The Statement calls for a holistic approach to planning and managing water resources necessary to safeguard aquatic ecosystems and make their benefits available to society on a sustainable basis.[287] The principles encompassed within the Statement provide guidance for States for the protection of water resources from over consumption, pollution, floods and droughts at the national and international levels.[288] The principles stress the basic right of all human beings to have access to clean water and sanitation at an affordable price as means for achieving equitable utilization and encouraging water conservation.[289] The principles also emphasize the vulnerability of water resources due to their finite nature.[290]

---

[284] UN Economic Commission for Africa (UNECA) and Department of Technical Cooperation for Development, Interregional Meeting on River and Lake Basin Development with Emphasis on the Africa Region, Addis Ababa, Ethiopia, 10-16 October 1988, Report of the Meeting at 36-38.

[285] Participants in the meeting disclosed many reservations regarding the duty to exchange data or to consult reciprocally and stated that even if there is a moral obligation to do that this must proceed on the basis of agreement.

[286] The Dublin Statement on Water and Sustainable Development 1992 *found at* http://www.gdrc.org/uem/water/dublin-statement.html (*last visited* April 2003)

[287] *Ibid*, Principle 1

"....Since water sustains life, effective management of water resources demands a holistic approach, linking social and economic development with protection of natural ecosystems. Effective management links land and water uses across the whole of a catchment area or groundwater aquifer."

[288] *Ibid,* Guiding principles.

[289] *Ibid,* Principle 4:

"Water has an economic value in all competing uses and should be recognized as an economic good."

[290] *Ibid,* Principle 1:

The Rio Declaration on Environment and Development[291] of 1992 recommends the governments to establish equitable partnership through cooperation with a vision to conclude international agreements which respect the interests of all and protect the global environment.[292] Similar to the Dublin Statement of 1992, the Rio Declaration refers to the sovereign rights of States to exploit their own resources based on international law and the Charter of the UN along with an obligation not to cause harm to other States.[293] There is a duty to cooperate between states in order to achieve the goals of sustainable development, eradicate poverty and decrease disparities in standards of living. Cooperation is also indispensable for States to improve the scientific knowledge and understanding over their shared environment and natural resources.[294] States are encouraged to solve their disputes amicably and peacefully in accordance to the principles of the UN Charter.[295]

Chapter 18 of Agenda 21 confirms what the Rio Declaration has already embraced. This chapter includes more elaboration on the integrated approaches to development including the integration of land and water activities, management and use of water resources and greater emphasis on the need to cooperate and establish mechanisms to solve transboundary issues.[296] This Chapter recommends the ecosystem approach referring to unitary nature of fresh water and the interconnectedness of the elements related to fresh water quality.[297] Additionally, reference was made to the importance of considering the human basic needs and need to safeguard the ecosystem when prioritizing uses.[298] The text includes recommendations and action plans pertinent to the implementation of the principles in specific areas.[299] There is a special consideration of the need to

---

"Fresh water is a finite and vulnerable resource, essential to sustain life, development and the environment…"

[291] Rio Declaration On the Environment and Development (3-14 June 1992) UN Doc. A/CONF.151/26 also found at http://www.unep.org/unep/rio.htm (last visited April 2003).

[292] *Ibid*, Annex I Preamble.

[293] *Ibid*, Principle 2

[294] *Ibid,* Principles 5-9.

[295] *Ibid,* Principle 26.

[296] Agenda 21, Chapter 18, *supra* note 88.

[297] *Ibid*, Para. C.

[298] *Ibid*, Para. 18.8 and 18.9.

[299] *Ibid*, the programme areas are: (a) Integrated water resources development and management, (b) Water resources assessment, (c) protection of water resources, water quality and aquatic ecosystems, (d) Drinking water supply and sanitation, (e)Water and sustainable development, (f) Water for sustainable food production and rural development, and (g) Impacts of climatic change on water resources.

understand the extent and severity of contamination of unsaturated zones and aquifers and to develop a good base of knowledge on aquifer systems.[300]

The Ministerial Declaration of The Hague on Water Security in the 21st Century adopts the ecosytemetic approach and advocates the integrated water resources management that takes account of economic and environmental factors and integrates surface and groundwater and the ecosystems through which they flow.[301] The declaration calls for equitable allocation of water for food production and the need to consider the principle of equity in assessing the economic, social, environmental and cultural values of water.[302] The principle of Polluter Pays, including the implementation of appropriate measures in the fields of liability and compensation is considered. There is a strong recommendation to develop collaboration and partnership in order to build a secure and sustainable water future.[303] This Declaration also refers to the challenges of water security, namely the human basic need to access safe and sufficient water and sanitation that are essential to health and well-being.[304]

In the Johannesburg Summit of 2002 the participants reaffirmed their commitment to sustainable development and to the building of a "humane, equitable and caring global society cognizant of the need for human dignity for all." The participants recognized that "overarching objectives of, and essential requirements" for sustainable development are poverty eradication, changing consumption and production patterns, and protecting and managing the natural resource base for economic and social development.[305]

Finally, in the Kyoto Ministerial Declaration[306] integrated management of water resources has been emphasized.[307] The Declaration further reaffirms the

---

[300] *Ibid*, item 18.37
"The extent and severity of contamination of unsaturated zones and aquifers have long been underestimated owing to the relative inaccessibility of aquifers and the lack of reliable information systems. The protection of groundwater is therefore an essential element of water resources management."

[301] The Ministerial Declaration of The Hague on Water Security in the 21st Century, 22 March 2000, the Hague. Para. 3. *found at* http://www.idi.or.jp/vision/declarat.htm (*last visited* May 2003).

[302] *Ibid.*

[303] *Ibid*, para.10.

[304] *Ibid*, para.3

[305] found at http://www.un.org/jsummit/html/documents/summit_docs/0409_l6rev2_pol_decl.pdf (last visited May 2003).

[306] The Third World Water Forum Ministerial Declaration, Ministry of Foreign Affairs, Kyoto 2003, *found at* http://www.mofa.go.jp/policy/environment/wwf/declaration.html (*last visited* April 2003).

[307] *Ibid*, Para. 2 and 10.

need for countries to enhance their monitoring and assessment capacities at local, basin and national levels and to develop the appropriate indicators at the national level.[308] States were encouraged to promote cooperation on transboundary and/or boundary watercourses confirming that such cooperation contributed to sustainable water management and mutual benefits.[309]

4.7.5 Relevant International Initiatives and Programmes

There are other initiatives of relevance to transboundary groundwater such as the UNECE groundwater programme.[310] This Programme started in November 1995 with the drafting of the guidelines on monitoring and assessment of transboundary groundwaters. The outcome of the work in phase one was the endorsement of the Guidelines on Monitoring and Assessment of Transboundary Groundwaters.[311] The groundwater guidelines are based on and supported by four technical background reports. The Implementation of the Groundwater Guidelines started in 2000. For this second phase an overall working programme on the implementation of the guidelines on monitoring and assessment of transboundary groundwaters has been drafted.[312] Another very significant initiative on transboundary groundwater aquifers is the one by the International Association of Hydrogeologists (IAH) who established a Commission to

---

[308] *Ibid* Para. 8.

[309] *Ibid*, Para. 11

[310] The working programme is based on the results of a preparatory study in which relevant developments have been investigated important for the implementation and testing of the groundwater guidelines. Relevant developments are the experiences and lessons learned from the pilot studies on the implementation of the river guidelines , the co-operation with UNESCO-IAH-FAO in the so-called ISARM/TARM project and the European Water Framework Directive (WFD). UNECE Task Force on Monitoring and Assessment, under the Convention on the Protection and Use of Transboundary Watercourses and International Lakes (Helsinki 1992), *Inventory of Transboundary Groundwaters* , Vol. 1, ISBN 9036952743. Lelystad, 1991, *see also* UNECE Task Force on Monitoring and Assessment, under the Convention on the Protection and Use of Transboundary Watercourses and International Lakes (Helsinki 1992), Problem-Oriented approach and the Use of Indicators, Vol. 2, ISBN 9036952751, Lelystad, 1991, *see also* UNECE Task Force on Monitoring and Assessment, under the Convention on the Protection and Use of Transboundary Watercourses and International Lakes (Helsinki 1992), Application of Models, Vol. 3, ISBN 90369527X, Lelystad, 1991, *see also* UNECE Task Force on Monitoring and Assessment, under the Convention on the Protection and Use of Transboundary Watercourses and International Lakes (Helsinki 1992), State of the Art on Monitoring and Assessment of Groundwaters, Vol. 4, ISBN 9036952778, Lelystad, 1991.

[311] *Ibid.*

[312] Dutch Institute for Inland Water Management and Waste Water Treatment (RIZA), *found at* http://www.iwac-riza.org/IWAC/iwacsite.nsf/0/a4d3390ec02e4e9bc1256aa0003eb929?OpenDocument *(last visited* May 2004.

investigate the issue in 1999. This initiative is known as the International Shared Aquifer Resources Management (ISARM).[313] The main objectives of the said initiative are to support co-operation among countries to develop their scientific knowledge and to eliminate potential for conflict, particularly where conceptual differences might create tensions. It aims to train, educate, inform and provide inputs for policies and decision-making, based on good technical and scientific understanding.[314]

4.7.6 Summary

Although international declarations and statements are considered as soft law with a non-binding nature, they present a crucial subsidiary source for the determination of the law. These instruments also appear to be more responsive to the requirements of equitable and sustainable regulation and management of transboundary groundwater. The "drainage basin" concept has been unanimously recommended as the proper hydrological management unit in relation to transboundary groundwater management and protection.

The concept of integrated water resources management with the possibility of conjunctive use with surface water has been promoted to optimally regulate manage and develop such resources. The principle of equitable and reasonable utilization has recurrently appeared in the outcomes of international conferences and meetings as the governing principles for transboundary groundwater utilization and development. Cooperation, including the exchange of appropriate

---

[313] TARM is an international initiative on Internationally Shared (Transboundary) Aquifer Resource Management. This initiative has been adopted by resolution XIV-12 during the 14th Session of the Council of UNESCO's International Hydrological Programme (Paris, 5-10 June 2000). For this programme co-operation has been established with UNECE for the implementation of guidelines on monitoring and assessment of transboundary groundwaters and the setting up of training and exchange of information with the other regions. The Internationally Shared Aquifer Resources Management Program ISARM operates through a co-ordination committee drawn from UNESCO, FAO, UNECE and the International Association of Hydrogeologists (IAH). Each of these organizations provides a specific contribution to this multidisciplinary initiative aiming to improve understanding of scientific, socio-economic, legal, institutional and environmental issues surrounding the management of transboundary aquifers. The growing concern of the international community for the development and proper management and legal frameworks governing international water resources has been manifested in numerous forums. This has been reflected in the work of international governmental and non-governmental organisations and in the writing of scholars and publicists as means of focusing the wide attention on the question of development and management of water resources.

[314] S. Puri (editor), B. Appelgren, G. Arnold, A. Aureli, *Internationally Shared Aquifer Resources Their Significance and Sustainable Management: A Framework Document, found at* http://www.fao.org/Legal/Prs-OL/isarm.pdf (*last visited* May 2003).

data and the establishment of proper institutions, has been promoted as the vehicle for achieving equitable and reasonable utilization.

4.8 CONCLUSIONS

The preceding analysis finds that international law provides the mechanisms for amicable and peaceful solutions over the utilization, development and protection of shared water resources. The modern development of the law in the field of transboundary groundwater demonstrates an increased awareness of the current and emerging water crises, of the risks from the uncontrolled use of waters that cross borders between two or more States, and the importance of international cooperation in resolving conflicts over international waters.

In the context of "scope and use of terms", and as reflected in the codification and progressive development of governmental and non-governmental organizations, there is a clear consensus to consider connected transboundary groundwater as an integral component of the overall hydrological cycle. However, this consideration varies substantially within the various legal instruments reviewed. On the one hand non-governmental organizations and groups have confirmed that the "drainage basin" is the most appropriate hydrogeological unit for the utilization, development and protection of transboundary groundwater. Accordingly, the 1961 IDI Salzburg Resolution, the 1966 ILA Helsinki Rules and its complementary Seoul Rules, the Bellagio Draft Treaty, and most declarations and conferences relating to groundwater, adopt the "drainage basin" approach which better reflects the interrelationship between groundwater and the surrounding environment. On the other hand, the ILC uses the term "international watercourse system", which was a compromise that satisfied the States' reluctance to accept the "drainage basin" approach.

Non-renewable groundwaters in confined aquifers were only covered in the scope of the Seoul Rules and the Bellagio Draft Treaty. The two instruments provide that the same rules and principles that apply to connected transboundary aquifers apply to those that are confined. Such resources, however, were completely excluded in the final text of the 1997 UN Watercourses Convention. In this regard, this research concludes that despite the obvious recognition by the ILC experts to the importance of including confined aquifers within the scope of the 1997 UN Watercourse Convention, as reflected in the ILC "Resolution on Confined Transboundary Aquifers", States played the decisive role in the final product. This ambiguity in the treatment of these resources calls for further study this type of transboundary groundwater and to elaborate rules and principles that govern their utilization, development, and protection which is currently taking place in the ILC.

As for the substantive rules, the above review confirms that there is a collective consensus that equitable and reasonable utilization is a fundamental principle governing the utilization, development, management and protection of transboundary groundwater. There is an inherent relationship between this principle and the "no harm" rule and there are enough grounds to conclude that the "no harm" rule is subordinate to the equitable and reasonable utilization principle. The operationalization of the latter principle requires that hydrologic, economic, social, and political factors are taken on the account with the final decision to be made based on the whole.

The procedural rules in the case of transboundary groundwater are the most distinct as compared to those relating to international surface water. The Bellagio Draft Treaty which presents the most advanced framework in relation to transboundary groundwater regulation, offers the best suited mechanisms and procedures for the protection, utilization development and management of such resources.

The provisions on dispute avoidance/settlement found within the Helsinki Rules and the Bellagio Draft Treaty, are innovative and therefore inspiring. The procedures can be modified and adapted to suit the specific groundwater course situation and be as well procedures for cooperation and collaboration. In summary, the Helsinki Rules and the Draft Treaty agree that the first step in dispute settlement is negotiation and consultation. As a second step and having failed to reach an agreement the intervention of a third party becomes essential. The proposal embodied within the 1997 UN Watercourses Convention and the Bellagio Draft Treaty to appoint a "Fact Finding Commission" or a "Commission of Inquiry", respectively is important for the verification of facts especially when one party has more control of relevant data and information. The two instruments agree that last resort in all cases would be mediation, conciliation and arbitration.

From this research perspective, it is proposed that the 1997 UN Watercourses Convention is the most authoritative statement developed so far in the field of non-navigational uses of international watercourse, including transboundary groundwaters. The 1997 UN Watercourses Convention, however, fails to address the specific needs of optimal and sustainable utilization and development of transboundary groundwaters. Additionally, the Convention does not adequately and comprehensively respond to management and regulation needs of transboundary groundwater. To compensate for this gap, it is suggested here to derive the detailed procedures, mechanisms and required institutions contained in the Bellagio Draft Treaty. It is therefore proposed that an agreement that draws lessons from the 1997 UN Watercourses Convention and the Bellagio Draft Treaty might provide an ideal alternative for the regulation of utilization of transboundary groundwater. If States intend to apply the above proposition, they have to resolve two important questions at the outset of their negotiations. The first relates to the scope and whether or not confined transboundary groundwaters are included, and second to decide on the geographic/hydrographic scope and how broad or narrow they are willing to be in their approach.

Finally, this chapter included some indications on what States accept and what they resist as far as transboundary groundwater law is concerned. It is observed, for example, that States accept, in general, the inclusion of connected transboundary groundwater aquifers within the scope of the legal arrangements, but not those which are confined. States also appear to resist concepts that have implications on their territorial sovereignty such as the “drainage basin” one. A more in-depth analysis of States’ positions and practice in that regard is reviewed in the next chapter.

5

Transboundary Groundwater in State Practice

5.1 INTRODUCTION

In Chapter Three an enquiry into the modern development of international law relating to transboundary groundwater was made guided by Article 38 of the ICJ Statute.[1] It was concluded that in accordance with the principles of international law, States have the

[1] Article 38 (1) of the Statute of International Court of Justice ICJ states:
"(1)The Court, whose function is to decide in accordance with international law such disputes as are submitted to it, shall apply: a. international conventions, whether general or particular, establishing rules expressly recognized by the contesting states; b. international custom, as evidence of a general practice accepted as law;......"
found at : http://www.un.org/Overview/Statute/chapter2.html, (*last visited* April 2003).

sovereign right to equitably and reasonably exploit shared resources.[2] According to Article 38 of the ICJ Statute, international treaties and conventions are considered among the primary sources of international law that merit the greatest consideration in the creation of the law.[3] This chapter partly investigates some selected State treaty practice in the field of transboundary groundwater. This is done with the view of assessing whether or not there is enough evidence derived from such practice of the existence of international customs that are considered as law. This part of the chapter also aims to assess the relevance of existing legal arrangements in offering guidance to the progressive development of the law governing transboundary groundwaters. The other part of the analysis focuses on some technical non-binding co-operative arrangements concerning, in particular, confined groundwaters in confined aquifers. Two questions are addressed in this particular context: firstly, why do States choose these types of arrangements and not a comprehensive binding treaty? And what best practice relating to transboundary groundwater, if any, might be derived from such arrangements to be applied for the purposes of this research?

5.1.1 Background

International treaties are the major instruments of cooperation in the international relations and often involve a change in the relative positions of the States concerned.[4] After the Second World War, the general trend has been to enhance the role of treaties in international law making, partly in response to increasing interdependence, and partly as a solution to the controversies that existed between diverse groups of States as to the content and validity of older customary rules. Malanczuk argues that to some extent treaties have begun to replace customary law.[5] However, treaties as a source of international law are different from custom. While the latter is able to lay down legal rules for an unlimited number of subjects, a treaty is characterized by the fact that the rules created under it are binding only upon those subjects which are party to it.[6]

---

[2] See *supra* Chapter 4.

[3] The Vienna Convention on the Law of Treaties of 1969 (Signed at Vienna, 23 May 1969, entered into force, 27 January 1980) refers to the increasing importance of treaties. Paragraph 2 of the preamble states: "Recognizing the ever-increasing importance of treaties as a source of international law and as a means of developing peaceful co-operation among nations, whatever their constitutional and social systems." *found at* http://www.taiwandocuments.org/glossary1.htm#exchange (*last visited* April 2003), *See also* Encyclopaedia of International Law, Volume IV, (Elsevier: the Netherlands 2000), at 469 and 471, *see also* P. Malanczuk at Modern Introduction to International Law, (Routledge: New York, 1997) at 39-41, *and* I. Brownlie, Principles of Public International Law (5th ed.), (Oxford University Press: New York, 1999) at 4.

[4] *See* P. Malanczuk, supra note 3, at 37

[5] *Ibid.*

[6] *See* Encyclopaedia of International Law, supra note 3, at 469 and 471.

Additionally, the fact that a treaty exists does not necessarily imply that it is in force.[7] Even if the latter condition applies, the implementation on the ground might be totally variant and contradictory. The analysis of the treaties within this chapter has been conducted with due attention given to the foregoing facts. Whilst thousands of treaties have been concluded on international watercourses, State treaty practice is still unclear concerning transboundary groundwater.[8] States can choose to be a party to a various range of arrangement, formal and informal.[9]

Additionally States can enter into non-binding cooperation "arrangements" This fact is increasingly confusing the task of assessing what State practice is. In the absence of a formal agreement States are satisfied to maintain less formal co-operative arrangements. Bilateral and multilateral agreements that are concluded between States are usually a good evidence of States' willingness and acceptance to be bound by the rules and principles embodied in these agreements.[10] The significance of treaties derives from the fact that they are not concluded in vacuum, rather they are the accumulation and conglomeration of norms, customs and general principles.[11]

Notwithstanding the foregoing, and given the wide range of arrangements that States can be party to, treaties do not necessarily always contribute to the

---

[7] The Vienna Convention on the Law of Treaties, *supra* note 3 Article 2(b) refers to the various of instruments of ratifying a treaty which consist of: 'ratification', 'acceptance', 'approval' and 'accession' mean in each case the international act so named whereby a State establishes on the international plane its consent to be bound by a treaty'. The United Nations Treaty collection includes a guiding glossary of terms for the various terms to 'treaty action'. *Found at* http://www.taiwandocuments.org/glossary2.htm (*last visited* May 2003).

[8] A collection of treaties can be *found at* http://www.dundee.ac.uk/law/iwlri/Research_Documents_International.php (*last visited* May 2003), and at the Transboundary Freshwater Dispute Database (TFDD) at the Department of Geosciences at Oregon State University 1820-2002 *found at* http://www.transboundarywaters.orst.edu/ (*last visited* April 2003) and UN Treaty Collection *found at* http://untreaty.un.org/English/treaty.asp (*last visited* April 2003) and FAO Legislative Studies No. 50, 55 and 61. The latter *found at* http://www.fao.org/docrep/W7414B/W7414B00.htm (last visited May 2003), see also United Nations, Office of Legal Affairs, Codification Division, *The Law of International Groundwater,* December 1990, and D.A. Caponera and D. Alheritiere, *Principles for International Groundwater Law,* in L. Teclaff, L. & A. Utton (eds.) 25 International Groundwater Law (1981) also found in 18 Natural Resources Journal 589-618 (1978), and L.A. Teclaff and A.E. Utton, International Groundwater Law(Oceana Publication: New York, 1981) at 189.

[9] *See* generally *supra* note 2.

[10] *Ibid.*

[11] R.D. Hayton, *The Formation Of The Customary Rules Of International Drainage Basin Law*, in Garretson, Hayton and Olmstead, The Law of International Drainage Basins, 834 Oceana Publications; New York (1967).

formulation of general international law rules or principles, nor do they always codify an existing custom.[12] Except for *jus cogens* matters, States might agree to replace or modify the recognized general rule with a particular rule by special agreement. [13] Other types of treaties aim to confirm general rules or to apply these rules. The goal of this research is to identify where treaties or other types of States cooperation could be of significant contribution to the codification and progressive development of international law relating to the use of shared transboundary groundwaters.[14]

5.1.2 Transboundary Groundwaters in State Treaty Practice

The analytical reviews by scholars of State treaty practice relating to groundwater indicate that the regulation of transboundary groundwater aquifers is very limited.[15] Within this research a range of treaties relating to international watercourses were analysed. In total, approximately 423 treaties related to international watercourses were found, of which approximately 402 treaties were reviewed.[16] Of the 402 treaties, 59 treaties mention groundwater either explicitly or implicitly in the three selected regions (detailed analysis found in Annex I to III). The results of the survey indicate that early agreements dealing with border/frontier questions addressed mainly international surface water, including rivers, lakes and canals, with some implicit reference to groundwater (springs, wells, subsoil, water system, basin system, bodies of waters...etc.). Such reference was mostly intended to deal with a single topic like protecting existing rights or uses,[17] or restricting pumping on one segment of the boundary.[18] Some of these treaties

---

[12] *Ibid.*

[13] Article 38 of the ICJ Statute, *supra* note 1.

[14] The Ministerial Declaration of the second World Water Forum held in The Hague in 2000 on Water Security in the 21st Century (The Hague 22 March 2000) draws attention to the problems of managing shared water resources to promote peaceful co-operation and develop synergies between different users. This includes co-operation between adjacent states in the case of shared groundwater resources, *found at* http://www.worldwaterforum.net/Ministerial/declaration.html *(last visited* July 2003).

[15] S. McCaffrey, International Groundwater Law: Evolution and Context, found in S. Salman (ed.) Groundwater: Legal and Policy Perspective (World Bank Technical Paper No. 456: Washington, 1999). *See also* R.D. Hayton, The Law of International Aquifers, 22 NRJ 71-92 (1982) *and* D.A. Caponera, *supra* note 9, L.A. Teclaff, *supra* note 8.

[16] The TFDD *supra* note 8 includes summaries of what these treaties are, those which were found relevant to the topic were translated. 35 Agreements were in languages other than English or Arabic.

[17] Illustrative of agreements of this category are the following: the Agreement of Good Neighbourly Relations concluded between the British and French Governments on behalf of the Territories of Palestine, on the one Part, and on behalf of Syria and Great Lebanon, on the other Part. Signed at Jerusalem, February 2, 1926, *found at* http://ocid.nacse.org/qml/research/tfdd/toTFDDdocs/36ENG.htm (*last visited* May 2003).

included groundwater within their scope and mostly addressed water quality protection,[19] water allocations, and cooperation.[20] As a result, the survey finds that very few treaties

---

See also Agreement between the Republic of Syria and the Hashemite Kingdom of Jordan concerning the utilization of the Yarmouk waters, 6/4/1953 (Came into force on 8 July 1953), *found at* http://ocid.nacse.org/qml/research/tfdd/toTFDDdocs/82ENG.htm (*last visited* May 2003), *and* the 1947 Treaty of Peace between the Allies and Italy which outlines guarantees between Italy and Yugoslavia concerning springs in the Commune of Gorizia, United Nations Legislative Texts and Treaty Provisions Creating the Utilization of International Rivers for Purposes Other than Navigation, Treaties No. 6 and 120 U.N. Doc. ST/LEG/SER.B/12 (1963) *also found at* http://ocid.nacse.org/qml/research/tfdd/toTFDDdocs/394ENG.htm (*last visited* May 2003).

[18] Convention on Water Distribution Of Al-Asi River Between Syria And Lebanon, signed on the 20 September 1994 and came into force December 11th 1994 (In Arabic), *found at* http://ocid.nacse.org/qml/research/tfdd/toTFDDdocs/273ARA.htm. (*last visited* April 2003).

See also Item 5 of the Minute No. 242 between the United States and Mexico on the Colorado and Tijuana Rivers which states:

"Pending the conclusion by the Governments of the United States and Mexico of a comprehensive agreement on groundwater in the border areas, each country shall limit pumping of groundwaters in its territory within five miles (eight kilometres) of the Arizona-Sonora boundary near San Luis to 160,000 acre-feet (197,358,000 cubic meters) annually" Minute No. 242 of the International Boundary Water Commission (30 September 1973) of the Treaty between the United States of America and Mexico Relating to the Waters of the Colorado and Tijuana Rivers, and of the Rio Grande, 1944, Treaty Series No.994 UNTS 313, United States Treaties and Other International Agreements, vol. 24, part 2 (1973) (Washington D.C. United States Government Printing Office, 1974).

[19] *See* the Convention Concerning the Protection of Italo-Swiss Waters against Pollution between Italy and Switzerland, signed at Rome on 20 April 1972, Article 1 states:

"The Contracting Governments agree to collaborate closely with a view to protecting against pollution the following Italo-Swiss surface and groundwater including surface and groundwater of their affluent…….".

found at http://ocid.nacse.org/qml/research/tfdd/toTFDDdocs/259ENG.pdf (*last visited* April 2003).

[20] Illustrative examples are: (a) Agreement between the Federal Republic of Nigeria and The Republic of Niger Concerning the Equitable Sharing in The Development, Conservation and Use of Their Common Water Resources Done At Maiduguri, 18 July 1990, *found at* http://www.fao.org/docrep/w7414b/w7414b05.htm#TopOfPage (*last visited* May 2003)[hereinafter referred to as the 1990 Niger Nigeria Agreement] (b) the Convention Creating the Niger Basin Authority 1980 Signatories: Benin, Cameroon, Côte d'Ivoire, Guinea, Burkina Faso, Mali, Niger, Nigeria, Chad, (entered into force 21 January 1990) *found at* http://ocid.nacse.org/qml/research/tfdd/toTFDDdocs/374ENG.htm (*last visited* May 2003). (c) the Treaty of Peace Between The State of Israel and The Hashemite Kingdom Of Jordan, Done at Arava/Araba Crossing Point On 26 October 1994 *found at* http://ocid/nacse.org/qml/research/tfdd/toTFDDdoc/168ENG.htm (*last visited* May 2003) (d) The Israeli-Palestinian Interim Agreement on the West Bank and Gaza Strip, Washington DC, 28th December 1995 *found at*

deal with the principles governing the utilization, management, development and protection of shared transboundary groundwaters.[21]

The statistical analysis of the reviewed treaties demonstrates that of the 72 treaties that cover connected groundwater either explicitly or implicitly, 32 are in Europe, followed by Africa, which has 19. Only few agreements were identified in Asia (5), the Middle East (8), and North America (8). The following sections analyze closely a selection of these treaties in representative regions. The selection criteria for the regions were based on the following: (i) importance of transboundary groundwater for the region (ii) availability of relevant State treaty practice (iii) availability of relevant information and data. Based on the selection criteria and the results of the treaty survey, Africa, Europe and the Middle East were selected as regions with relevant treaty practice requiring closer analysis and scrutiny.

The choice of the Middle East is particularly important to this work for many reasons, including: (i) the importance of groundwater resources in the region; (ii) the existing conflicts between many neighbours in the region on the shared groundwater resources and the potential for further escalation; (iii) the general lack of sound State treaty practice for cooperation and dispute settlement and avoidance in the region; (iv) the vulnerability of the water situation, which is highly influenced by the prevailing politics; and (v) The Middle East is the region of the case study.

---

http://www.ejil.org/journal/Vol4/No4/art8.html (*last visited* May 2003), (e) The Convention regarding the Water Supply of Aden between Great Britain and the Sultan of Abdali, April 1910 which deals with digging of wells and substitutes in case of failure in addition to obligations to safeguard the quality of the well, *found at* ocid.nacse.org/qml/research/tfdd/toTFDDdocs/21ENG.htm (*last visited* May 2003).

[21] The Convention between the Canton of Geneva in Switzerland and the Department of Upper Savoy in France in 1978 is one good example on such agreements. The Convention deals primarily with artificial recharge of the Geneva Aquifer and the regulation of its management protection and recharging. (Copy received from *Service Cantonal de Géologie in Geneva* through Mr. Michel Meyer), [hereinafter referred to as the 1978 Franco-Suisse Convention]. The Franco-Swiss transboundary Genevese Aquifer (See : Figure 1) extends over 19 km between the southern extremity of Lake Geneva and its effluent, the Rhone river which crosses the Canton of Geneva westwards. The width of the aquifer varies between 1 km and 3.5 km. It is located partly on the southern border of the Canton of Geneva with the French Department of Haute-Savoy.

5.2 SIGNIFICANCE OF GROUNDWATER IN THE SELECTED REGIONS

5.2.1 Africa

Water in Africa is abundant on a regional scale but is unevenly distributed.[22] Although a few African countries have high annual averages of rainfall, many already or soon will face water-stress or scarcity conditions where the population cannot be sustained with available water resources. Given current population projections, over 400 million people are expected to be living in at least 17 water-scarce African countries by the year 2010. Their lack of water will severely constrain food production, ecosystem protection and socio-economic development.[23] Africa has over 50 significant water basins covering nearly all the countries.[24] In Sub-Saharan Africa (SSA), international river basins constitute the principal source of water resources.[25] Groundwater, on the other hand, is extremely important in Africa. It is estimated that more than 75% of the African population uses groundwater as the main source of drinking water supply,[26] while groundwater presents 15% of Africa's water resources.[27] In Algeria, for example, more than 60% of all water withdrawals are from groundwater, and in Libya 95 % of all withdrawals are from groundwater.[28] Increases in demand for freshwater are anticipated in all African countries, in all sectors over the next 10 years. In some countries, demand is projected to double within the next 30 years.[29] In the parts of Africa, where recharge is poor, much of the water is fossilized and cannot be used sustainably. In many areas, withdrawals from the non-renewable groundwater aquifers (confined aquifers) are dramatically increasing.[30]

---

[22] African Ministerial Declaration at the International Conference on Freshwater in Bonn, Germany, 3 to 7 December 2001, *found at* http://www.thewaterpage.com/Documents/bonn_africa.doc. (*last visited* May 2003).

[23] *Ibid.*

[24] S.M.K. Donkor and E. Wolde, *Integrated Water Resources Management In Africa: Issues And Options* United Nations Economic Commission for Africa, Addis Ababa 1997, *found at* http://www.ramsar.org/wurchbk4cs1.doc (*last visited* June 2003).

[25] *Ibid.*

[26] G. Khroda Strain, Social and Environmental Consequences, and Water Management in the Most Stressed Water Systems in Africa, *found at* http://www.idrc.ca/books/focus/804/chap7.html (*last visited* May 2003).

[27] UNEP, Africa Environment Outlook: Past, Present and Future Perspectives, 2002 *found at* http://www.unep.org/aeo/153.htm (*last visited* June 2003).

[28] *Ibid.*

[29] *Ibid.*

[30] UNEP, *supra* note 27. For example current annual rates of groundwater withdrawal in the sub-region are 407 per cent of the recharge rate in Egypt, and 560 per cent in Libya.

The severity of the water problem varies from region to region and accordingly the importance of groundwater is not the same across the African countries. For example, the Northern African sub-region is mostly arid and deserts are dominating.[31] Groundwater is an important source of water supply in this region. Renewable groundwater resources are in the form of shallow alluvial aquifers, recharged from the main rivers (for example, the alluvial aquifer beneath the Nile delta in Egypt) or from precipitation (along the north African Mediterranean coast). In the Sahara desert, the major water resources are the combined Nubian Sandstone Aquifer System (NSAS) and the Kalahari non-renewable confined aquifer, which extend from Egypt to Mauritania. These aquifers are mostly non-renewable (confined aquifer) whose exploitation over the past ten years has led to a reduction in water pressure levels at the oasis of the western desert. Over-extraction from the delta shallow aquifer has led to increased water salinization and a rapid inland advance of the saltwater interface.[32]

Eastern Africa experiences high variability in rainfall over time and space, including frequent episodes of flooding or drought. Eastern Africa, on the whole, is fairly well endowed with freshwater, with total average renewable freshwater resources amounting to 187 km^3/yr. Uganda has the largest share of this, with 39 km^3/yr, whilst Eritrea has the least, with 2.8 km^3/yr. Surface water dominates freshwater resources in Eastern Africa (the groundwater resources of Ethiopia and Eritrea, for example, are just 2.6% of the total resources for these countries).[33]

Southern Africa is mostly semi-arid, and experiences variation in rainfall, both over time and between countries. Groundwater is the main source of water for about 60 per cent of both rural and urban residents throughout Southern Africa. [34]

---

[31] With the exception of parts of southern Sudan and an intermittent narrow strip along the Mediterranean shoreline, where the climate is more humid. *See generally supra* note 25.

[32] *Ibid.*

[33] M. Ramazzotto, Readings in African Customary Water Law, (FAO Legislative Study No. 58, Rome, 1996). The drier countries in the Horn of Africa (Ethiopia, Eritrea and Somalia) frequently experience drought, and have been devastated by drought-induced famine on several occasions over the past 30 years. The largest freshwater source in eastern Africa is Lake Victoria, the second largest lake in the world. Lake Victoria provides freshwater to the populations of Uganda, Kenya and Tanzania directly and, through the Nile River, to Sudan and Egypt. It is also the life and livelihood support of millions of people living around the lake, providing: fish; irrigation water; tourism and recreation; communications; and transport. Other major freshwater lakes in eastern Africa include: Lake Tanganyika in Tanzania; Lake Edward, Lake George, Lake Kyoga and Lake Albert in Uganda; Lake Turkana in Kenya; and eleven freshwater lakes in Ethiopia.

[34] United Nations Environmental Program (UNEP), *Regional Challenges for Sustainability*, 2000, *found at* http://www.uneptie.org/media/review/vol25no1/IEvol25no1.pdf (*last visited* June 2003). For example, Angola, the wettest country in the sub-region, has average annual internal water resources of 184 km^3/yr (14 000 m^3/capita/yr), and Mozambique and Zambia

A large part of the sub-region is characterized by small towns, villages and dispersed rural settlements. Thus, access to surface water resources has been limited because of the high costs and long distances that need to be covered in order to establish infrastructure for formal water services. The sustainable management of groundwater is, therefore, important for rural livelihoods in the sub-region, even though the proportion of water coming from groundwater sources is relatively small.

Central Africa rarely experiences problems of water availability, because rainfall is high and generally predictable. With the exceptions of the deserts of northern Chad and the Sahelian parts of northern Cameroon and central Chad, the Central Africa region is well-endowed with water resources. Lake Chad is an important source of water and economic activities, including agriculture and fisheries. Satellite images show that the lake has shrunk over the past 30 years, and is now 5 per cent of its former size, due to persistent low rainfall in the region.[35] The Intergovernmental Panel for Climate Change (IPCC) predicts reduced rainfall and run-off, and increased desertification, risks in the Sahelian belt which could mean further reductions in the size of the lake.[36] This would threaten the livelihoods of fishermen and farmers in the region, and poses a challenge for Integrated Water Resources Management and international cooperation to ensure sustainability of the resource.[37]

Most Western African countries are well-endowed with freshwater, except for those bordering on the Sahel, which frequently experience drought. Three major types of groundwater aquifers are observed in the region, namely: basement aquifers; deep coastal sedimentary aquifers; and superficial aquifers.[38] The availability of groundwater varies considerably from one type of substrate to another, and depends on the local levels of precipitation and infiltration, which determine the actual recharge. In Mauritania, for example, internal renewable groundwater resources are estimated at 0.3 km^3/yr, and these are important sources of water for domestic use, irrigation and livestock watering. About 400 000 people

---

have 100 km^3/yr and 80 km^3/yr respectively (5 000 m^3/capita/yr and 8 700m^3/capita/yr). By contrast, the driest countries, Botswana and Namibia, have just 2.9 km^3/yr. and 6.2 km^3/yr respectively (1 700m^3/capita/yr and 3 500m^3/capita/yr respectively).

[35] The Lake Chad basin is a depression of the seven countries grouped around it, forming a freshwater lake (the Conventional Basin), which is shared by Cameroon (9 per cent), Chad (42 per cent), Niger (28 per cent) and Nigeria (21 per cent), *see generally supra* note 27.

[36] International Program for Climate Change (IPCC), *Africa to suffer most under Global Warming,* 2001, *found at* http://eces.org/articles/static/98264880059304.shtml (*last visited* May 2003).

[37] *Ibid.*

[38] *Ibid.*

live in the 218 oases, and are dependent on 31 400 wells, extracting the water manually.[39]

The treaty survey that was conducted within this research revealed that 19 of the total 64 treaties mentioning connected groundwater were concluded in Africa (5 implicitly and 14 explicitly). The survey reveals that the treaties that were signed after 1970 in Africa demonstrate greater awareness to the question of transboundary groundwater utilisation and development. The scope of earlier treaties and arrangements covered navigation, protection of the resources and co-operation. Non-renewable confined groundwater aquifers were not included in any of the treaties reviewed. Of the 19 treaties, six relate to international water resources in Central Africa (five in the Niger Basin and one in the Chad Basin) which, as was described earlier, is region with great reliance on groundwater. Therefore, connected groundwater constituted part of the water systems was dealt with in all these agreements. In the Southern African sub-region 10 Agreements mentioning connected groundwater were identified.

5.2.2 Europe

Water resources range from 3000 mm (annual average runoff) in western Norway to 100-400 mm in much of Central Europe and less than 25 mm in central and southern Spain.[40] Groundwater is a major source for drinking water around Europe. For Austria, Denmark and Portugal it represents more than 75% of the public water supply, and in Belgium (Flanders), Finland, France, Germany, Luxembourg, UK and the Netherlands between 50 and 75 % of the public water supply is abstracted from groundwater.[41] The status of groundwater in terms of quality and quantity is therefore essential. Groundwater also plays an important role within the environment, both for aquatic and terrestrial ecosystems. Some countries depend heavily on water coming from beyond their borders. Water pollution is a serious issue; some progress has been made in Western Europe, but less in Central and Eastern Europe.[42]

The water situation in Europe is alarming.[43] The groundwater lying under 87 per cent of the agricultural area in the EU contains more than 25 milligrams of

---

[39] FAO, Le foncier-environnement - Fondements juridico-institutionnels pour une gestion viable des ressources naturelles renouvelables au Sahel, 1997, (on file with author).

[40] UNEP, *supra* note 27.

[41] See generally www.eeb.org/activities/water/EAF%Groundwater%issues%20%paper.pdf. (*last visited* May 2003).

[42] *Ibid.*

[43] B. Bosnjakovic, *The Regulation of International Watercourses Under the UNECE Regional Agreements*, 25 Water International 544-552 (2000), between 1990 and 1995 the membership of the UNECE increased from 34 to 55 countries including 27 countries in transition from a centrally planned to a market economy. Not much knowledge is available

nitrates per litre.[44] For a quarter of the agricultural land, the level has even risen above the drinking water limit for nitrate of 50 milligrams per litre.[45] Nitrate pollution is worst in the heavily farmed regions of North-western Europe where groundwater quality is also damaged by high pesticide inputs. Nitrate and pesticide levels are also rapidly rising in Southern European states. At least two thirds of all drinking water in the EU comes from groundwater reserves. These valuable underground resources are not only jeopardized by pollutants, but also by excessive abstraction.[46]

In the context of international watercourses the "precautionary principle" appears to have a unique importance in Europe. The precautionary principle was originally adopted by the Swedish and German environmental law. Some States prefer the term "precautionary approach" over "precautionary principle", with the former offering greater flexibility and will be less restrictive than principle'. In their perspective precaution is as an obligation of diligent prevention and control that depends on the magnitude and probability of harm.[47]

---

on the extent of dependency of Europe on this resource. Quoting the Environment Commissioner Margot Wallström who said in September 2003: "Water is a divine gift, but much abused. We depend on groundwater, and we must safeguard it. At the moment, we do not even have enough data about the quality of this vitally important resource in Europe...."

[44] K. Lanz, and S. Stefan, J. Hontelez (ed.), EEB Handbook on EU Water Policy under the Water Framework Directive (EEB: Brussels, 2001), found at http://www.rivernet.org/general/handbook.pdf (*last visited* November 2003).

[45] *Ibid.*

[46] *Ibid.*

[47] States adopting this interpretation consider that this obligation must take into account their capabilities, their economic and social priorities, the cost effectiveness of preventive measures and the nature and degree of environmental risk. The USA supported the precautionary 'approach' rather than 'principle' which was globally endorsed in the 1992 Rio Declaration In sum the precautionary principle is not globally applied. State practice has demonstrated some hesitance in adopting and applying this concept equally in all fields. Within this research these two principles shall not be considered as substantive rules rather they will be treated as procedural rules since they entail activities and measures that need to be done in order to guarantee the basic rule which is the equitable and reasonable utilization principle. For details *See* P. Bernie and A. Boyle, International Law and the Environment, (Oxford University Press: New York, 2002), Also See J. Tickner and N. Myers, *Precautionary Principle: Current Status and Implementation* Centre for Sustainable Production, University of Massachusetts, 2000 *found at* http://web.greens.org/s-r/23/23-17.html (*last visited* May 2003). Tickler explains that the "Precautionary Principle" was first employed internationally in the North Sea Conference in 1984 and later supported by the 1990 Bergen Ministerial Declaration on Sustainable Development and by the 1992 Rio Declaration on Environment and Development. European treaties in general refer to 'precautionary principle' and it is perceived that each EU member state accepts 'precaution' as a general principle of environmental policy. The Danish Environmental Protection Agency held a national conference in 1998 to examine how the principle should be

5.2.2.1 Groundwater Law Evolution under the Auspices of the EU

Water is one of the most comprehensively regulated areas of EU environmental legislation.[48] The process of developing European Union water policy began in the 1970s with the First Environmental Action Programme in 1973. This effort was followed by a first wave of legislation, starting with the 1975 Surface Water Directive and culminating in the 1980 Drinking Water Directive. The produced water legislation included bathing waters (1976), water quality standard legislation on fish waters (1978), shellfish waters (1979), and groundwaters (1980). In the field of emission limit value legislation the Dangerous Substances Directive (1976) and its daughter directives on various individual substances were adopted.[49]

In the early 1990's a review of existing legislation was conducted and as a consequence the gaps and essential improvements were identified. This phase of water legislation included the Urban Waste Water Treatment Directive (1991) and the Nitrates Directive (1991). Other elements identified were revisions of the Drinking Water and Bathing Water Directives to bring them up to date, the development of a Groundwater Action Programme and a 1994 proposal for an Ecological Quality of Water Directive.[50] Also, for large industrial installations, the integrated Pollution Prevention and Control (IPPC) Directive that was adopted in 1996 covered water pollution as well.[51] During this time it had become, after extended discussion at Member States' and Community level, increasingly clear that an efficient protection of water needed emission limit value legislation as well as water quality standards legislation[52], i.e. a so-called "combined approach". This combined approach is also in accordance with principles established in the

---

implemented. The British, Scottish, and Swedish governments have been undergoing similar exercises. The European Environment Agency will soon release a report examining several case studies of where precaution was and was not taken and offer recommendations for its implementation. Several other countries including Hungary and Brazil have adopted precaution as a guiding principle.

[48] Guide to the Approximation of EU Environmental Legislation: *found at* http://europa.eu.int/comm/environment/guide/contents.htm (*last visited* June 2003).

[49] *Ibid.*

[50] European Environment Agency Report (EEA), found at http://reports.eea.eu.int/92-9167-025-1/en/page005.html (*last visited* October 2003).

[51] Council Directive 96/61/EC of 24 September 1996 Concerning Integrated Pollution Prevention and control Official Journal L 257, 10/10/1996, P.0026–0040, *found at* http://europa.eu.int/smartapi/cgi/sga_doc?smartapi!celexapi!prod!CELEXnumdoc&lg=en&numdoc=31996L0061&model=guichett *(last visited* June 2003).

[52] H. Bloch, *European Water Policy Facing the New Millennium: The EU Water Framework Directive*, *found at* www.ecologic.de/interaction/CaseStudies/Water_Framework_Directive.pdf (*last visited* June 2003).

Treaty[53] -the precautionary principle and the principle that environmental damage should as a priority be rectified at the source, as well as the principle that environmental conditions in the various regions shall be taken into consideration. On 26 February 1997, the EC published a proposal for a Council Directive establishing a Framework for Community Action in the field of water policy. The aim of this proposed directive is twofold: to offer an integrated framework for the EU water policy;[54] and to improve the protection of surface and groundwater, in particular to reach a 'good water status' within 16 years after adoption.[55]In Europe the WFD[56] presents a good example on modern trends in legal treatment of water resources including groundwater. It is meant to be a State and interstate European Union (EU) related legal instrument which aims at improving the protection of surface and groundwater.[57] In September 2003, the EC adopted a proposal for a

---

[53] *Ibid,* Community legislation and the jurisprudence of the Court of Justice are the sources of law. The Single European Act (SEA) came into force in 1987, followed by the Treaty on European Union TEU, which entered into force in 1993, and most rntly the Treaty of Amsterdam adopted in 1997 but not yet in force.

[54] The European Union WFD 2000/60/EC, published in the Official Journal of the European Communities on 22 Dmber 2000, (entered into force on that date). The fifteen Members States of the Community are: Austria, Belgium, Denmark, Finland, France, Germany, Greece, Ireland, Italy, Luxembourg, The Netherlands, Portugal, Spain, Sweden, and the United Kingdom: *found at* http://www.europa.eu.int/eur-lex (*last visited* July 2003).

[55] *Ibid.*

[56] WFD, *supra* note 54. The WFD sets out the following basic provisions in Article 4, 7 and 11. Member States shall implement the measures nssary to prevent or limit the input of pollutants into groundwater and to prevent deterioration of the status of all bodies of groundwater.(i) Member States shall also protect, enhance and restore all bodies of groundwater with the aim of achieving good status by 2015. (ii) Member States shall implement the measures nssary to reverse any significant and sustained upward trends in the concentration of any pollutant resulting from the impact of human activity in order to progressively reduce pollution of groundwater. (iii) Member States shall ensure the nssary protection for bodies of water used for the abstraction of water intended for human consumption or intended for such future use and they shall take measures to safeguard water quality in order to reduce the level of purification treatment required for the protection of drinking water, (iv) Member States have to adopt a set of basic measures that shall include a prohibition of direct discharges of pollutants into groundwater. The basic definitions for good groundwater status are set out in Annex V of the WFD. Good groundwater chemical status is considered when the chemical composition of the groundwater body is such that the concentration of pollutants do not exhibit the effects of saline or other intrusions, and do not exceed the quality standards related to Article 17. The chemical compound of the groundwater body must not be such that would result in failure to achieve the good chemical or ecological status in associated surface waters nor any significant damage to terrestrial ecosystems depending on the groundwater body."

[57] European Freshwater Program, *found at* www.wwfreshwater.org/initiative/wfd.html (last visited March 2003). The WFD has been included under the discussion of Modern Trends

new Directive to protect groundwater from pollution. The proposed Directive is aimed at decisively improving the quality of Europe's groundwater. It introduces monitoring requirements and quality objectives, obliging Member States to monitor and assess groundwater quality and to identify and reverse trends in groundwater pollution.[58]

5.2.2.2 Groundwater Law Evolution under the Auspices of the UNECE

The increase in the membership of the UNECE and the emergence of new countries with new economies, and frontiers has created more transboundary situations and issues. Between 1980 and the present, the UNECE has been active in adopting declarations and decisions concerning the protection of and international cooperation over transboundary watercourses including connected groundwater. In 1980, for example, the UNECE adopted a Declaration of Policy on Prevention and Control of Water Pollution, including Transboundary Pollution, [59] which stresses the need for the conservation of water resources and the prevention and control of water pollution.[60] The Declaration recognizes the interaction between the various component of the environment, including air land and water, and points out the need to protect them from pollution in an integrated manner.[61] Groundwater is explicitly considered as part of the water system whose quality and quantity protection has to be undertaken alongside with the other components of the environment.[62] In 1982 the UNECE adopted the Decision on International Co-operation on Shared Water Resources,[63] which includes connected groundwater within its scope.[64]

---

in national regulations in Chapter 2 section 1.4. It is considered to be an interstate legal instrument that regulates interstate watercourses rather than an international agreement. Nevertheless it is a good guidance for States in their practice at the international level.

[58] EU Document IP/03/1280 *found at* http://www.eu-oplysningen.dk/emner/politikker/miljoe/RAPID/ (*last visited* May 2004).

[59] Declaration of Policy on Prevention and Control of Water Pollution, including Transboundary Pollution (Decision B (XXV))- Geneva, 1980, *found in* FAO, Sources of International Water Law, Rome, 1998.

[60] *Ibid,* Principle 1.

[61] *Ibid,* Principle 2: "Water pollution control should be handled taking account of possible interactions of pollutants on air, land and water."

[62] *Ibid,* Principle 3: "The aim of water pollution control is to preserve, as far as possible, the natural quality of surface and ground water, to protect the environment which depends on such water, and to decrease existing levels of water pollution in order to protect public health and to allow the satisfaction of the needs of such water, under the best economic conditions and in sufficient quantity.."

[63] Decision on International Co-operation on Shared Water Resources (D (XXXVII))- Geneva, 1982, *found in* FAO, Sources of International Water Law, Rome, 1998.

[64] *Ibid,* Preamble para 1: "Recognizing the growing significance of economic, environmental and physical interrelationships between UNECE countries, in particular

The Decision encourages the member Governments to strengthen their efforts in the field of international cooperation on the utilization of international watercourses and to effectively make use of administrative and institutional arrangements.[65] The Protocol on Civil Liability and Compensation for Damage Caused by the Transboundary Effects of Industrial Accidents on Transboundary Waters[66] to the Helsinki Conventions[67] was formally adopted and signed by 22 countries[68] at the Ministerial Conference in Kiev. The Protocol is a legally binding instrument that must be ratified by the parliaments of at least 16 countries before it becomes enforceable.[69] As this instrument is linked to the Helsinki Conventions, the term "transboundary waters" include connected transboundary groundwater.[70] The Protocol provides a comprehensive regime for civil liability and for adequate and prompt compensation for damage caused by the transboundary effects of industrial accidents on transboundary waters.[71]

Finally, in an attempt to enhance Europe's knowledge concerning its transboundary aquifers, and under the auspices of the UNECE, a report for transboundary Aquifers (1999) has documented over 100 aquifers that are shared by two or more states. The report demonstrates that the UNECE countries have concluded regional watercourses agreements, which promote conflict prevention and dispute settlement in the field of transboundary watercourses.[72] However, only

---

where streams or lakes and related ground water aquifers cross or are located on international boundaries".

[65] *Ibid,* Principle 3: "Encourages member Governments to continue their efforts to extend already existing international arrangements in the light of changing socio-economic requirements or of changing priorities in the utilization of shared water resources and in particular to speed up procedures within international river commissions and competent governmental bodies to cope efficiently with emergency situations and/or to embody sufficient power to existing legal and administrative arrangements to deal with such situations."

[66] The Protocol on Civil Liability and Compensation for Damage Caused by the Transboundary Effects of Industrial Accidents on Transboundary Waters, Kiev, Ukraine 21 May 2003, *found at* http://untreaty.un.org/English/notpubl/27_16E.doc (*last visited* May 2004).

[67] The 1992 Helsinki Convention on the Protection and Use of Transboundary Watercourses and International Lakes *infra* note 110, and to the 1992 Convention on the Transboundary Effects of Industrial Accidents, *found at* http://www.unece.org/env/teia/text.htm (*last visited* May 2004).

[68] Armenia, Belgium, Bosnia and Herzegovina, Bulgaria, Cyprus, Denmark, Estonia, Finland, Georgia, Greece, Hungary, Latvia, Lithuania, Luxembourg, Monaco, Norway, Portugal, the Republic of Moldova, Romania, Sweden, Ukraine and the United Kingdom.

[69] *Supra* note 66, Article 29.

[70] *See* relevant discussion *infra* section 4.3.3.

[71] *Supra* note 66, Article 1.

[72] B. Bosnjakovic, *supra* note 42.

a few cooperative arrangements have been identified in relation to the utilization and development of these aquifers. [73]

5.2.3 The Middle East

The area of the Middle East within this research encompasses the region between the Nile and the Tigris-Euphrates river systems and extending southward to encompass the Arabian Peninsula and the Gulf Islands, a little bigger than what is sometimes called the Mashrek. For most countries in the Middle East, water is the limiting resource for development. Iran, Iraq, Lebanon, Sudan, Syria, and Turkey [74] Israel, Egypt,[75] Jordan, and Palestine are in the Middle group, while the countries of the Arabian Peninsula are the least well endowed with water. The Middle East region depends upon two main sources of water: rivers and aquifers. Rivers are the best-known sources of water, and the rivers in this region include two of the greatest in the world, the Nile and the Euphrates. There are many short streams or ephemeral wadis, typically being fed by springs in the mountains and spilling into the sea. Aquifers of various types are also common. Some are replenished regularly by rainfall and constitute renewable; others contain water buried in deep confined aquifers and therefore non-renewable; a few others occur along fracture zones.

Many of the aquifers found in the Middle East region are shared between two or more States. Jordan, for example, shares some of its most important water resources with its neighbouring countries. [76] The significant international watercourse in the region is the Jordan River, which has an average annual flow of 1300 mcm.[77] One of the most important shared water resources includes the groundwater resources of the north Jordan "Basalt Aquifer", where a large percentage of the natural recharge occurs in the Syrian territories. The Qa Disi

---

[73] Illustrative Examples are: the aquifer system of the Vechte Catchment that is shared by Germany and the Netherlands and the alluvial aquifer system is shared by the Czech Republic and Poland, see in E. Almássy, and Z Buzás, Inventory of Transboundary Groundwaters (UNECE Task Force on Monitoring and Assessment, Lelystad, 1999).

[74] D. Brooks, *Between the Great Rivers: Water in the Heart of the Middle East*, found in : Eglal Rached et al (eds.), Water Management in Africa and the Middle East: Challenges and Opportunities (International Development Research Centre (IDRC): Ottawa, Canada, 1996), *Also found at* http://www.idrc.ca/books/focus/804/chap4.html (*last visited* April 2003)

[75] Egypt is sometimes classified as a North African Country as well as a Middle Eastern Country.

[76] M. Abu-Taleb, *Environmental management in Jordan: Problems and Recommendations*, 21 Environmental Conservation 35-40 (1994).

[77] *Ibid.*

Aquifer is a non-renewable aquifer shared with Saudi Arabia,.[78] One current question on sustainable water development in Jordan is whether it can afford to continue to develop fossil groundwater in deep sandstone aquifers, such as the Disi, and if so then for what purposes and for how long.[79] Disputes are appearing to arise between Saudi Arabia and Jordan over their shared aquifer. Although currently utilised almost exclusively by Saudi Arabia, Jordan's vested interest in the aquifer, which runs beneath both countries, has increased in recent years, with Jordan's Minister of Agriculture accusing Saudi Arabia of overuse of the aquifer as far back as November 1992. [80] Saudi Arabia is rapidly approaching a dramatic crisis over water where the problem stems from the country's lack of rivers and permanent bodies of water, as a result of which it relies heavily upon underground water sources for its agricultural and potable water supply.

Lebanon, Syria, Israel, Jordan and Palestine share the Jordan River and its tributary the Yarmouk. Most of the Jordan River flow originates from outflow from springs and other groundwater sources. Accordingly, the management and utilization of the River Basin cannot overlook the fact that connected groundwaters have to be managed too.[81] All international efforts to reach a settlement on the Jordan River Basin failed, partly, because the Arab States, especially Jordan, did not need a comprehensive water development plan directly involving Israel in order to achieve their immediate development goals.[82]

The Mountain Aquifer shared between the Palestinians and Israel can also be cited as a significant source of water supply for Israel and Palestine with disputes arising around its utilization. The complex natural, political, economic and social circumstances associated with this aquifer system were discussed in chapter 3. This chapter analyzes the signed treaties relating to this aquifer. The following sections employ the same analytical framework that was used in chapter 3 to analyse the manner in which States in the selected regions have dealt with their transboundary groundwater. The justification and relevance of each element in the analytical framework were explained earlier in Introduction.

---

[78] Water in Jordan, *found at* http://www.emwis-jo.org/Intro_wtr_Resources.htm#Introduction (*last visited* May 2003).

[79] A. Wolf and M. Murakami, *Techno-political Decision Making for Water Resources Development: The Jordan River Watershed, found at* http://www.oieau.fr/semidoc/SEMCOM/TEXTE1.htm. (*last visited* May 2003).

[80] L. Karim, *Wet Conflicts, found at* http://www.alshindagah.com/may2000/specialfeature.htm. (*last visited* May 2003).

[81] J. Dellapenna, Water in the Jordan Valley: The Potential and Limits of Law, V PYIL 15-49 (1989).

[82] Dating back to 1913, many plans relating to the Jordan River Basin were initiated. Starting with the Franghia plan in 1913 and ending with the Johnston Jordan Valley Plan in 1955.

5.3 SCOPE AND USE OF TERMS

5.3.1 Introduction

Annexes I to III include an analysis of all the treaties in Africa, Europe and the Middle East that were considered to directly or indirectly cover connected transboundary groundwaters within their scope. Some of the treaties employ general terms like "River Basin", the "waters" or the "resources of the River Basin" with no explicit reference to groundwater.[83] The terms can be subject to varying interpretations by the various parties to the agreements, which might provoke a dispute. Within this research, agreements which do not exclude groundwaters from their scope were considered to cover connected transboundary groundwater. The scope of most of the treaties in Africa covers three main categories: (i) conservation and development of the water resources (eight treaties), (ii) planning and study (two),[84] and (iii) setting up institutional mechanisms (nine treaties). In Europe, however, the scope of the earlier bilateral

---

[71] The Act Regarding Navigation and Economic Co-Operation between the States of the Niger Basin of 1963 refers to the River Niger Basin term without specifying what this term embraces, Article 5 states:

"In order to further their co-operation for the attainment of the objectives of this Act, the riparian States ....the exploitation of the resources of the 'River Niger Basin....'".

Entered into Force in February 1966, Parties Cameroon, the Republic of Chad, the Republic of Dahomey, the Republic of Guinea, the Republic of the Ivory Coast, the Republic of Mali, the Republic of Niger, the Federal Republic of Nigeria, the Republic of Upper Volta. (United Nations, Treaty Series, vol. 587, No. 8506, p. 9, *also found at* FAO Treaty Collection No. 2 *found at* http://www.fao.org/docrep/W7414B/w7414b04.htm#TopOfPage (*last visited* April 2003), (d) the Agreement for the Establishment of the Organization for Management and Development of the Kagera River Basin of 1977, which refers to the general term 'River Basin and considers this unit as a geographical unit which is interpreted here to include groundwater, the Preamble of the Agreement Para 4 states:

" Recognizing that the Kagera River Basin constitutes a geographical unit offering a valuable base for fruitful cooperation between Riparian Countries"

Entered Into Force 5th February 1978, Signatories: Tanzania, Rwanda, Burundi, FAO Treaty Collection No. 8. *found at* http://www.fao.org/docrep/W7414B/w7414b0a.htm#TopOfPage (*last visited* May 2003).

[84] Illustrative examples on such Agreements are (a) the Action Plan for the Environmentally Sound Management of the Common Zambezi River System Signed at Harare, 28th May 1987, which offers a plan for integrated management for all the resources in the region without specifically refer to groundwater *found at* ocid.nacse.org/qml/research/tfdd/toTFDDdocs/177ENG.htm (*last visited* June 2003) (b) the Tripartite Ministerial Meeting Of Ministers Responsible for Water Affairs held on the 15th February, 1991 in Swaziland which binds the Parties to conduct a joint study of the water resources, demands and development potential on the whole Inkomati River Basin, *found at* ocid.nacse.org/qml/research/tfdd/toTFDDdocs/234ENG.htm (*last visited* June 2003).

treaties covers mainly the protection of the water quality of the resources. The analysis of these treaties reveals that water quantity was not of particular importance. The treaty survey in the Middle East is rather scanty and inconsistent.

5.3.2 Implicit Consideration of Transboundary Groundwater

Of the 19 treaties identified in Africa, seven treaties deal implicitly with transboundary groundwater. The treaty survey conducted within this research revealed that 32 Agreements in Europe mention groundwater (12 implicitly and 20 explicitly, 14 bilateral and 18 multilateral). The majority of those treaties have the notion of groundwater embedded in the term "river basin" or "water system", "bodies of waters", "waters"...etc.[85] This research reveals that the Middle East treaty practice relating to transboundary groundwater is very limited. If some of these arrangements deal specifically with the management of springs and wells located in border areas, most of them regulate surface waters and only incidentally connected groundwaters. Groundwater was found to be regulated in six treaties, three of which mention groundwater explicitly.[86]

5.3.3 Explicit Consideration of Transboundary Groundwater

A first good example of treaties explicitly including connected groundwater within its scope is the Convention and Statutes Relating to the Development of the

---

[85] Agreement between the Federal People's Republic of Yugoslavia and the Republic of Austria Concerning Water Economy Questions in Respect of the Frontier Sector of the Mura and the Frontier Waters of the Mura (Mura Agreement), signed at Vienna on 16 Dmber 1954 and Protocol signed on November 1954.(Came Into Force On 9 Feb 1956) *found at* http://ocid.nacse.org/qml/research/tfdd/toTFDDdocs/88ENG.pdf (*last visited* November 2003) and Treaty between the Hungarian People's Republic and the Republic of Austria Concerning the Regulation of Water Economy Questions in The Frontier Region, Signed At Vienna, On 9 April 1956. Came into force On 31 July 1956 by the exchange of the instruments of ratification at Budapest, in accordance with Article 22. http://ocid.nacse.org/qml/research/tfdd/toTFDDdocs/94ENG.htm Article 1 provides :
"1) Frontier waters i.e. Frontier waters, i.e. : (a)Those sectors of bodies of water which form the frontier between Hungary and Austria or along which the frontier runs (frontier waters in the restricted sense);(b) Bodies of water which intersect the frontier and those situated in the frontier region, i.e., in the territory of either country within six kilometres of the frontier.2) Hydraulic structures and installations situated in the frontier region (sub-paragraph 1). 3)Those sectors of watercourses and those hydraulic structures and installations which are enumerated in annex 1 of this Treaty or are designated at some future time by the Hungarian-Austrian Water Commission (article 12).

[86] The Convention between Syria and Lebanon, *supra* note 18, the Israeli/Jordanian Peace Treaty, *supra* note 20 and the Israeli Palestinian Interim Agreement, *supra* note 20.

Chad Basin,[87] which regulates the utilization of the water resources of the Chad Basin. Under this Agreement the "Chad Basin" includes surface and connected groundwaters.[88] The Statutes attached to the Agreement introduce the "water system" concept thus focusing on the physical relationship between surface and underground environment to ensure that any development undertaken takes into account this interaction.[89] In another Agreement between the Federal Republic of Nigeria and the Republic of Niger of 18 July 1990,[90] the parties have explicitly declared the important role of orderly management of the shared water resources.[91] They also acknowledged the significant influence of these resources on the economic and social development and well-being of their respective populations living on both sides of the frontier.[92] This Agreement includes connected underground water within its scope.[93] It is stipulated that such waters must contribute to the flow of surface waters, a requirement that emphasizes the

---

[87] Article 4 of the Statutes Relating to the Development of the Chad Basin, signed at Fort Lamy on the 22nd of March 1964 states:

"The exploitation of the Chad Basin and especially the utilization of surface and underground waters has the widest meaning and refers in particular to the needs of domestic and industrial and agricultural development and the collecting of its fauna and flora products."

Entered into force 22 May 1964), FAO Legislative Study, FAO Treaty Collection No. 3, *found at* http://www.fao.org/Legal/treaties/treaty-e.htm (*last visited* May2003).[hereinafter referred to as the 1964 Chad Convention and Statutes].

[87] *Ibid,* Preamble of the Convention.

[88] *Ibid*, Chapter I article 4.

[89] *Ibid* Chapter II Article 5 Para. 2

"In particular, the Member States agree not to undertake in that part of the Basin falling within their jurisdiction any work in connection with the development of water resources or the soil likely to have a marked influence upon the system of the water courses and levels of the Basin ......".

[90] The 1990 Niger Nigeria Agreement, *supra* note 20.

[91] *Ibid,* Preamble.

[92] *Ibid.*

[93] *Ibid*, Part I: Article 1 Para. 3 states:

"Subject to the provisions of Article 9, a reference to the shared river basins shall include a reference to underground waters contributing to the flow of surface waters".

hydrologic unity of the hydrologic system.[94] Rules and principles embodied within this Agreement apply to surface and underground waters without any distinction.[95]

Most recently, in 2000, the Southern African Development Community (SADC)[96] adopted the Revised Protocol on Shared Watercourse Systems[97] which provides a set of substantive and procedural rules that apply equally to surface and groundwater within the same "drainage basin".[98] The Protocol derives its rules

---

[94] *Ibid*, See also Part III Article 9 which confirms that only those interrelated groundwaters are considered within the river basin concept:

"Groundwater resources shall not be accounted for the purpose of equitable sharing determination unless: (a) such resources are part of shared river basins within the meaning of Article 1, paragraph (3); or (b) such resources lie in whole or only in part within the shared river basins and are bi-sected by the common frontier between the Contracting Parties".

[95] *Ibid*, Part I: Article 1 Para. 1.

[96] The SADC was originally known as the Southern African Development Co-ordination Conference (SADCC), the organisation was formed in Lusaka, Zambia on 1 April 1980, following the adoption of the Lusaka Declaration. The Declaration and Treaty establishing the Southern African Development Community (SADC) which has replaced the Co-ordination Conference was signed at the Summit of Heads of State or Government on 17 August 1992, in Windhoek, Namibia. Member states of SADC are Angola, Botswana, Democratic Republic of the Congo, Lesotho, Malawi, Mauritius, Mozambique, Namibia, Seychelles, South Africa, Swaziland, Tanzania, Zambia and Zimbabwe. The Treaty of 1992 establishing the SADC is based on principles of international law including mutual understanding good neighbourliness, meaningful co-operation. The substantive principles embodied within the Treaty hinges on the following: (a) sovereign equality of all Member States; (b) solidarity, (c)peace and security; (d) human rights, democracy, and the rule of law; (e) equity, balance and mutual benefit; (f) peaceful settlement of disputes (Art. 4). Among the objectives of the SADC is promoting and maximise productive employment and utilisation of resources of the Region and achieving sustainable utilisation of natural resources and effective protection of the environment (Article 5 f and g), *found at* http://www.itcilo.it/english/actrav/telearn/global/ilo/blokit/sadc.htm *(last visited* July 2003).

[97] Revised Protocol on Shared Watercourses in the Southern African Development Community (SADC), done at Windhoek on the 7th August 2000, replacing the Protocol that entered into force on 29 September 1998 and signed in 1995. (entered into force on the 7 Sep. 2000) Signatories: Republic of Angola, the Republic of Botswana, the Kingdom of Lesotho, the Republic of Malawi, the Republic of Mauritius, the Republic of Mozambique, the Republic of Seychelles, the Republic of South Africa, the Kingdom of Swaziland, the Republic of Tanzania, the Republic of Zambia and the Republic of Zimbabwe, *found at:* http://www.dundee.ac.uk/law/iwlri/Documents/Treaties/Multilateral%20Treaties/2000%20 Revised%20SADC%20Protocol.pdf *(last visited* April 2003).

[98] *Ibid*, Article 1 Para. 1 identified "Drainage Basin" as:

"a geographical area, determined by the watershed limits of a system of waters including underground waters flowing into a common terminus". Which resembles the definition found in the 1966 Helsinki Rules, International Association of the 52nd Conference xi, 447-

from the customary international law, the restatement of international water law by the International Law Association [99], the relevant provisions of Agenda 21 of the United Nations Conference on Environment and Development,[100] and the concepts of environmentally sound management, sustainable development and equitable utilization of shared watercourse systems in the SADC Region.[101] The influence of the 1997 UN Watercourses is also evident, although the Protocol does not explicitly refer to it in the Preamble.[102] With regard to the identification the meaning of the term "watercourse", the Protocol confirms the physical unity and coherence of each shared watercourse.[103] This is apparent from the definition of the "Watercourse System" that clearly includes underground water that contributes to the hydrologic system.[104] The Protocol imposes stricter restrictions on its parties within a wider basin concept to maintain a proper balance between resource development and conservation at the same time maintaining the environment for sustainable development.[105] Finally, the majority of African States have concluded a legally binding instrument, which has a framework character and include groundwater within its scope. The African Convention on the Conservation of Nature and Natural Resources[106] is a general framework Convention covering all natural resources including soil, water flora and fauna.

---

533 (1967), *found at* http://www.internationalwaterlaw.org/IntlDocs/Helsinki_Rules.htm. (*last visited* May 2003), *also see supra* Chapter Three.

[99] *Ibid.*

[100] Agenda 21, Chapter 18, UNECOD Report Annex II Agenda 21, June 1992, UN Doc. A/CONF.151/26 Vol. 1,14 and Vol. 2, 167,206, *found at* http://www.igc.org/habitat/agenda21/a21-18.htm (*last visited* March 2003).

[101] Revised SADC Protocol, *supra* note 97, Preamble.

[102] The United Nations Convention on the law of Non-Navigational Uses of International Watercourses 21 May 1997, U.N.Doc.A/RES/51/229, July, 8, 1997 *found at* http://www.un.org/law/ilc/texts/nonnav.htm#abstract (*last visited* May 2003) [hereinafter referred to as the 1997 UN Watercourses Convention].

[103] Revised SADC Protocol, *supra* note 97.

[104] *Ibid*, Article 1:

"watercourse means a system of surface and ground waters consisting by virtue of their physical relationship a unitary whole normally flowing into a common terminus such as the sea, lake or an aquifer".

[105] *Ibid*, Article 3 Para. 4.

[106] The African Convention on the Conservation of Nature and Natural Resources, (Entered into force 16th June 1969), Signatories: Algeria, Cameroon, Central African Republic, Congo, Cote D'Ivoire, Djibouti, Egypt, Ghana, Kenya, Madagascar, Malawi, Mali, Morocco, Mozambique, Niger, Nigeria, Rwanda, Senegal, Seychelles, Sudan, Swaziland, Togo, Tunisia, Uganda, Tanzania, Zaire, and Liberia. FAO Treaty Collection No. 1, found at http://www.fao.org/docrep/W7414B/W7414B00.htm#Contents (*last visited* May 2003), *also found at* http://ocid.nacse.org/qml/research/tfdd/toTFDDdocs/341ENG.htm (*last visited* May 2003)[hereinafter referred to as the 1968 African Convention].

Groundwater is not explicitly mentioned in the definition of "natural resources", [107] however, from the analysis of the text it can be concluded that groundwater is implicitly included in the term "water". The latter is confirmed in the special section on water that explicitly mentions groundwaters to be treated equally as surface water. [108]

In Europe, the period after 1970, presents a transition in the manner in which connected transboundary groundwater is treated, although some earlier treaties explicitly included groundwater within their scope. [109] The recently adopted multilateral conventions within the UNECE present good examples of treaties explicitly dealing with connected transboundary groundwater. The 1992 UNECE

---

[107] *Ibid*, Article III Definition:
"(1) Natural Resources mean renewable resources that is soil, water, flora and fauna".
[108] *Ibid,* Article V:
"Water: (1) The Contracting States shall establish policies for conservation, utilization and development of underground and surface water, and shall endeavour to guarantee for their populations a sufficient and continuous supply of suitable water.....(2) Where surface or underground water resources are shared by two or more of the Contracting States, the latter shall act in consultation, and if the need arises, set up inter-State Commissions to study and resolve problems arising from the joint use of these resources, and for the joint development and conservation thereof".

[109] *See the* Agreement between the Government of The Federal People's Republic of Yugoslavia and the Government of the People's Republic of Albania Concerning Water Economy Questions, together with the Statute of the Yugoslav-Albanian Water Economic Commission and with the Protocol Concerning Fishing In Frontier Lakes And Rivers,12 May 1956 (entered into Force on 6 August 1957) *found at* http://ocid.nacse.org/qml/research/tfdd/toTFDDdocs/97ENG.htm (*last visited* May 2003) Article 1:

" The provisions of this Agreement shall apply to all water economy questions, measures and works on watercourses which form the State frontier and watercourses, lakes and water systems which are intersected by the State frontier......., and which are of interest to both Contracting Parties, and in particular to:(h) Groundwater (3) The expression "water system" shall mean, in this Agreement, all watercourses (surface or underground, natural or artificial), installations, measures and works which may affect watercourses from the standpoint of water economy, and installations forming or intersected by the State frontier".

Also see the Agreement between the Government of the Czechoslovak Republic and the Government of the Polish People's Republic Concerning the Use of Water Resources in Frontier Waters, Signed at Prague, On 21 March 1958, (entered into force on 7 August 1958, by an exchange of diplomatic notes signifying the approval of the agreement by both contracting Parties pursuant to their legislation, in accordance with Article 14), Article 2 of the Agreement states: "For the purposes of this Agreement, the term "frontier waters" mean: (b) Surface and ground waters flowing from the territory of one State to the territory of the other, at those places where they are intersected by the State frontier".
found at http://ocid.nacse.org/qml/research/tfdd/toTFDDdocs/104ENG.htm (*last visited* May 2003).

Helsinki Convention on the Protection and the Use of Transboundary Watercourses and International Lakes explicitly deals with the utilization and protection of transboundary groundwater.[110] This Convention was signed in 1992 by 25 countries and by September 16 2003 ratified by 34 countries including the European Community. It is a multilateral treaty governing environmental protection of European watercourses and the first regional framework convention dealing with international watercourses. The 1992 Helsinki Convention emphasizes the need to apply the precautionary principle[111] and the polluter pays principle, [112] and encourages the achievement of the goals of sustainability.[113] The Convention primarily addresses the issue of protection of transboundary waters including connected groundwaters.[114] The UNECE Protocol on Water and Health[115] was inspired by the 1992 Helsinki Convention and therefore includes transboundary groundwater within its scope.[116] The UNECE Protocol on Health

---

[110] UNECE Convention on the Protection and the Use of Transboundary Watercourses and International Lakes, Helsinki, 1992, 31 ILMN 1312 (entered into force on 6 October 1996). Parties: Albania, Bulgaria, Estonia, France, Hungary, Liechtenstein, Helsinki Netherlands, Portugal, Russian Federation, Spain and United Kingdom. [hereinafter referred to as the 1992 UNECE Helsinki Convention].

[111] *Ibid.*

[112] *Ibid*, Preamble, para (b).

[113] *Ibid*, Art. 2 Para 3

"(c)Water resources shall be managed so that the needs of the present generation are met without compromising the ability of future generations to meet their needs."

[114] *Ibid*, Article 1 Para. 1 identifies transboundary waters as:

"any surface or groundwaters which mark, cross or are located on the boundaries between two or more States; wherever transboundary waters flow directly into the sea, these transboundary waters end at a straight line across their respective mouths between points on the low water line of their banks".

[115] UNECE Protocol on Water and Health to the 1992 Convention and Use of Transboundary Watercourses on International Lakes, London 1999, *found at* (www.who.dk/london99) *and* http://ocid.nacse.org/qml/research/tfdd/toTFDDdocs/338ENG.htm (*last visited* May 03), [hereinafter referred to as the UNECE Protocol on Health]. The primary objective of the Protocol is to promote the protection of human health and well being at the national, transboundary and international levels. The objectives of sustainable development as provided by the protocol would be achieved by improving water management, including the protection of water ecosystems and through preventing, controlling and reducing water-related diseases.

[116] *Ibid*, Preamble:

"Aware that surface waters and groundwater are renewable resources with a limited capacity to recover from adverse impacts from human activities on their quantity and quality, that any failure to respect those limits may result in adverse effects, in both the short and long terms, on the health and well-being of those who rely on those resources and

applies to (a) surface freshwater (b) groundwater; (c) estuaries, (d) coastal waters which are used for recreation or for the production of fish by aquaculture or for the production or harvesting of shellfish (e) enclosed waters generally available for bathing, (f) water in the course of abstraction, transport, treatment or supply and (g) waste water throughout the course of collection, transport, treatment and discharge or reuse.[117]

The Danube River Protection Convention has been signed in June 1994 in Sofia. [118] With its entry into force the 1998 Danube Convention became the principal legal instrument for co-operation and transboundary water management in the Danube River Basin. The main objective of the Convention is the protection and sustainable use of ground and surface waters and riverine ecology, at basin-wide and sub-basin-wide co-operation with transboundary relevance. [119] The

---

their quality, and that in consequence sustainable management of the hydrological cycle is essential for both meeting human needs and protecting the environment".
and Article 2 (3)
"Groundwater" means all water which is below the surface of the ground in the saturation zone and in direct contact with the ground or subsoil"
and Article 2(5) defines "Transboundary waters" as:
"any surface or ground waters which mark, cross or are located on boundaries between two or more States; wherever transboundary waters flow directly into the sea, these transboundary waters end at a straight line across their respective mouths between points on the low-water line of their banks".

[117] *Ibid,* Article 3.

[118] Convention on Co-operation for the Protection and Sustainable Use of the River Danube, Sofia 1994, (entered into force October 22nd 1998) *found at* http://ksh.fgg.uni-lj.si/danube/envconv) *also found at* http://ocid.nacse.org/qml/research/tfdd/toTFDDdocs/331ENG.htm (*last visited* April 2003). [hereinafter referred to as the 1994 Danube Convention]. Contracting parties are: Austria, Bulgaria, Croatia, the Czech Republic, the European Commission, the Federal Republic of Germany, Hungary, Moldova, Romania, the Slovak Republic and Slovenia. The Danube River Basin is in the heartland of Central Europe. The main river is 2,857 km long and drains 817,000 km^2 including all of Hungary; most part of Romania, Austria, Slovenia, Croatia and Slovakia; and significant parts of Bulgaria, Germany, the Czech Republic, Moldova and Ukraine. Territories of the federal Republic of Yugoslavia, Bosnia and Herzegovina and small parts of Italy, Switzerland, Albania and Poland are also included in the basin.

[119] *Ibid,* Article 1: Definitions
"(g) Water balance means the relationship characterizing the natural water household of an entire river basin as to its components (precipitation, evaporation, surface and underground run-off. In addition a component of current man-made effects originating from water use and influencing water quantity is included."
Article 2: (1)

Convention adopts the ecosystem approach thus confirming the integrity of the Danube in addition to applying the polluter pays principle, the precautionary principle and the principle of preventive action.[120] These principles constitute the principal elements of the optimal utilization and protection of groundwater.

The Convention on the Protection of the Rhine[121] is based on the ecosystem approach and aims at ensuring the sustainable development of the Rhine by maintaining the quality and quantity of the river basin waters, including connected transboundary groundwater.[122] The waters covered within this convention comprises the river, the groundwater interacting with it, the aquatic and terrestrial ecosystems interacting with the Rhine and the Rhine Catchment.[123] The Convention embodies the principles of prevention, precaution, polluter pays, fighting further degradation of the environment, sustainable development, and other technical principles that are emphasized within this research.[124]

"The Contracting Parties shall strive at achieving the goals of a sustainable and equitable water management, including the conservation, improvement and the rational use of surface waters and ground water in the catchment area as far as possible. ....".

[120] *Ibid*, Art. 2 Para.

"(4) The Polluter pays principle and the Precautionary principle constitutes a basis for all measures aiming at the protection of the Danube River and of the waters within its catchment area. And (5) Water management cooperation shall be oriented on sustainable water management, that means on the criteria of a stable, environmentally sound development, which are at the same time directed to."

[121] Convention on the Protection of the Rhine Signed on January 22 1998 by the European Union, France, Germany and Luxemburg., Rotterdam the Netherlands, *also found* at: http://www.internationalwaterlaw.org/RegionalDocs/Rhine_River.htm (*last visited* May 2003), [hereinafter referred to as the 1998 Rhine Convention]. The river Rhine flows from its source in Switzerland through Germany, France and the Netherlands to the North Sea. It is the 3rd biggest river of Europe. The total length is 1,320 km and the catchment area is 185,000 km^2. The average discharge is 2,300 m^3/sec and total volume per year is in average 73 km^3. At present, more than 50 million people live in the catchment in 9 countries (Switzerland, Austria, Liechtenstein, Italy, France, Germany, Luxembourg, Belgium and the Netherlands). It provides the source for drinking-water production for more than 20 million people. The Convention stresses some principles mainly derived from international environmental law like the procedural principles that guide the implementation of the Convention are the prevention principle, the precautionary principle, and the principle of polluter pays.

[122] *Ibid,* Article 3 provides a list of the targets of the Convention which emphasizes sustainable development, the protection of quantity and quality of the waters for drinking water purposes, the safe disposal of wastes to ensure harm prevention the holistic flood prevention and protection and the protection of marine waters.

[123] *Ibid*, Art.2.

"The scope of this Convention comprises: 1. the Rhine 2. the ground-water interacting with the Rhine,........."

[124] *Ibid,* Article 4 provides a list of these principles.

The Convention between Switzerland and France relative to the Protection Utilization and Recharging of the 1978 Franco-Suisse Convention[125] is an international agreement that deals solely with connected groundwater, which is rarely found elsewhere in the world.[126] The Agreement establishes institutional mechanisms and sets forth procedures for the protection, utilization and recharging of the Geneva aquifer. The Convention's scope is mainly to regulate artificial recharge.[127]

The adopted EU WFD[128] establishes a framework for the right conditions to encourage efficient and effective water protection at the State and Interstate levels, by providing for a common approach and common objectives, principles, definitions and basic measures.[129] The hydrological unit of measurement for water

---

[125] *See* the 1978 Franco Suisse Convention, *supra* note 21.

[126] *Ibid,* this agreement is in French but due to its relevance to this thesis it was translated. Direct links was made with the Commission in Geneva that is responsible for the implementation of the agreement, Mr. De Los Cobos who kindly answered a questionnaire that was designed and sent for him. The questions included: 1-How many years it took you to be able to agree on this convention 2- did it enter into force on the date of signing? 3- Are the parties fulfilling their obligations? 4- do you have joint monitoring? 5- did you have any dispute on the interpretations of provisions of the Convention? 6-Any critical issues regarding the co-operation that you think should have been included in the Conventions? 7- Any particular problems you are encountering in implementing the convention?

[127] Reply to the questionnaire by Mr. De Los Cobos from the Service Cantonal de Géologie Geneva, in the reply it was explained that during the 1960's and 1970's, over pumping lowered the transboundary groundwater level by more than 7m. Artificial recharge has been carried out from the Arve River into the Genevese aquifer in order to maintain the groundwater level and enable water resources management. With the French part, it was estimated that the natural groundwater input coming from the French border was around 2millions m^3/year. This explains the free limit (quota) for French part (5 wells). If the total French pumping exceeds this limit of 2 million m^3/year, they have to pay to the State of Geneva the surplus calculated with the formula described in Article 14 of the convention. The administrative and technical management of the convention was prepared by Service cantonal de Géologie. The first documents found dealing with these works are dated back to 1968, it took 10 years of negotiations to sign the Convention. The treaty entered in force on the date of signing, but in fact the artificial recharge started to work in 1980.The information rived from the Commission indicate that there are no problems encountered in the implementation. During the last years no dispute arose between the parties. However, some discussion took place concerning an increase the minimal quota for the French part. The latter had to prove that the 2M m^3/year limit is too low by hydrogeological investigations. The first that were studies developed in the south limit of the aquifer (on the French border) demonstrated nothing positive as the potential aquifers linked to the main aquifer were all dry. Accordingly the French quota remained the same.

[128] The WFD, *supra* note 54. The relevance of the WFD at the State level was discussed in Chapter 1.

[129] *Ibid,* paragraph 23 of the WFD Preamble states that:

resources management within the WFD is the "River Basin". The WFD deals with water as an integral part of the ecosystem. It provides that measures taken for the integrated management of ground and surface water belonging to the same hydrogeological ecological, hydrological should be coordinated.[130] The objective of achieving "good water status" [131] should be pursued for each river basin, so that measures with respect to surface and groundwaters that belong to the same ecological, hydrological and hydrogeological system are integrated.[132] River Basin Management is to be coordinated across administrative and political borders with coordinated (national) programs of measures to achieve objectives set forth in the context of River Basin Management Plans.[133]

In the Middle East three Agreements that explicitly regulate transboundary groundwater were identified. These are: (a) the Convention between Syria and Lebanon on the Distribution of Al 'Asi River of 1994,[134] (b) the Jordan Israel Peace Treaty of 1994[135] and the (c) Interim Agreement between the PLO and Israel of 1995.[136] The Israeli/Palestinian Interim Agreement devotes an entire section[137] to water and wastewater cooperation and coordination. Israel, and for the first time, recognised Palestinian water rights to these waters; the nature of which

---

" Common principles are needed in order to coordinate Member States' efforts to improve the protection of Community waters in terms of quantity and quality, to promote sustainable water use, to contribute to the control of transboundary water problems, to protect aquatic ecosystems, and terrestrial ecosystems and wetlands directly depending on them, and to safeguard and develop the potential uses of Community waters".

[130] *Ibid,* paragraph 33 of the WFD Preamble.

[131] *Ibid,* Article 4 of the WFD paragraph 1-9 identifies the environmental objectives that have to be achieved within the River Basin.

[140] *Ibid,* Article 33 of the WFD Preamble.

[133] *Ibid,* Paragraph 33 of the WFD Preamble states that:

"the objective of achieving good water status should be pursued for each river basin, so that measures in respect of surface water and groundwaters belonging to the same ecological, hydrological and hydrogeological system are coordinated".

[134] Convention No. 15 between Lebanon and Syria Concerning the Distribution of Al Asi River, *supra* note 18. Article 4 of the Agreement sets restrictions on the withdrawal from spring waters in the border region and specified a circle of 1500 meters radius from those springs where withdrawals will be prohibited.

[135] Peace Treaty between The Hashemite Kingdom of Jordan and the State of Israel, *supra* note 20.

[136]The 1995 Israeli-Palestinian Interim Agreement, *supra* note 20. This treaty can be classified as a 'modus vivendi' which is an instrument recording an international agreement of temporary or provisional nature intended to be replaced by an arrangement of a more permanent and detailed character, see generally *supra* note 3.

[137] The 1995 Israeli-Palestinian Interim Agreement, *supra* note 20, Appendix II Annex III Article 40.

was not identified neither the criteria for their determination was set.[138] The scope of the Agreement covers transboundary groundwater and sewage. [139] Groundwater aquifers that are included are those renewable groundwater aquifers underlying the West Bank in Palestine and extending to Israel from the north-east and west. In the Treaty of Peace between the State of Israel and the Hashemite Kingdom of Jordan, [140] there is a section on the Israeli/Jordan cooperation concerning the waters of the Arava Valley Groundwater.[141] Both Parties agreed mutually to recognize the rightful allocations of both of them in the Araba/Arava groundwater in accordance with the agreed acceptable principles, quantities and quality that shall be fully respected and complied with.[142]

Also relevant to this discussion are a number of recent regional conventions such as the 1970 Convention on Establishing the Arab Centre for Studies of Dry Areas and Arid Lands which is applicable, among others, to Iraq, Jordan, Kuwait, Lebanon, Saudi Arabia, Syria and Yemen in the ESCWA[143] Region and which promotes the investigation of both surface and underground water resources and the irrigation of arid lands;[144] the 1976 Convention for the Protection of the Mediterranean Sea Against Pollution and the 1980 Protocol for the Protection of the Mediterranean Sea Against Pollution from Land-Based Sources applicable, among others, to Egypt, Israel, Lebanon and Syria (for the Convention but only Israel and Lebanon for the Protocol);[145] the 1978 Regional Convention between

---

[138] *Ibid*, Principle 1 of Article 40 of Annex III regarding water and sewage states:
"Israel recognizes the Palestinian water rights in the West Bank. These will be negotiated in the permanent status negotiations and settled in the Permanent Status Agreement relating to the various water resources."

[139] *Ibid*, Appendix II of Annex III Article 40 in the Interim Agreement is the section dealing with water and sewage for the transitional period until permanent status negotiations commence, which was planned not to be later than 4th May 1994.

[140]Peace Treaty between The Hashemite Kingdom of Jordan and the State of Israel, *supra* note 19 done At Arava/Araba Crossing Point On 26 October 1994.

[141]*Ibid,* Art. 6 Para. IV.

[142] *Ibid*, Para. 1:
"The Parties agree mutually to recognise the rightful allocations of both of them in Jordan River and Yarmouk River waters and Araba/Arava ground water in accordance with the agreed acceptable principles, quantities and quality as set out in Annex II, which shall be fully respected and complied with".

[143] Economic and Social Commission for Western Asia.

[144] Convention Establishing the Arab Centre for Studies of Dry Areas and Arid Lands signed in Cairo on 9 September 1970 (regional studies relating to water resources, geological and geomorphological characteristics, economics of exploiting arid lands, irrigation and drainage), in The Weekly Gazette of the Republic of Iraq, No. 36, 8September 1971, p. 3.

[145] ESCWA, Expert Working Group on Legal Aspects of the Management of Shared Water Resources, Sharm El-Sheikh, 8-11 June 2000, ESCWA/ENR/2000/WG.1/6, 29 May 2000.

Bahrain, Iran, Iraq, Kuwait, Oman, Qatar, Saudi Arabia, and the United Arab Emirates for Co-operation in the Protection of the Marine Environment from Pollution which regulates the prevention and abatement of marine pollution in the Persian Gulf from land-based sources and establishes the principle of civil liability and of compensation for damages; and the 1982 Regional Convention between Jordan, Saudi Arabia, Somalia, Sudan, the Arab Republic of Yemen and the People's Republic of Yemen for the Conservation of the Red Sea and of the Gulf of Aden Environment which enjoins contracting states to take all reasonable measures to prevent pollution originating from land-based sources, all regional conventions which, whether formally or informally, recognize the "Drainage Basin" concept as well as that of the inter-connection of surface and underground water resources.[146] This fact is very relevant for the current research in terms of demonstrating States' acceptance to the "Drainage Basin" concept as an appropriate unit for regulating international watercourses.

5.3.4 Summary

The preceding discussion on "scope and use of terms" in Africa, Europe and the Middle East demonstrates a greater recognition to connected transboundary groundwater as an integral part of the international water system. In Africa, the use of terms within the treaties is inconsistent and somehow misleading to whether or not groundwater is included. By and large, there is some evidence that supports the proposition that groundwater is treated as part of the international water system and the "river basin" concept is the preferred concept of the region. The scope of most of the existing treaties covers the utilization, protection management and cooperation over these resources. Transboundary Confined groundwaters, which constitute a major source of the water supply in the Northern and Southern regions remain outside the scope of these treaties.

In Europe, on the other hand, even under earlier practice, there appears to be a general uniformity among the treaties to include groundwater as part of the "River Basin". Many of the earlier treaties dealing with frontier waters focus on the protection against pollution, whilst recent treaties within the EU tend to relate to utilization and development. Only one treaty is found to solely deal with transboundary groundwater in Europe; the Geneva Aquifer.

In the Middle East- apart from the Israeli Palestinian unsettled relationship, cooperative arrangements between the ESCWA countries has been established to protect and develop shared surface and groundwater resources. In the context of pollution many of these countries have adopted the "Drainage Basin" concept

[146] *Ibid.*

which is a very appropriate management and regulation unit in the case of transboundary groundwater.

5.4 SUBSTANTIVE RULES

5.4.1 Introduction

Recalling the definition made for "substantive rules" earlier in this book[147] this section builds on the treaty survey undertaken within this research to analyze State practice concerning the substantive rules and principles governing the utilization, development and protection of transboundary groundwater.

5.4.2 Equitable and Reasonable Utilization Principle Vs. the No Harm Rule

The analysis of the treaties in Africa reveals that governing rule under the different treaties varies from one treaty to the other. In most of these treaties, however, the protection of the territorial sovereignty and existing uses of States are emphasized.[148] For example, the 1964 Chad Basin Convention and Statutes[149] recognise that the exploitation of the Chad Basin resources is based on the sovereign rights of each of the Member State sharing the Basin. [150] The Statute calls upon the Member States to utilize the waters of the Basin without causing marked influence on the basin which is considered a particular requirement of the "equitable and reasonable utilization" principle.[151] The Niger Nigeria

[147] *See supra* Introduction (Section 8) and Chapter 3 (Section 3.4).

[148] Illustrative examples are the 1964 Chad Basin Convention, *supra* note 86, the 1990 Niger Nigeria Agreement, *supra* note 20, and the Revised SADC Protocol *supra* note 97.

[149] The 1964 Chad Basin Convention, *supra* note 86, Chapter I, Article 3:

"The Chad Basin shall be open to the exploitation of all Member States who are parties to the Convention, in respect for the sovereign rights, of each of them and in accordance with the terms and conditions of the present Statutes, subsequent revisions or regulations or special agreements".

[150] *Ibid,* Chapter I, Article 3:

"The Chad Basin shall be open to the exploitation of all Member States who are parties to the Convention, in respect for the sovereign rights, of each of them and in accordance with the terms and conditions of the present Statutes, subsequent revisions or regulations or special agreements".

[151] *Ibid,* Chapter II Article 5 provides:

"The Member States undertake a refrain from adopting, without referring to the Commission beforehand, any measures likely to exert a marked influence either upon the extent of water losses, or upon the form of the annual hydrograph and limnograph and certain other characteristics of the Lake, upon the conditions of their exploitation by other

Agreement[152] explicitly calls upon the parties to develop, conserve and use these resources in an equitable manner.[153] According to the provisions of the 1990 Niger Nigeria Agreement, each Party is entitled to an equitable share in the development, conservation and use of the water resources in the shared river basins. [154] This Agreement also establishes that existing uses have inherent priority over future or existing uses later in time unless otherwise determined by the requirement of equitable sharing.[155] Any use that existed at the time the determination of equitable sharing is made has priority over any future uses and competing existing uses later in time. [156] The Revised SADC Protocol emphasises the principle of sovereign rights of States and calls for the participation of all watercourse States in the utilization of these water resources.[157] The Revised SADC Protocol calls for respecting and applying the existing rules of general or customary international law relating to the utilization and management of the resources of shared watercourse systems.[158] The utilization of the watercourses including groundwater is governed by the principle of equitable and reasonable utilization and participation which includes a degree of factor of harm.[159] The

---

bordering States, upon the sanitary condition of the water resources or upon the biological characteristics of the fauna and the flora of the Basin. In particular, the Member States agree not to undertake in that part of the Basin falling within their jurisdiction any work in connection with the development of water resources or the soil likely to have a marked influence upon the system of the water courses and levels of the Basin."

[152] The 1990 Niger Nigeria Agreement, *supra* note 20.

[153] *Ibid.*

[154] *Ibid,* Part I: Article 2.

[155] *Ibid,* Part III: Articles 6 and 7.

[156] *Ibid.* "Article 6 :
"A water use existing at the time an equitable sharing determination is made shall take prdence over a future use, provided the existing use is beneficial to both Contracting Parties and reasonable under the circumstances."
And Article 7
"A water use existing at the time an equitable sharing determination is made shall take precedence over a competing existing use which came into being later in time, provided that: (a) the use prior in time is beneficial to both Contracting Parties and reasonable under the circumstances; and (b) the weighting of the factors under Article 5 does not warrant accommodation of the later use, in whole or in part".

[157] SADC Revised Protocol, *supra* note 97, Article 3 para. 2
"2. The utilisation of shared watercourses within the SADC Region shall be open to each Watercourse State, in respect of the watercourses within its territory and without prejudice to its sovereign rights, in accordance with the principles contained in this Protocol. The utilisation of the resources of the watercourses shall include agricultural, domestic, industrial, navigational and environmental uses."

[158] *Ibid*, para. 3

[159] *Ibid*, para. 7 (a and b). 7.

Revised SADC Protocol allows for "significant harm" to occur provided that consultation is properly made and adequate appropriate measures are undertaken. In such situations, there is an obligation on the State causing the harm to eliminate or mitigate such harm and where appropriate discuss the question of compensation.[160]

Treaty practice in Europe relating to international watercourses is mostly concerned with limiting transboundary impact.[161] The 1992 UNECE Helsinki Convention embraces two general principles of international law; reciprocity and equality, which govern the relation between Riparian Parties. In the context of international watercourse the substantive rule within this Convention is the equitable and reasonable utilization.[162] In order to avoid or limit the transboundary impact Article 2(2) calls for the equitable and reasonable utilization of the transboundary resources.[163] The substantive rules embodied within the UNECE Protocol on Health[164] build on the sovereign rights of States to exploit their resources in accordance with their policies, and the responsibility to ensure that activities within their jurisdiction do not cause damage to other State's beyond

---

"a) Watercourse States shall in their respective territories utilise a shared watercourse in an equitable and reasonable manner. In particular, a shared watercourse shall be used and developed by Watercourse States with a view to attain optimal and sustainable utilization thereof and benefits therefrom, taking into account the interests of the Watercourse States concerned, consistent with adequate protection of the watercourse for the benefit of current and future generations b) Watercourse States shall participate in the use, development and protection of a shared watercourse in an equitable and reasonable manner. Such participation, includes both the right to utilize the watercourse and the duty to co-operate in the protection and development thereof, as provided in this Protocol."

[160] *Ibid*, para. 10 (a, b and c).

[161] See generally the 1992 UNECE Helsinki Convention *supra* note 110, the UNECE Protocol on Health *supra* note 115, the 1994 Danube Convention *supra* note 118, and the 1998 Rhine Convention *supra* note 121.

[162] The 1992 UNECE Helsinki Convention, Article 1(2) *supra* note 110, identifies "Transboundary impact" as:

"Any significant adverse effect on the environment resulting from a change in the conditions of transboundary waters caused by a human activity, the physical origin of which is situated wholly or in part within an area under the jurisdiction of a Party, within an area under the jurisdiction of another Party. Such effects on the environment include effects on human health and safety, flora, fauna, soil, air, water, climate, landscape and historical monuments or other physical structures or the interaction among these factors; they also include effects on the cultural heritage or socio-economic conditions resulting from alterations to those factors".

[163] Article. 2 para. 2 (c), *ibid.*

"To ensure that transboundary waters are used in a reasonable and equitable way, taking into particular account their transboundary character, in the case of activities which cause or are likely to cause transboundary impact".

[164] UNECE Protocol on Health, *supra* note 115.

their jurisdiction.[165] The goals of the 1994 Danube focus on achieving the sustainable and equitable water management of the Danube River through the conservation, improvement, and the rational use of surface and groundwater.[166]

Existing State treaty practice in the Middle East is very limited and scanty. The analysis of substantive rules governing the existing treaties between Israel and its neighbours is conducted in view of the prevailing political conflicts between the parties. Current allocations among watercourse States from their transboundary watercourses are not equitably distributed, therefore, the equitable and reasonable utilization principle is not governing within the existing treaties. Under Israeli Palestinian Interim Agreement the primary substantive rule that this agreement emphasizes is the "No Harm" to existing uses.[167] There is also a duty on both parties to protect the environment and utilise the natural resources on a sustainable basis.[168] "Powers and responsibilities" and not "sovereign rights" of each side in their respective territories over water and sewage are confirmed. [169] In the Israeli Jordanian Agreement the no harm rule is also the governing principle within this Agreement. Each party is also obliged to use the waters within their territory without jeopardising the quality and quantity of water in the other party's

---

[165] Article 5 paragraph c, *Ibid.*

[166] *Ibid*, Art. 2 Para. (1).

"The Contracting Parties shall strive at achieving the goals of a sustainable and equitable water management, including the conservation, improvement and the rational use of surface waters and ground water in the catchment area as far as possible. ....".

And Article 6:

"Specific Water Resources Protection Measures: The Contracting Parties shall take appropriate measures aiming at the prevention or reduction of transboundary impacts and at a sustainable and equitable use of water resources as well as at the conservation of ecological resources, especially: (a) enumerate groundwater resources subject to a long-term protection as well as protection zones valuable for existing or future drinking water supply purposes;….."

[167] Israel PLO Interim Agreement, *supra* note 20, Appendix II Annex III Article 40 paragraph

"Each side shall take all nssary measures to prevent any harm to the water and sewage systems in their respective areas" .

[168] *Ibid*, Article 12 of Annex III 'Civil Affairs states that :

"The Palestinian side and Israel, recognizing the need to protect the environment and to utilize natural resources on a sustainable basis, agreed upon the following: This sphere includes, inter alia, licensing for crafts and industry, and environmental aspects of the following: sewage, solid waste, water, ...".

[169] *Ibid*, Para. 3,

"While respecting each side's powers and responsibilities in the sphere of water and sewage in their respective areas, both sides agree to coordinate the management of water and sewage resources and systems in the West Bank during the interim period."

territory.[170] Practice in the Middle East apart from the Israeli related relations has in most of the legal arrangements reduced the principle of equitable utilization to an effective quantitative apportionment of the waters shared among their contracting Basin States.[171]

5.4.3 Determination of Equitable and Reasonable Utilization

The treaty survey made within this research demonstrates that where equitable and reasonable utilization principle is the governing principle within a treaty, the Parties have a free hand to determine what equitable and reasonable utilization is. Under some of these agreements, the Parties agree to benefit from a certain proportion of the annual renewable flow of the shared resource.[172] On the other hand, some treaties have identified criteria for the determination of equitable and reasonable utilization, which takes natural and socio-economical factors into account.[173] The Revised SADC Protocol, for example, provides guidance on the determination of equitable and reasonable utilisation. The list of factors provided in Article 3 paragraph 8 (a&b) resemble to a great extent those factors embodied within the UN Convention. In addition to promoting co-operation the Protocol SADC provides guidance to the operationalization of equitable and reasonable

[170] Israel Jordan Peace Agreement, *supra* note 20, Para. 1.
"Neither country shall take, nor cause to be taken, any measure that may appreciably reduce the yields or quality of these wells and systems".

[171] ESCWA, *supra* note 145. See for example the 1994 Syrian Lebanese Convention, *supra* note 18, The Agreement between the Republic of Syria and the Hashemite Kingdom of Jordan Concerning the Utilization of the Yarmuk Waters. Signed At Damascus, On 4 June 1953.(entry into force on 8 July 1953),[hereinafter referred to as the 1953 Syrian Jordanian Agreement], *found at* http://ocid.nacse.org/qml/research/tfdd/toTFDDdocs/82ENG.htm (*last visited* May 2003).

[172] For example, *see* the 1978 Franco-Suisse Convention, Article 9.1, *supra* note 21 which provides that:
"until the commissioning of the groundwater recharge installation, and subject to possible derogations by the Commission, the French authorities and local communities would see to it that the total annual volume of water extractions within their territory would not exceed 5 million cubic meters, of which 2 million cubic meters corresponding to existing uses were recognized as a free allocation in favour of the French users".
See also the 1994 Syria Lebanese Convention, Article 3, *supra* note 18, which divides the waters between the Lebanese and Syrians. These quantities are changeable and depend on the abundance of water in the specific year, *See also* the 1953 Syrian Jordanian Agreement, Article 8, *supra* note 152, which also divides the quantities of water used for irrigation in addition to dividing the power generated by the installations on the river (Article 9).

[173] See the 1990 Niger Nigeria Agreement *supra* note 20, and the Revised SADC Protocol *supra* note 97.

utilization.[174] In the Niger Nigeria Agreement Article 5 Part III of the Agreement lists twelve factors that help determine the equitable and reasonable sharing. [175]

5.4.4 Summary

The preceding analysis reveals that in Africa, where scarcity of resources is the major problem, the principle of equitable utilization has been reiterated in most of the existing treaties. The emphasis by States on territorial sovereignty is interpreted to emphasise the co-relative rights of States from the shared water resources under the principle of "equitable and reasonable utilization". In the context of Africa, this research therefore finds that the principle of equitable and reasonable utilization principle is considered the governing rule in most of the

---

[174] Revised SADC Protocol, *supra* note 97, Article 3 paragraph 8
"Utilisation of a shared watercourse in an equitable and reasonable manner within the meaning of Article 7(a) and (b) requires taking into account all relevant factors and circumstances including: (i) geographical, hydrographical, hydrological, climatic, ecological and other factors of a natural character; (ii) the social, economic and environmental needs of the Watercourse States concerned; (iii) the population dependent on the shared watercourse in each Watercourse State; (iv) the effects of the use or uses of a shared watercourse in one Watercourse State on other Watercourse States; (v) existing and potential uses of the watercourse; (vi) conservation, protection, development and economy of use of the water resources of the shared watercourse and the costs of measures taken to that effect; and (vii) the availability of alternatives, of comparable value, to a particular planned or existing use.(b) The weight to be given to each factor is to be determined by its importance in comparison with that of other relevant factors. In determining what is an equitable and reasonable use, all relevant factors are to be considered together and a conclusion reached on the basis of the whole."

[175] The 1990 Niger Nigeria Agreement, *supra* note 20, Article 5 Part III:
"1. In determining the equitable share to which each Contracting Party is entitled pursuant to Article 2, the following factors shall be taken into account: (a) the climate of the region, and its influence on rainfall patterns; (b) rainfall patterns, and their influence on surface hydrology, and related hydro-geology; (c) surface hydrology and related hydro-geology; (d) existing uses of the waters; (e) reasonable planned water development requirements; (f) the economic and social development needs of the Contracting Parties; (g) the dependence of local populations on the waters in question for their own livelihood and welfare; (h) the availability of alternative sources of water to satisfy competing water demands; (i) the practicability of compensations either in cash or in kind one or the other Contracting Party as a means of adjusting competing water demands; (j) maintaining an acceptable environmental balance in and around a particular body of water; (k) the avoidance of unnssary waste in the utilization of waters, with due regard for the technological and financial capabilities of each Contracting Party; (l) the proportion in which each Contracting Party contributes to the water balance of the basin. 2. Each factor is to be given the weight warranted by the circumstances peculiar to each individual river basin, or group of basins, and all factors so weighted are to be considered together and a determination arrived at on the basis of the whole".

reviewed treaties. In most of Europe, on the other hand, where quality related problems constitute the major challenge, the limited transboundary impact is the rule. Recently, the increased awareness of the need for integration of quality and quantity and the drastic deterioration in the water quality shifted the attention in Europe to the principle of equitable and reasonable utilization. Recent UNECE treaties, therefore, consider equitable and reasonable utilization principle as the governing principle.

In the context of limiting transboundary impact it is understood that equitable and reasonable utilization is regarded as essential for its achievement. The situation in the Middle East is completely different and incomparable with the other two regions. The impact of politics in the formulation of existing treaties is very evident. This is demonstrated in the Israeli/Palestinian Interim Agreement of 1995 and the Jordanian/Israeli Peace Treaty of 1994 where the no harm rule has been emphasized mainly to protect the status quo of the existing uses by the Israelis. The absolute prohibition of harm within these Agreements is not realistic and cannot be enforceable. The emphasis on powers and responsibilities rather than sovereign rights of the Parties in the Israeli Palestinian Interim Agreement is a confirmation Israel's rejection to the sovereign right of the Palestinians over the shared water resources. Existing treaty practice in the Middle East provides minimum guidance on what the substantive rules governing the utilization, development and management of transboundary groundwaters are.

5.5 PROCEDURAL RULE

5.5.1 Introduction

The importance of "procedural rules" has been described earlier in this research.[176] The sections below seek to identify the basic procedural rules relating to transboundary groundwater in State treaty practice. In most of the reviewed treaties the requirement of cooperation is translated into three main procedural mechanisms: (i) the duty to exchange information; (ii) duty to notify; and (iii) consultation and negotiations. The following analysis includes an assessment to the binding nature of these rules under the selected treaty practice.

5.5.2 The Duty to Exchange Information

The requirement to exchange reliable and updated information and data relating to the shared water resources is found to be an obligation under most of the instruments reviewed within this research. To that end, some of these legal

[176] *See supra* Introduction (Section 8), *and* Chapter 4 (Section 4.5).

instruments require the Parties, on regular basis, to collect, process and assess these information and data[177], while some other arrangements restrict this procedure to the duty of exchange data that is requested by any of the Parties.[178] For example, in Africa, the 1990 Niger-Nigeria Agreement[179] requires the Parties to collect, process and provide at regular intervals to the appropriate joint institution all information and data necessary for the determination of equitable sharing.[180]

The Parties are as well responsible each within its territory to install and protect the required monitoring equipment.[181] The Revised SADC Protocol requires the exchange of the available data that is necessary for the assessment of the conditions of the shared watercourses.[182] The Protocol further delegates the power of ensuring exchange of data to the joint water institutions that are established within the Protocol.[183] The 1992 UNECE Helsinki Convention obliges the Parties

---

[177] Illustrative examples: the 1990 Niger Nigeria Agreement, *supra* note 20 and the UNECE Helsinki Convention *supra* note 110 and the 1978 Franco-Suisse Convention *supra* note 21.

[178] The 1968 African Convention, *supra* note 106, under "Inter-State Cooperation" in Article XVII (c) States are obliged to provide the African Unity with:

"all the information nssary for the complete documentation of matters dealt with by this Convention if requested".

Also see Article 11 of the Agreement for the establishment of the organization for Management and Development of the Kagera River Basin, Concluded At Rusomo, Rwanda 24 August 1977:

"Whenever possible, technical information on national projects, works and activities in the Basin or affecting the Basin, shall be communicated to the Secretariat by the respective governments…".

Signatories: Tanzania, Rwanda, Burundi.(Entry into force 5th February 1978), *found at* TFDD treaty collection at http://ocid.nacse.org/qml/research/tfdd/toTFDDdocs/200ENG.pdf (*last visited* May 2003).

[179] The 1990 Niger Nigeria Agreement, *supra* note 20.

[180] *Ibid,* Part II Article 3 para. 1:

"1. The Contracting Parties undertake to collect, process and provide at regular intervals the Nigeria-Niger Joint Commission for Co-operation[60] with all the data and information which, in the opinion of the Commission, are needed to arrive at equitable sharing determinations, and to monitor the continued viability thereof"..

[181] *Ibid,* Article 3 para 2.

"At the request of the Commission, the Contracting Parties shall: (a) install in their territory the required measuring equipment, and protect such equipment from interference; and (b) permit and facilitate inspections by the Commission of such equipment"..

[182] The Revised SADC Protocol, *supra* note 97, Article 2 para. 6 provides that:

"State Parties shall exchange available information and data regarding the hydrological, hydro geological, water quality, meteorological and environmental condition of shared watercourses."

[183] *Ibid,* Article 5

to allow for the provision of wide range of information in relation to the fulfilment of the objectives of the Convention.[184] For the purpose of protection of the shared water resources Article 9 of the Convention encourages the establishment of joint bodies that are responsible for (a) collecting, compiling and evaluating data in order to identify pollution sources likely to cause transboundary impact; (b) elaborating joint monitoring programmes concerning water quality and quantity; (c) drawing up inventories and exchange information on the pollution sources (d) elaborating emission limits for waste water and evaluate the effectiveness of control programmes; (f) developing concerted action programmes for the reduction of pollution loads from both point sources (e.g. municipal and industrial sources) and diffuse sources (particularly from agriculture); (g) establishing warning and alarm procedures.[185] Exchange of information under the Protocol on Health is of high priority in decision-making in order to enhance the quality and implementation of the decisions, to contribute to public awareness. [186] In transboundary context the exchange of information and sharing knowledge supports the Parties to assess problems and risks in relation to their shared water resources.[187] This might include the establishment of joint management plans with early warning systems.[188] In the Middle East and under the Israeli/Palestinian Interim Agreement,[189] the Israeli side agreed to make available all relevant data and to determining the appropriate locations for drilling of wells.[190]

5.5.3 The Duty to Prior Notification

State practice in the selected regions has also demonstrated a wide acceptance to the duty to notify other States on measures that might have a negative impact on

---

"c) Shared Watercourse Institutions shall provide on a regular basis or as required by the Water Sector Co-ordinating Unit, all the information nssary to assess progress on the implementation of the provisions of this Protocol, including the development of their respective agreements."

166The 1992 UNECE Helsinki Convention, Article 6, *supra* note 110, provides:

"The Parties shall provide for the widest exchange of information, as early as possible, on issues covered by the provisions of this Convention."

185 *Ibid,* Article 9.

186 UNECE Protocol on Health, Article 13, *supra* note 115, provides that:

"(a) Exchange information and share knowledge about the transboundary waters and the problems and risks which they present with the other Parties bordering the same waters;…………..".

187 *Ibid,* Article 12 and as part of the "Joint and International Actions' the Parties are obliged to develop integrated information systems and databases, exchange of information and sharing of technical and legal knowledge and experience;…"

188 *Ibid,* Article 13 (b).

189 Israeli/Palestinian Interim Agreement, *supra* note 20.

190 These are the historical data that the Palestinians were prohibited access to.

the shared water resources. The level of detail of such requirement varies among the instruments. The 1964 Chad Basin Convention and Statutes require the Parties to inform the Commission of any studies and schemes which they propose to undertake, as from their initial stage especially if such measures are expected to cause marked influence on the shared resources.[191] Under the Revised SADC Protocol prior notification is essential when planned measures might have an adverse impact on the shared watercourses.[192] The Protocol includes mechanisms for reply to notification, duration for reply and procedures in absence of notification.[193] Any Member State can proceed with planned measures of the utmost urgency, [194] provided that a declaration of what these measures are and the justification of their urgency is issued. Additionally, Member States are obliged to cooperate on the regulation of flow and have the equal rights to participate in the construction and maintenance of necessary installations and consequently bear the cost for such works.[195]

---

[191] The 1964 Chad Basin Convention and Statutes, Article 5, *supra* note 86, provides
"The Member States undertake a refrain from adopting, without referring to the Commission beforehand, any measures likely to exert a marked influence either upon the extent of water losses, or upon the form of the annual hydro graph and limnograph and certain other characteristics of the Lake, ... In particular, the Member States agree not to undertake in that part of the Basin falling within their jurisdiction any work in connection with the development of water resources or the soil likely to have a marked influence upon the system of the water courses and levels of the Basin without adequate notice and prior consultations with the Commission, provided always that the Member States shall retain the liberty of completing any plans and schemes in the course of execution or such plans and schemes as may be initiated over a period of 3 years to run from the signature of the present Convention."
Article 6 provides:
"With a view to obtaining the closest possible co-operation on the matters specified in article 5, the Member States agree to inform the Commission of any studies and schemes which they propose to undertake, as from their initial stage".
Also within the responsibilities of the Chad Basin Commission Article 9 confirm the need for prior information and notification:
"(d) Following up the execution of works and studies in the Chad Basin falling within the present Convention and keeping Member States informed at least once a year through the exploitation of systematic periodic reports which each State undertakes to address to it;..."

[192] The Revised SADC Protocol, Article 4, *supra* note 97.

[193] *Ibid,* paras (a-h).

[194] UNECE Protocol on Health, *supra* note 115.
"These urgent planned measures must be of the utmost urgency in order to protect public health, public safety or other equally important interests".
In this case article 4 (i) Para.
"(i-iii) can immediately proceed with the planned measure, and shall at the request of any of the States enter into consultations and negotiations".

[195] *Ibid,* Article 4 Para 3 (b).

5.5.4 Negotiations and Consultation

Most of the reviewed instruments demonstrate that the requirement for negotiations and consultation is favourable and recommended but is not regarded as an obligation. There is nearly always a reference to general principles of international law as the basis for undertaking such procedures. For example under the Revised SADC Protocol Member States are encouraged on the basis of good faith to enter into consultation and establish joint mechanisms for shared water resources management.[196] Article 4 of the `1990 Niger Nigeria Agreement supports the same procedure.[197] The 1992 Helsinki Convention recommends the Parties on the basis of reciprocity, good faith and neighbourliness to undertake consultations in matters relating to the provisions of the Convention.[198] The UNECE Protocol on Health recommends the Parties to consult each other on issues pertaining to adverse impacts on human health.[199]

5.5.5 Specific Procedural Rules

Some of the reviewed instruments were found to embody more detailed procedural rules which are considered highly relevant in the case of transboundary groundwater. These instruments include the African Convention that introduces the concepts of "conservation areas", "restricted areas", where activities that might cause harm to the natural resources are either restricted or fully prohibited.[200] Procedural measures to protect water might

---

[196] *Ibid*, Article 4 Para 3 (a).

"(i) If a communication is made under paragraph (e) that implementation of the planned measures would be inconsistent with the provisions of Article 3 (7) or (10), the notifying State and the State making the communication shall enter into consultations and, if nssary, negotiations with a view to arriving at an equitable resolution of the situation.(ii) The consultations and negotiations shall be conducted on the basis that each State must in good faith pay reasonable regard to the rights and legitimate interests of the other States...."

[197] The 1990 Niger Nigeria Agreement, Article 4, *supra* note 20, provides:

"The Contracting Parties undertake to inform and consult with the Nigeria-Niger Joint Commission for Co-operation in advance of undertaking any project, programme or plan for the implementation of agreed upon equitable sharing determinations, or likely to have an appreciable impact on any such determination."

[198] The 1992 Helsinki Convention, Article 10, *supra* note 110,

"Consultations shall be held between the Riparian Parties on the basis of reciprocity, good faith and good-neighbourliness, at the request of any such Party. Such consultations shall aim at cooperation regarding the issues covered by the provisions of this Convention. Any such consultations shall be conducted through a joint body established under article 9 of this Convention, where one exists."

[199] *Ibid*, Arts. 11 and 12.

[200] The 1968 African Convention, supra note 106, Art. III (4) defines Conservation Area as:

include the study of water cycles and the investigation of each catchment area, the co-ordination and planning of water resources development projects, the administration and control of all water utilization, and prevention and control of water pollution.[201] The 1994 Danube Convention provides a strong legal instrument in relation to connected groundwater management and protection. It gives adequate reference to groundwater under very important provisions including the definition of terms, the objectives and specific measures to protect the resource by introducing protection areas and zones.[202] Under the 1978 Franco-Suisse Convention,[203] a detailed Technical Regulation for the operation of the groundwater recharge system has been drafted jointly and adopted

---

"any protected natural resource area, whether it be a strict natural reserve, a national park or a special reserve; a) "strict nature reserve" means an area: i) under State control and the boundaries of which may not be altered nor any portion alienated except by the competent legislative authority, ii) throughout which any form of hunting or fishing, any undertaking connected with forestry, agriculture or mining, any grazing, any excavation or prospecting, drilling, levelling of the ground or construction, any work tending to alter the configuration of the soil or the character of the vegetation, any water pollution and, generally, any act likely to harm or disturb the fauna or flora, including introduction of zoological or botanical species, whether indigenous or imported, wild or domesticated, are strictly forbidden".
Article X. justifies the need for conservation areas:
"The Contracting States shall maintain and extend where appropriate, within their territory and where applicable in their territorial waters, the Conservation areas existing at the time of entry into force of the present Convention and, preferably within the framework of land-use planning programmes, assess the nssity of establishing additional conservation areas in order to: i) protect those ecosystems which are most representative of and particularly those which are in any respect peculiar to their territories, ii) ensure the conservation of all species and more particularly of those listed or which may be listed in the annex to this Convention; 2. The Contracting States shall establish, where nssary, around the borders of conservation areas, zones within which the competent authorities shall control activities detrimental to the protected natural resources".
This procedural rule within the African Convention is very relevant in the case of groundwater and can be looked at as best practice in the context of protecting groundwater from depletion and pollution. The declaration of "Conservation Areas" is becoming a key measure at national and international levels to protect groundwater. The prohibition or restriction of certain activities in certain area around the groundwater recharge area is very essential for the optimal and sustainable protection of groundwater.

[201] *Ibid*, Article V(1).

[202] *Ibid*, Article 6:
"Specific Water Resources Protection Measures: The Contracting Parties shall take appropriate measures aiming at the prevention or reduction of transboundary impacts and at a sustainable and equitable use of water resources as well as at the conservation of ecological resources, especially: (a) enumerate groundwater resources subject to a long-term protection as well as protection zones valuable for existing or future drinking water supply purposes;"

[203] The 1978 Franco-Suisse Convention, Article 21, *supra* note 21.

following the commissioning of the installation. This procedure is vitally relevant and important as far as transboundary groundwater is concerned.

5.5.6 Summary

The preceding analysis reveals that the treaty practice in Africa Europe and the Middle East reveal a satisfactory level of consistency and uniformity in the context of procedural rules. This research finds that there is adequate evidence that cooperation is collectively considered the means for achieving successful implementation of the Agreements. The modes of cooperation, however, vary from region to another with some procedures common to all. The common denominators found in the three regions include the duty for exchange of information, notification and to a certain extent the need for consultation. In Africa some of the legal instruments namely the Revised SADC Protocol of 2000 and the Niger Nigeria Agreement of 1990 provide guidance on how equitable and reasonable utilization can be operationalised. This guidance resembles that provided by the ILA Helsinki Rules and the 1997 UN Convention. The declaration of conservation areas found in the 1968 African Convention, is one important rule of procedure that is relevant for the protection of the quality and quantity groundwater resources namely the declaration of conservation area. In Europe the development of the appropriate legal and technical measures for the abatement of pollution are common to most treaties.

5.6 INSTITUTIONAL AND DISPUTE SETTLEMENT MECHANISMS

5.6.1 Institutional Mechanisms in State Practice

In Africa the 1990 Niger-Nigeria Agreement[204] employ an exiting institutional mechanism as the machinery for the implementation of the present Agreement- the Niger-Nigeria Joint Commission.[205] The Niger Nigeria Commission has an authoritative status as it can take decisions on behalf of the Contracting Parties regarding the provisions of the Agreement.[206] The 1964 Chad Basin Convention and Statutes establish the Chad Basin Commission for the purpose of implementing the Agreement with scope covering the whole Basin.[207] The Commission's functions cover a broad range of activities including regulatory,

---

[204] The 1990 Niger Nigeria Agreement, *supra* note 20.

[205] Created by the Convention done at Niamey on 3 March 1971, amended by Protocol of 22 December 1973.

[206] *See* Part IV "Institutional Arrangements".

[207] The 1964 Chad Convention and Statutes, *supra* note 86 Article 9 (a-h)

technical and dispute settlement.[208] However, the Chad Commission has limited authority compared to the Niger Nigeria Joint Commission as all rules and regulations prepared by the Chad Commission have to be approved by the governments. Disputes are first referred to the Commission for its review and possible settlement.[209] The 1968 African Convention[210] encourages States that are sharing surface or groundwater to set up inter-State Commissions to study and resolve problems arising from the joint use of these resources, and for the joint development and conservation.[211] The Revised SADC Protocol encourages the Member States to establish of a number of institutions that are responsible for the implementation of the Protocol. The Protocol identifies the functions of each institution in order to eliminate any overlap or conflict in their roles and responsibilities.[212]

In Europe the 1992 UNECE Helsinki Convention calls for at least annual meetings between the parties during which they will review the implementation of the Convention and discuss viability of procedures adopted and seek assistance of UNECE bodies and other competent international institution as deemed appropriate.[213] With the entry into force of the Convention on the Protection of the Danube in 1998[214] the International Commission for the Protection of the Danube River was legally established.[215] The Danube International Commission facilitates a "common platform" for the sustainable use of the basin's resources in relation to its aquatic ecology and coherent and integrated river basin management.[216] Decisions and recommendations are adopted by consensus of the delegations to the Danube Commission. Such decisions become effective upon the formal adoption of all Contracting Parties voting for it.[217] The first institutional structure to manage the river Rhine dates back to 1449: the Strasbourg regulations with the objectives to protect the river against pollution and over fishing. [218] In 1950 the riparian states established the International Commission for the Protection of the

---

[208] *Ibid,* Article 9.

[209] The 1990 Niger Nigeria Agreement, *supra* note 20.

[210] The 1968 African Convention, *supra* note 106.

[211] *Ibid,* Article V(2)

[212] *Ibid,* Article 5 sets forth the institutional mechanisms and establishes the following institutions 1) Water Sector Organs constituting from :the Committee of Water Ministers, the Committee of Water Senior Officials, the Water Sector Coordinating Unit, the Water Resources Technical Committee and Subcommittees; 2) the Shared Water Institutions.

[213] The 1992 Helsinki Convention, Article 17, *supra* note 110.

[214] The 1994 Danube Convention, *supra* note 118.

[215] *Ibid,* Part III Article 18.

[216] *Ibid,* paragraph 1.

[217] *Ibid,* Annex IV Article 5

[218] K. Wieriks, and A. Schulte-Wulwer-Leidig, Integrated water management for the Rhine river basin, from pollution prevention to ecosystem improvement 21 NRF no. 2 (1997).

Rhine (ICPR).[219] The most powerful institutional mechanism the International Commission on the Protection of the Rhine is found in the 1998 Convention.[220] The Rhine Commission has almost the same functions and authority as the Danube Commission.[221] The 1978 Franco-Suisse Convention[222] establishes a Joint Commission for Exploitation composed of six members (three Swiss, three French), of who at least four (two of each State) are experts in water matters.[223] The Commission gives its technical opinion on new water extraction works and utilizations as well as on the modification of existing ones and audits the construction and operation costs of the groundwater recharge installation. Each delegation reports to its parent authority.[224] The Commission is purely consultative. The Commission convenes periodically, in practice twice a year, alternatively in France and in Switzerland.[225] The recommendations reached by the Commission are jointly submitted to the consideration of the contracting parties which make sovereign decisions on the implementation thereof within their respective territory.[226]

In the Middle East, the Palestinian and Israeli sides agreed within the Interim Agreement[227] to establish a permanent Joint Water Committee (JWC) for the interim period, the function of which is to deal with all water and sewage related issues in the West Bank including coordinated management of water resources.[228] In order to ensure compliance with the provisions of Article 40 both sides agreed to establish, a Joint Supervision and Enforcement Teams (JSETs) for the West Bank, under the control and supervision of the JWC, which shall commence operation immediately.[229] Currently, and due to the political tense situation

---

[219] *Ibid.*

[220] The 1998 Rhine Convention, *supra* note 121, Art. 5-12.

[221] *Ibid,* Articles 10 and 11.

[222] The 1978 Franco-Suisse Convention, *supra* note 21.

[223] *Ibid,* Article 1 of the Agreement paragraphs 1 and 2.

[224] *Ibid,* Article 2.

[225] *Ibid,* Article 3.

[226] *Ibid,* Article 5.2.

[227] The 1995 Israeli-Palestinian Interim Agreement, *supra* note 20.

[228] *Ibid,* Appendix II Annex III Article 40 Schedule 8.

[229] *Ibid,* Schedule 9. The functions of these teams are:

"JSETs shall operate, in the field, to monitor, supervise and enforce the implementation of Article 40 and this Schedule, and to rectify the situation whenever an infringement has been detected, concerning the following: a. Extraction from water resources in accordance with the decisions of the JWC, and the Schedule to be prepared by it in accordance with sub paragraph 1 (e) of Schedule 8. b. Unauthorized connections to the supply systems and unauthorized water uses; c. Drilling of wells and development of new projects for water supply from all sources; d. Prevention of contamination and pollution of water resources and systems; e. Ensuring the execution of the instructions of the JWC on the operation of monitoring and measurement systems; f. Operation and maintenance of systems for

between Israel and Palestine, the work of the JWC is almost frozen.[230] Israel and Jordan within their 1994 Peace Agreement[231] established a Joint Water Committee for the implementation. The Joint Committee comprised of three members from each country. The Joint Water Committee has the power to specify its work procedures, the frequency of its meetings, and the details of its scope of work. [232]

5.6.2 Dispute Settlement Mechanisms in State Practice

In Africa the first resort for dispute settlement within most of the analysed treaties is through amicable means and negotiation. In some of the treaties the first resort is the joint commission.[233] As a second resort these agreements refer to the Commission of Mediation, Conciliation and Arbitration of the Organization of African Unity for the purposes of decision.[234] The Revised SADC Protocol limits

---

collection, treatment, disposal and reuse, of domestic and industrial sewage, of urban and agricultural runoff, and of urban and agricultural drainage systems; g. The electric and energy systems which provide power to all the above systems; h. The Supervisory Control and Data Acquisition (SCADA) systems for all the above systems; i. Water and sewage quality analyses carried out in approved laboratories, to ascertain that these laboratories operate according to accepted standards and practices, as agreed by the JWC. A list of the approved laboratories will be developed by the JWC; j. any other task, as instructed by the JWC".

[230] *See infra* Chapter Five.

[231] The 1994 Israeli-Jordanian Peace Treaty, *supra* note 20.

[232] *Ibid*, Art. VII.

[233] The 1968 African Convention *supra* note 106 Article XVIII
"Any dispute between the Contracting States relating to the interpretation or application of this Convention which cannot be settled by negotiation, shall at the request of any party be submitted to the Commission of Mediation, Conciliation and Arbitration of the Organization of African Unity".
The 1990 Niger Nigeria Agreement *supra* note 20, Article 17:
"1. Any difference concerning the interpretation or implementation of this Agreement, or of any equitable sharing determination made there under, shall be referred to the Nigeria-Niger Joint Commission for Co-operation. 2. Any such difference which cannot be settled by the Commission within six months after a reference to it under the last prding paragraph, shall be regarded as a dispute and shall, at the request of both Contracting Parties, be referred to the Commission of Mediation, Conciliation and Arbitration of the Organization of African Unity for binding determination".
The SADC Revised Protocol *supra* note 97 Article 7 Para. 1,2 and 3.encourages the Member States to follow customary rules and resolve the conflicts agreeably. The second step within the Protocol is to refer the case to the tribunal.

[234] The 1964 Chad Basin Convention and Statutes, Article 7, *supra* note 86,
"Any dispute concerning the interpretation of implementation of the present Convention, which has not been determined by the Commission, shall be submitted to the Commission

itself to encouraging Member States in situations of dispute to follow the customary procedure conflict agreeably; if this fails the dispute is referred to the tribunal.[235] Under the 1964 Chad Basin Convention and Statutes, failure to settle disputes amicably, the case shall be referred to mediation, conciliation and finally to the arbitration of the Organisation of the African Unity for their decision.[236]

In Europe and under the 1992 Helsinki Convention disputes are firstly to be resolved by negotiation or by any other means of dispute settlement acceptable to the parties to the dispute.[237] Upon failure to agree, the 1992 Helsinki Convention offers two alternative compulsory procedures.[238] These are the submission of the dispute to the International Court of Justice or arbitration in accordance with the procedure set out in annex IV. The UNECE Protocol on Health of 1992 provides that settlement of dispute between parties shall follow the normal procedures of negotiations to solve conflicts amicably or by any other means acceptable to the disputants.[239] The 1994 Danube Convention encourages the Parties to seek settlement of dispute through negotiation or any other acceptable means. Disputes not solved though the aforementioned means, are either submitted to the ICJ or to arbitration in accordance with the provisions of the Convention.[240] Under the 1998 Rhine Convention the parties concerned with a certain dispute shall negotiate and strive to solve the conflicts amicably whereas if they fail the Convention provides arbitration procedures to be carried out upon the demand of one of the parties to the dispute.[241] Any differences arising out of the implementation of the 1978 Franco-Suisse Convention[242] are to be referred to the conciliation of the Regional Franco-Genevese Committee, failing which, the matter is to be decided by the Mixed Consultative Commission for problems of neighbourliness.[243]

---

of Mediation, Conciliation and Arbitration of the Organisation of African Unity for the purposes of decision".

[235] *Ibid.*

[236] *Ibid.*

[237] *Ibid,* Article 22 paragraph 1

[238] *Ibid*, Para 2.

[239] *Ibid*, Art. 20.

[240] *Ibid*, Art. 24. Annex V details the procedure of the arbitration.

[241] *Ibid*, Art. 16.

[242] The 1978 Franco-Suisse Convention, *supra* note 21.

[243] *Ibid,* Article 20, The first Committee is the "*Comité Regional Franco-Genevese*" which was established in 1974 as one of the permanent committees of the Mixed Consultative Commission .and the second Committee is the "*Commission Mixte Cnsultative pour les Poblèmes de Visinage entre Genève et les Départements de l'Ain et de la Haute-Savoie*" which was established by an Exchange of Letters dated 12 July 1973 (info rived from the Service Cantonal de Géologie in Geneva).

5.6.3 Summary

Most of the reviewed legal instruments agreements that were analyzed recognized the need for an appropriate institutional mechanism for the implementation of desired cooperation. To that end, States in the three selected regions resorted to different types of institutional mechanisms. Such mechanisms are classified into four main categories (i) establishing mechanisms for regular meetings between the Parties, (ii) making use of existing institutions, (iii) establishing new institutions and (iv) encouraging the set up the proper mechanisms. The powers, structure and scope of work of such institutions range from one instrument to the other. While some institutions have pure technical and supervisory functions[244] others have complete authority on the decision making with regard to the implementation of the legal arrangement.[245] Additionally, some of the legal instruments employ the available institutional mechanisms as the first resort for dispute settlement.[246] As for disputes that are not determined by the respective institutional mechanisms the

---

[244] Illustrative examples are: (a) the Joint Commission established under the Treaty between the Hungarian People's Republic and the Republic of Austria concerning the Regulation of Water Economy Questions in The Frontier Region, Signed At Vienna, On 9 April 1956, (entry into force on 31 July 1956), *found at* http://ocid.nacse.org/qml/research/tfdd/toTFDDdocs/94ENG.htm (*last visited* May 2003) and (b) the Joint Technical Committee established under the Israeli Palestinian Interim Agreement, *supra* note 20.

[245] Illustrative examples are: (a) the Boundary Commission established under the Treaty between Germany and Poland for the Settlement of Frontier Questions, signed At Poznan, January 27, 1926. http://ocid.nacse.org/qml/research/tfdd/toTFDDdocs/35ENG.htm Article 39 provides:

"The decisions of the Boundary Commission in connection with the present Treaty are enumerated in a special list and are hereby recognised by both Parties as binding. These decisions shall be executed by the Committees provided for in Article 6 in so far as the latter still regard this as necessary."

And (b) the Tripartite Standing Committee under the Protocol To Establish Tripartite Standing Committee on Polluted Waters between Luxemburg, France And Belgium, Brussels 8 April 1950 (Came Into Force 8 April 1950) http://ocid.nacse.org/qml/research/tfdd/toTFDDdocs/284ENG.pdf (*last visited* May 2003), and the Niger Basin Authority under Convention creating the Niger Basin Authority. 21 January 1980 Signatories: Benin, Cameroon, Chad, Ivory Coast, Guinea, Mali, Niger, Nigeria, Upper Volta. ocid.nacse.org/qml/research/tfdd/toTFDDdocs/374ENG.htm (*last visited* May 2003) and the 1992 UNECE Helsinki Convention, *supra* note 110.

[246] The 1964 Chad Basin Convention and Statutes, *supra* note 86. Article 9 which lists the functions of the Chad Commission item (g) Examining complaints and assisting in settling disputes;

Also under the 1990 Niger Nigeria Agreement, *supra* note 20, settlements m of disputes are first referred to the Joint Commission (Art. 17).

legal instruments mostly offer one or a combination of three stepwise mechanism for dispute settlement-namely, mediation conciliation and arbitration.

5.7 INFORMAL COOPERATIVE ARRANGEMENTS

5.7.1 Introduction

In the absence of regional treaties regulating transboundary groundwater aquifers, especially if States are hesitant to conclude a special treaty relating to transboundary groundwater, the resort to informal arrangements is increasingly growing. The circumstances, types and character of such State practice are described in the following sections.

5.7.2 Cooperative Frameworks on Transboundary Confined Aquifers

It was observed in this research that there is an emerging trend to establish networks for regional cooperation on the development utilization and management of transboundary aquifers. In Africa, for example, and in order to better understand confined transboundary aquifers, geohydrologists from more than 20 States have recently drawn up the first continental survey of transboundary aquifers in Africa.[247] The Survey has uncovered about 20 transboundary aquifers, five of which have never been identified before.[248] Geohydrologists from Benin learned that the aquifer providing water for the capital city, Cotonou, extends across the border into Togo.[249] Both countries have expressed interest in developing a framework to jointly manage this aquifer, which will become increasingly important in the future as rainfall declines with climate change. [250] The survey also uncovered a major aquifer between Ghana and Côte D'Ivoire. Hydrologists from Côte D'Ivoire have now begun studying the aquifer, which provides 80% of the local population's water needs.[251] In Europe, on the other hand there is an emerging trend in developing networks and frameworks for cooperation concerning unregulated transboundary groundwater. In Africa the cooperative frameworks include cooperation on the Nubian Sandstone Aquifer

---

[247] Africa's Hidden resources: UNESCO Water Portal, found at http://www.unesco.org/water/unescopress_05_06.shtml (last visited April 2003). UNESCO Press Release No.2002-35, found at http://www.unesco.org/bpi/eng/unescopress/2002/02-39e.shtml. *(last visited* April 2003).

[248] *Ibid.*

[249] *Ibid.*

[250] *Ibid.*

[251] *Ibid.*

System NSAS [252] which is a huge confined aquifer, located in the eastern Sahara desert in north-eastern Africa.[253] In the 1970's, the States sharing the NSAS expressed their interest in regional cooperation to share experience and to study and develop this regional aquifer.[254] The recent initiative to formulate a regional strategy for the utilization of the NSAS is a step towards the sustainable management of the aquifer, through the sharing of information, monitoring of the aquifer and the exchange of updated information on the behaviour of that shared resources.[255] The four States sharing the NSAS have signed two agreements: one instrument related to the sharing of data incorporated in the regional information system. The second agreement updated this information by continuous monitoring of the aquifer.[256] The most important achievements of these agreements were the revitalization of the Joint Authority on the NSAS that was previously formed between Egypt, Libya, and later Sudan. Chad was invited to attend as an observer at the first meeting (during the project) of the revitalized Joint Authority in February 1998. The Joint Authority acted as the Regional Programme Steering Committee. Currently the efforts are concentrated on developing a regional mathematical model for simulating scenarios for the utilisation of the NSAS, hence developing the regional strategy for the sustainable management of the aquifer.[257] Another example of cooperative framework is the one on the North Western Sahara Aquifer System shared by Libya Tunisia and Algeria. The increased utilization of the North-Western Sahara Aquifer System NWSAS [258] since 1982 has led to this system being identified as a high national priority.[259] The parties sharing the aquifer agreed to undertake a project that determined the best scenario for exploiting the water resources in a manner that ensures sustainable water utilization in both quantitative and qualitative terms, particularly focusing on the protection of the recharge areas and the wetland.[260] The FAO has funded the Watercourse States to undertake a project entitled "Establishment of a mechanism for consultation at the level of the North-Western Sahara Aquifer

---

[252] Nubian Sandstone Aquifer System Programme, *found at* http://isu2.cedare.org.eg/nubian/ (*last visited* April 2003). See *supra* Chapter One.

[253] *Ibid.*

[254] *See* Africa's Hidden Resources, *supra* note 247.

[255] *Ibid.*

[256] *Ibid.*

[257] Nubian Sandstone Aquifer System Programme, http://isu2.cedare.org.eg/nubian/project/activities.htm (*last visited* November 2003).

[258] Shared by Libya Algeria and Tunisia and it covers well over 1,000,000 km^2 of which 700,000 are in Algeria, 80,000 in Tunisia and 250,000 in Libya.

[259] Global Environmental Facility, *found at* http://www.gefweb.org/Documents/Medium-Sized_Project_Proposals/MSP_Proposals/Regional_-_North_West_Sahara_Aquifer_System.pdf (*last visited* May 2003)

[260] *Ibid.*

System", establishing a "consultation mechanism", which consists of a Coordination Unit or Secretariat, guided by a Steering Committee of representatives of the water resources agencies of the member States.[261]

The cooperation on The Kalahari/Karoo Multi-layered Aquifer multi-layered aquifer that is shared by Botswana, Namibia and South Africa should also be examined. The aquifer is a key component of future human and economic development in the arid environment of the Kalahari.[262] Groundwater abstraction for irrigation, which depends highly on recharge from flash floods in the river systems, is a matter of concern in the upper Molopo River, in South Africa. The largest portion of the aquifer falls within Namibia, which is expected to have the largest demand from the system and where needs are expected to rise in the future. [263] The major issue at this stage is for all 3 countries to obtain a proper understanding of all aspects of the aquifer, for joint assessment, monitoring and management of the resource.[264] The States can then work out a legal framework as to a common abstraction policy. This could be done through the existing International River Commissions, bilateral technical committees and institutions of the Southern African Development Community (SADC).

5.7.3 Technical Cooperation Concerning Connected Transboundary Groundwater

A good example that can be cited cooperation on the Vechte catchment area between the Netherlands and Germany. On the initiative of the German and Dutch water management authorities the Regional Geohydrological Information System (REGIS) is being applied for the River Vechte cross-border basin.[265] In Europe the system REGIS-Vechte has been developed in the framework of the Digital Waterway Vechte project in which water managers from both Germany and the Netherlands were participating. The information system which is now operational within the water management institutions on both sides of the border is a first important step towards a sustainable groundwater data exchange between Germany and the Netherlands. The institutions have now access to the groundwater data, the subsurface data and the geohydrological subsurface models

---

[261] FAO Legal Office website, *found at* http://www.fao.org/legal/advserv/waternews.htm (*last visited* June 2003)

[262] Internationally Shared Aquifer Resources Management (ISARM) Framework Document, *found at* http://www.iah.org/News/2001/isarm.pdf (*last visited* June 2003).

[263] *Ibid.*

[264] *Ibid.*

[265] Email communications with Geo Arnold geo-Hydrologist, Senior Project Engineer at the Dutch Institute for Inland Water Management and Waste Water Treatment (RIZA) and Programme Coordinator for groundwater at IWAC (UNECE).

of the whole catchment. To enable cross-border geohydrological characterisation and modelling of the subsurface the relevant German data (e.g. bore hole descriptions) had to be processed and interpreted in a uniform manner that would be comparable to the set up of the Dutch subsurface model. The three geological surveys involved agreed upon the geohydrological layer schematisation to be applied. Accordingly about five thousand bore holes have been interpreted and entered into the REGIS Database. The geometry and hydraulic properties of the geohydrological units of the different subsurface models have been ascertained. All these characteristics can be consulted, evaluated and compared with the information system. The catchment wide approach and the cross border collection, compilation and evaluation of data is in accordance with the WFD. The information system can provide the required information on groundwater characteristics, monitoring programs, the compliance testing of the environmental objectives and on the anthropogenic impact on groundwater. Missing information can easily be identified and the effect of water management measures verified.[266] This cooperation includes the so-called "Permanent Commission" that is charged with groundwater issues (state level) and Commissions at regional (provincial) level. Different Commissions have been organized in relation to groundwater basins.[267]

Apart from the signed Agreements in the Middle East, and in the context of Multilateral Peace Process,[268] a Group on Water Resources was established.[269]

---

[266] H. Uil, *The Transboundary Information System For The River Vechte Catchment In Accordance With The Water Framework Directive found at* http://Www.Copernicus.Org/Egs/Egsga/Nice01/Programme/Abstracts/Aac7322.Pdf (*last visited* May 2003).

[267] Email communications with Geo Arnold geo-Hydrologist, Senior Project Engineer at the Dutch Institute for Inland Water Management and Waste Water Treatment (RIZA) and Programme Coordinator for groundwater at IWAC (UNECE).

[268] The Middle East peace process and its bilateral track began with the Madrid Conference of October 1991. Subsequently, peace process partners agreed to establish a multilateral track, which began with an organizational meeting in Moscow in January 1992. The broad goal of the multilateral track is to focus on issues of common interest and importance throughout the region that can best be addressed on a regional basis. The multilateral track consists of five working groups: (1) Working Group on Water Resources, (2) Working Group on the Environment, (3) Working Group on Regional Economic Development, (4) Working Group on Refugees, and (5) Working Group on Arms Control and Regional Security.

[269] The Palestinians Israelis and Jordanians agreed For the Middle East, most of which has semi-arid to arid climatic conditions, the water problems are myriad. The Working Group on Water Resources (WGWR), for which the United States serves as Gavel holder and Japan and the European Union serve as co-organizers, established the following The Waternet project was the first joint initiative by the parties to implement parts of this declaration. The overall vision of Waternet is to stimulate cooperation between the parties

This group had four broad agenda items to address some of the critical water issues which include the enhancement of water data availability, water management practices, including conservation, enhancement of water supply and concepts of regional water management and cooperation. [270] As an outcome of these efforts, which was facilitated by the Government of Norway and the Office of Gavel holder, the core parties have identified common denominators[271] in their water resources management systems and consequently adopted a "Declaration on Principles for Cooperation".[272] These common denominators on water include (i) the need for modified water legislations, which apply to all types of water resources including wastewater and desalinated water; (ii) ownership and administration of water resources, which are publicly owned and/or centrally

---

on water-related issues. One of the first objectives to be achieved was the development of a common information system for water-related matters – the Waternet Information System. By facilitating regional cooperation on water issues and the development of mutual trust between the parties, the system benefits all parties. Currently, the status of all these projects is unclear as the political situation has hindered all progress in that endeavour.

[270] This study was undertaken by the Centre for Environmental Studies and Resource Management (CESAR) in Norway.

[271] The Core Parties promote public participation in water resources management. Well drilling, water production and supply are allowed only by permit or license. A Central Water Authority/ Government Agency exists in each of the Core Parties, and exercises effective control over water resources. 3. Allocation: Domestic uses occupy the first priority in the allocation of water resources. 4. Drought Measures: The Core Parties will take appropriate measures in periods of drought and water scarcity. 5. Water Quality and Protection Water quality standards for various water uses have been adopted by each of the Core Parties; water preservation is an overriding concern, and enforcement powers exist in the hands of competent authorities to prevent water pollution, and to mitigate any negative environmental impacts on them at the expense of the polluters. 6. Data and Record Keeping Obligations exist, pursuant to the legislation of each Core Party, to measure, monitor and keep proper record of all water production, supplies and consumption. 7. Compliance and Enforcement Proper sanctions against non-compliance are explicit in the respective legislation of each of the Core Parties. Enforcement of the water legislation is the norm. 8. Water Charges: Water is not supplied free of charge in any of the Core Parties. Tariff structures, taking into account different extents of cost recovery, apply to domestic, industrial and agricultural sectors. These tariffs are periodically reviewed and adjusted. Declaration on Principles for Cooperation on Water-Related Matters and New and Additional Water Resources Multilateral Peace Process in the Middle East Multilateral Working Group on Water Resources, Oslo, February 13, 1996, *found at* http://www.israel-mfa.gov.il/mfa/go.asp?MFAH00tr0 (*last visited* June 2000). The author of this thesis believes that the signing of this Declaration is merely to express willingness to cooperate, as the Declaration does not conflict with the Palestinians' demands for equitable and reasonable utilization of the shared water resources.

[272] *Ibid.*

controlled, used for the benefit of their respective societies.[273] The Core Parties in the Middle East Peace Process are considered to be Jordan, Syria, Israel, Lebanon and the PLO for the benefit of the Palestinian Authority. The signatories were Jordan Israel and the PLO. The Core Parties confirmed that this Declaration will not affect or alter in any form or manner any for the bilateral or other agreements or undertakings among them, nor does it prohibit or constrain any bilateral arrangements, understandings or agreements aimed at enhancing cooperation in water-related matters.

5.7.4 Summary

In the absence of formal treaties it was found that there is an emerging trend for the establishment of networks and cooperative frameworks that are characterized to be technical, informal and non-binding. Such arrangements rely primarily on the morals and ethics of States and their recognition of the need for such cooperation. In Africa there is a trend to resort for such arrangements in the context of confined transboundary groundwaters. In Europe however, these cooperative frameworks are emerging in the field of connected transboundary groundwater. Europe is increasingly realizing the need to build knowledge and awareness on their shared transboundary groundwaters, however, not yet in the form of a binding treaty. As for the Middle East, especially in the countries where the transboundary water resources are highly disputed such as in Israel, Jordan and Palestine, informal regional cooperation are considered to be more viable and preferable in the short term as the current prospects for concluding obligatory commitments on the side of the Parties are limited.

5.8 CONCLUSIONS

The preceding discussion demonstrates that States appear to rarely dedicate a special attention to the regulation of their transboundary groundwater. It is also evident that such resources, if included within the scope of international arrangements, are treated similarly to surface water. The increased knowledge and awareness of the special characteristics and behaviours of groundwater have not yet triggered the formulation of international treaties that specifically regulate the utilization, development and protection of these resources. This chapter also finds that only connected transboundary groundwaters are included within the scope of the existing treaties. Confined transboundary aquifers remain formally unregulated.

---

[273] *Ibid.*

In Africa, as the water situation is not uniform, State practice relating to transboundary groundwater is accordingly variant. By and large, it is concluded that the inclusion of connected transboundary groundwater within the scope of international watercourse treaties is widely recognized. The adoption of the Revised SADC Protocol in year 2000 is a clear recognition of this region to the significance of transboundary groundwater, yet treated as surface water. As for Europe recent treaty practice relating to international watercourses is increasingly recognising that groundwater is an integral component of the whole water system. Accordingly, the integrated ecosystem approach to the management of these resources is becoming common. Although more specific responses relating to transboundary groundwater are still required, the guidance that is offered by some of the multilateral conventions such as the 1994 Danube, the 1998 Rhine Conventions and the EU WFD is satisfactory. By and large it was found that there is a unique richness in the State treaty practice concerning transboundary water resources in Europe. In the Middle East although groundwater constitutes a major water supply source, many of aquifers that are shared by two States or more are not yet regulated. Apart from the recent Agreements signed between Israel and its neighbours (Palestine and Jordan), which merely serve political purposes- transboundary groundwaters in the region are not subject to any formal arrangements.

It appears from this research that the principle of territorial sovereignty is an important notion that affects State Practice. From this researcher's perspective this reflects the correlative rights of States under the equitable and reasonable utilization principle. In Africa, the balance between the equitable and reasonable utilization and the obligation not to cause harm has been explicitly in favour of the equitable and reasonable principle. This is demonstrated in the 1990 Niger Nigeria Agreement and the Revised SADC Protocol. In Europe the equitable and reasonable utilization principle began to appear after the 1990's. Prior that period, the focus was on transboundary impact and the obligation to limit this impact. In the Middle East, the few examples reviewed illustrate a unique mode of cooperation due to the tense political situation. The "no harm" principle appears to be emphasized, while the principle of equitable and reasonable utilization is almost absent under these arrangements.

The present chapter demonstrates that State treaty practice provides some common denominators with regard to procedural rules; at the heart of which lies cooperation. State treaty practice also includes a range of mechanisms for the successful implementation of desired cooperation. These include the duty to exchange information, prior notification and consultation. Consultation as referred to under the various legal arrangements is dependent on general principles of international law including good faith, neighbourliness and reciprocity. In Europe, existing treaties provide a significant guidance for the progressive development of the law relating to transboundary groundwater. The concepts of protection or restricted zones, the joint monitoring and control on abstraction, human activity and pollution constitute state of the art that can be borrowed and adapted to suit the local conditions. The "precautionary" and "preventive" approaches that are common to most of the legal instruments and agreements address the most critical problem associated with groundwater; that is irreversibility of damages and losses. These approaches are considered key for the efficient and optimum protection of transboundary groundwater.

Most of the reviewed legal instruments and agreements recognize the need for an appropriate institutional mechanism for the implementation of desired cooperation. To that end, States in the three selected regions resorted to different types of institutional mechanisms. Such mechanisms are classified into four main categories (i) establishing mechanisms for regular meetings between the Parties, (ii) making use of existing institutions, (iii) establishing new institutions and (iv) encouraging the set up the proper mechanisms. The powers, structure and scope of work of such institutions vary from one instrument to the other. While some institutions have pure technical and supervisory functions[274] others have complete authority on the decision making with regard to the implementation of the legal arrangement.[275] Additionally, some of the legal instruments employ the available

---

[274] Illustrative examples are: (a) the Joint Commission established under the Treaty between the Hungarian People's Republic and the Republic of Austria concerning the Regulation of Water Economy Questions in The Frontier Region, Signed At Vienna, On 9 April 1956, (entry into force on 31 July 1956), *found at* http://ocid.nacse.org/qml/research/tfdd/toTFDDdocs/94ENG.htm (*last visited* May 2003) and (b) the Joint Technical Committee established under the Israeli Palestinian Interim Agreement, *supra* note 20.

[275] Illustrative examples are: (a) the Boundary Commission established under the Treaty between Germany and Poland for the Settlement of Frontier Questions, signed At Poznan, January 27, 1926. http://ocid.nacse.org/qml/research/tfdd/toTFDDdocs/35ENG.htm Article 39 provides:

"The decisions of the Boundary Commission in connection with the present Treaty are enumerated in a special list and are hereby recognised by both Parties as binding. These decisions shall be executed by the Committees provided for in Article 6 in so far as the latter still regard this as nssary."

And (b) the Tripartite Standing Committee under the Protocol To Establish Tripartite Standing Committee on Polluted Waters between Luxemburg, France And Belgium, Brussels 8 April 1950 (Came Into Force 8 April 1950) http://ocid.nacse.org/qml/research/tfdd/toTFDDdocs/284ENG.pdf (*last visited* May 2003), and the Niger Basin Authority under Convention Creating the Niger Basin Authority, 21 January 1980 Signatories: Benin, Cameroon, Chad, Ivory Coast, Guinea, Mali, Niger, Nigeria, Upper Volta. ocid.nacse.org/qml/research/tfdd/toTFDDdocs/374ENG.htm (*last visited* May 2003) and the 1992 UNECE Helsinki Convention, *supra* note 98.

institutional mechanisms as the first resort for dispute settlement.[276] As for disputes that are not determined by the respective institutional mechanisms the legal instruments mostly offer one or a combination of three stepwise mechanism for dispute settlement-namely, mediation conciliation and arbitration.

In the absence of formal treaties it was also found that there is an emerging trend for the establishment of networks and cooperative frameworks that are characterized to be informal and technical. In Africa the latter trend is primarily observed in the context of transboundary confined groundwaters, which is a highly valuable resource subject to no legally binding arrangement. In Europe however, these cooperative frameworks are emerging in the field of connected transboundary groundwater which are unregulated. Europe is increasingly realizing the need to build knowledge and awareness on their shared transboundary groundwaters, however, not yet in the form of a binding treaty.

As for the Middle East, especially in countries where the transboundary water resources are highly disputed such as in Israel Jordan and Palestine, informal regional cooperation is considered to be more viable and preferable as the current prospects for concluding obligatory commitments relating to the shared transboundary groundwaters are limited. To that end, the next chapter draws lessons from the research undertaken so far within this work to propose the most viable approach for solving the Palestinian Israeli conflict over transboundary groundwater resources.

---

[276] The 1964 Chad Basin Convention and Statutes, *supra* note 86. Article 9 which lists the functions of the Chad Commission item (g) Examining complaints and assisting in settling disputes;
Also under the 1990 Niger Nigeria Agreement, *supra* note 20, settlements of disputes are first referred to the Joint Commission (Art. 17).

6

A Progressive Approach for Resolving the Israeli- Palestinian Conflict on the Mountain Aquifer

6.1 INTRODUCTION

This chapter builds on the conclusions and findings resulting from the research undertaken within this research. The current chapter intends to attain a politically, technically and financially feasible solution for the Israeli-Palestinian water conflict over their shared groundwater resources. This will be done in light of (i) the physical, natural characteristics of the shared transboundary groundwater; (ii) the tense political situation and its impacts on the current water conflict and future prospects in that regard; (iii) the lessons learnt from international law in the field of transboundary groundwater resources; and (iv) the guidance provided by formal and informal relevant State practice.

6.1.1 The Palestinian-Israeli Water Conflict: Current Issues, Concerns and Challenges

6.1.1.1 Introduction

The research undertaken within this book raises many issues, concerns and challenges associated with the Palestinian-Israeli conflict over the Mountain Aquifer. On the one hand, these challenges are considered opportunities, if properly managed in the process of the solution identification. On the other hand, if neglected and renounced, these same challenges might intimidate all efforts towards achieving of workable arrangement concerning the Mountain Aquifer. Therefore, any proposal or attempt for resolving the present water conflict, must adequately and appropriately respond to the special requirements of these challenges.

6.1.1.2 Asymmetry among the Parties[1]

Asymmetry appears to be in the favour of Israel, and it exists at many levels and in different forms. The first level of asymmetry concerns the utilizations from the Mountain Aquifer. Whilst around 3 million Palestinians have access to 16% of the Mountain Aquifer's renewable waters, Israel, - besides all the other surface and groundwater resources available for its use - has control over the remaining 84% of these waters.[2] The major challenge is how to achieve an equitable mutually beneficial arrangement for the Parties in conflict despite the current relations and inequity in the power structures.

Asymmetry of information and knowledge regarding the transboundary groundwater resources is another important problem. At the one end; it is an abundance of data and information on the Israeli side, and at the other, a paucity of data and information on the Palestinian side.[3] By and large, the Palestinians were and are still recipients of data from the various Israeli or international organizations, due to the unilateral control over research and development in the field of hydrology water resources development. This fact is serious, and would require great efforts in order to verify existing information and reproduce national Palestinian figures and statistics.[4]

[1] Chapter 3 described the deep rooted history of political conflicts in the area of the case study, whereby it was demonstrated that Palestinians were underprivileged in terms of access to adequate resources, access to data and information, opportunities for development...etc.

[2] See discussion *supra* Chapter 3.

[3] See discussion *supra* Chapter 3.

[4] Palestinian Water Authority Archives, 2000.

The third level of asymmetry concerns the parties' interests in resolving the conflict. Israel, being the powerful party, does not appear to envision the benefits from achieving an equitable and reasonable solution based on the rules of international law.[5] However, the Palestinians have failed so far in building confidence with the Israeli politicians and negotiators concerning their intentions towards the shared transboundary groundwater resources. The challenge is how to reach a balance between the parties' interests at the table of negotiations to ensure that the solutions pursued are equitable.

The level of development in the field of water infrastructure and services is far less developed in the Palestine Territories compared with Israel.[6] After the establishment of the PA and especially after the take-over by the Palestinian Water Authority many projects have been implemented to construct new water networks or to rehabilitate existing ones. This is considered to be a time consuming task that requires commitments from the Palestinians and Israelis supported by the international community.

6.1.1.3 Parties' Emphasis on Positions

Earlier in this book, the positions by Israel and the Palestinians regarding the water conflict were described.[7] The Israeli position not to give up a drop of its existing uses and its persistent arguments concerning the need to develop new and additional water resources including desalinated water and reused wastewater, need serious reconsideration.[8] The Palestinians, on the other hand, confirm their rights to permanent sovereignty over their natural resources and rely on the principles of international law as the basis for resolving the conflict.

6.1.1.4 Resistance to International Law

To date international law has had limited role in resolving the water conflict between Israel and Palestine.[9] The existing inequitable utilization of the international watercourses –including transboundary groundwater - has been considered "*de facto*" as establishing water rights and the "no harm" rule appears

---

[5] *See supra* Chapter 3 (Section 2.4)

[6] *See supra* Chapter 3.

[7] *See* Chapter 3 pages 68-70.

[8] This argument is based on the Israeli statements and official positions in the issue of water. Recently, some joint Israeli Palestinian studies have been supporting this argument (e.g. the Regional Water Supply and Development Study that was funded by the German Government under the multilateral peace talks on water, 1998). Also reference is made to the Declaration of Principles for Cooperation *supra* Chapter 5, whose scope covers only non-conventional resources.

[9] *See supra* note 7.

to be the overarching principle embraced by at least the Israeli negotiators.[10] From this researcher's perspective, it is established that Israel is unlikely to be ready for a binding agreement relating to transboundary groundwater on the basis of international law. This justifies their hesitance to conclude the issue of water in the 1995 Israeli-Palestinian Interim Agreement.[11] The resistance on the side of Israel to conclude a legally binding agreement on the Mountain Aquifer is linked to three main reasons: (i) the great dependence of Israel on its waters;[12] (ii) Israel's undeclared recognition that their current utilizations of the shared groundwater resources violate the principles and rules of international water law,[13] (iii) any future legal arrangement that builds on the rules and principles of international law and relevant best practice threatens Israel's absolute control over the shared groundwaters.[14] As for the Palestinians, it is believed by this researcher that they have accepted the Interim arrangement on water for two main reasons. Firstly, given the transitional nature of these arrangements, the Palestinians expected that the final status negotiations will bring equitable solutions for them; secondly, the Interim Agreement is the only official document whereby Israel recognized the Palestinian water rights and therefore constituted - they hope - the real first step towards cooperation and agreement in the field of water.[15] The realities proved that the signed agreements do not provide an adequate framework for the regulating the relationship of parties as far as the Mountain Aquifer is concerned. The next section further elaborates on the question whether or not, in the context of the Mountain Aquifer, the signed treaties brought mutual benefits to the Parties.

---

[10] *See* discussion *supra* Chapter 5.

[11] Interim Agreement, Israeli-Palestinian Interim Agreement on the West Bank and Gaza Strip, Washington DC, 28th December 1995. *found at* http://www.miftah.org/Documents/documents/piia/interim.html *and* http://www.ariga.com/treaties/interim.html (*last visited* June 2003). *See also* relevant discussion in Chapter 4.

[12] *See supra* Chapter 3 (Section 2.4).

[13] The analysis undertaken in Chapter 4 on the rules and principles governing the utilization and development of transboundary groundwater confirm that the fundamental rule is equitable and reasonable utilization which includes the factor of degree of harm. The quantities of water currently used by Israel from the shared Mountain Aquifer are far from being equitable or reasonable and undoubtedly Israeli politicians realize this fact.

[14] *Ibid.*

[15] This view is based on the researcher's 6 years (1995-2000) of experience with the Palestinian Water Authority. Additionally, the researcher worked on the preparation of the technical files for the negotiations in addition to preparing documents and files for the Joint Technical Water Committee.

6.1.1.5 Relevance of Signed Agreements: A Mutual Benefit?

The obligation to cooperate contained under the relevant regional treaties appears to serve mainly the objectives of the Israelis - namely protecting their existing uses. The Interim Agreement of 1995 provided a merely temporary solution for this eternally complicated conflict.[16] The Agreement emphasizes the Israeli recognition of Palestinian water rights in the West Bank, but gives no definition of these rights.[17] Only additional supplies to serve the urgent water needs were allocated for the Palestinians under Article 40; these are to be developed from the Eastern Aquifer Basin and any other agreed sources.[18] Furthermore, there is no agreement on the overarching legal principles that will govern the rights and obligations of both parties. The negotiations on these rights were postponed for the permanent status negotiations. According to the Interim Agreement these should have commenced in 1998, three years after the signing. However, to date, the unstable political environment within the region has hindered the commencement of any serious negotiations on water.[19]If one compares the agreements with what has actually been achieved, there is a strong indication of the complexity of the situation and the inequality of the power structures that appears to favour the Israelis. The existing joint mechanisms and institutions - namely the Joint Water Committee- completely failed to fulfil their obligations.

---

[16]. See *supra* note 11.

[17] *Ibid*, Annex III Article 40 paragraph 1 of the Agreement states:
"Israel recognizes the Palestinian Water rights in the West Bank. These will be negotiated in the permanent status negotiations and settled in the Permanent Status Agreement relating to the various resources."

[18] *Ibid*, Annex III Article 40 paragraph 6 of the Interim Agreement states:" Both sides have agreed that the future water needs of the Palestinians in the West Bank are estimated between 70-80mcm/yr". Of this amount 28.6 mcm/yr were considered urgent and a plan to how to develop it was included in paragraph 7(a and b). While the remainder estimated quantity of the Palestinian needs mentioned in paragraph 6 (41.4-51.4) shall be developed by the Palestinians from the Eastern Aquifer and other agreed sources in the West Bank), *see* paragraph 7.

[19] As stipulated from the Declaration of Principles, the aim of the Israeli-Palestinian negotiations within the Middle East peace process is among other things to establish a Palestinian Interim Self-Government Authority and the elected Council for the Palestinian people in the West Bank and the Gaza Strip. This arrangement was agreed upon for a transitional period not exceeding five years, leading to a permanent settlement based on Security Council Resolutions 242 and 338. In the field of water the Declaration included commitment from both sides to cooperate in the field of water and the protection of the environment including the initiation of studies and research to strengthen this cooperation.
The Israel-PLO Agreement and Self-Determination Declaration of Principles on Interim Self-Government Arrangements, September 1993. *found at* http://ocid.nacse.org/qml/research/tfdd/toTFDDdocs/88ENG.pdf (*last visited* May 2003),..

On a technical level, the essential projects for development were delayed, rejected or put on hold due to unjustified reasons.[20]

6.1.1.6 Lack of Cooperation and Coordination

Coordination between Israel and Palestine on the transboundary groundwater resources has been absent so far because the magnitude of the problems has outpaced efforts for comprehensive management. In effect, during the period of occupation the Palestinians have had limited -if any- control over these resources. Any further delays in taking serious steps towards genuine co-operation will lead to a more deteriorated water situation and to inflation in the water crises, thus causing harm to the present and future generations and to the groundwater resources themselves. The current situation of no co-operation has widened the gap between the parties and deepened the mistrust in all matters and at all levels. As was shown in Chapter Two, the continuation of the status quo in relation to the imbalance in utilization and the uncoordinated management of transboundary groundwater are not in the interest of both parties.

6.1.1.7 Palestine's Water Institutions and Good Water Governance

Good governance is only realistic with a political will and courage to make the right decisions. At the national level, the newly established Palestinian Water Authority (PWA),[21] is empowered to regulate and manage all water resources of Palestine including transboundary water resources.[22] The mandates of PWA are not fully achieved, and will not be, unless the sovereign rights of the Palestinians over the transboundary groundwaters are respected and recognized.[23] The PWA is facing many obstacles in its efforts to rehabilitate, build and operate new systems and structures. There is a need for investment in the water sector. Building the water institutions together with the required infrastructure will require huge investments to accomplish. This need calls for a genuine commitment from the international community and Israel to help the establishment of strong institutions

---

[20] Researcher's interpretation based on the Joint Water Committee (JWC) minutes of meeting from 1996-1998, (On file with the author).

[21] The Palestinian Water Authority PWA was established by the Presidential Order No. 90 for 1995.

[22] Palestine Water Law No. 3/2002. (on file with the author). The Palestinian National Authority has established the Palestinian Water Authority, as a statutory institution with its own budget, and under the authority of the Palestinian Cabinet of Ministers.

[23] As the water problem was delayed to the permanent status negotiations Israel continues to maintain control over all the resources in the West Bank and Gaza Strip. The PWA decisions even within the Palestinian Territories to develop new water resources are dependent on decisions taken by the Israeli Authorities. *See supra* Chapter 2 pages: 64, 65 and 66.

of comparable capacities to those in Israel in order to pave the way for co-operation although the Parties were expected to start cooperating in 1993, 1994 and 1995 when they signed three consecutive formal legal arrangements.[24] Presently, the emphasis on positions rather than interests from both sides continues to intensify the problem and amplify the complexity of the water conflicts.

6.2 THE REQUIRED RESPONSE: WHAT LESSONS FROM INTERNATIONAL LAW

6.2.1 Introduction

In light of the historical and current political background, there are valid reasons to suggest that in the short-term at least, the attaining of a binding arrangement is impossible. The issues, concerns and challenges listed above, require an innovative approach that draws lessons from international law and relevant State practice in order to systematically address them following a stepwise manner.[25] This approach, in light of the highly stressed political environment that exists in this region, is advisable for many reasons, including the fact that it has the potential to aid the decision-making process by clarifying the various steps that states should take before committing themselves to implementing international law. The option of a binding treaty is only realistic when the parties collectively agree to commit themselves entirely to the rigors of a formal treaty regime.[26] Under such circumstances, it can be shown that international cooperation efforts are more efficient and allow for greater options.

If the Palestinians and Israelis start with such an informal arrangement and demonstrate tangible benefits and progress, they will hopefully find themselves in

---

[24] Declaration on Principles, *supra* note 20, and the Gaza Jericho Agreement between Israel and the PLO May 4, 1994 *found at* http://www.us-israel.org/jsource/Peace/gazajer.html (*last visited* May 2003) , *see also* the 1995 Palestinian-Israeli Interim Agreement on the West Bank and Gaza Strip See *supra* note 12.

[25] *See* Chapters 4 and 5. The efforts that were facilitated by Norway and the United States in the Water Working Group on Water, which were referred to in Chapter 5, followed a project-based approach which ignored the issues concerns and challenges listed in Section 6.1.1 of this Chapter. These projects are semi frozen now due to the unstable political situation described in Chapter 3.

[26]The review undertaken in Chapter 5 was based on a treaty survey by this researcher of around 400 treaties concerning international watercourses (*See* Annexes I to III). Most of these treaties include a mutual recognition by States to each other's sovereign rights to the use of a portion of the shared transboundary water. The absence of such recognition, as in the case of Israel, continues to hinder the development of any equitable arrangement on the Mountain Aquifer.

a position to generate the required national willingness to enter into arrangements that are more formal. This approach, which relies on meeting the underlying legal concept of "equity" at all times, in all circumstances, is by no means a repetition of earlier studies and researches done on the topic in the area of the case study.[27] It is, however, non-deniable that earlier studies had their influence.

The following sections consolidate an approach that is believed to adequately responds to the challenges raised with due attention to what lessons learnt from international law including but not restricted to the 1997 UN Watercourses Convention, the Bellagio Draft Treaty, and relevant State Practice.[28]

6.2.2 Introducing the Progressive Approach

From this research opinion, without co-operation, each entity's pursuit for more water supplies or less demand will take place unilaterally, with the probability of duplicate efforts and foregone opportunities.[29] In the initial stages at least, it is preferable, without all the formalities of a binding treaty, to reach understandings or administrative arrangements that are sufficient to allow the desired work to proceed and to bear fruit.[30] The delay and difficulties usually encountered in the more cautious drafting and required approval of a formal treaty are thus avoided.

---

[27] The only relevant study that can be cited here is one done by an academic Palestinian team. The study, which presented a stepwise open ended approach for the identification of joint management structures for shared aquifers, falls short in terms of its application of International Law. In fact, the study does not deal with international law is completely absent, as evidenced by the fact that at the outset of the research, the "equity criterion" the researchers referred to was dropped from their list of criteria for the analysis. This might have been justified in this particular case, given the political sensitivities under which the study had to be undertaken. To date, the study has not been brought to the implementation level due to many political reasons. M. Haddad and E. Feitleson, Management of Shared Groundwater Resources: The Israeli-Palestinian Case with an International Perspective (Kluwer Academic Publishers, December 2000).

[28] *See* relevant discussions, *supra* Chapter 4.

[29] *See supra* Chapters 4 and 5, which demonstrate that "cooperation" is a requirement under international law, which derives from States sovereign rights and equality. Most of the reviewed State practice in the field of international watercourses provide evidence that States are required to cooperate in order to fulfil their rights and obligations under the legal arrangements. Even informal arrangements are based on the concept of cooperation. The core elements of cooperation as concluded from this research are: the exchange of data and information, the prior notification and consultation and negotiations.

[30] The research on State Practice *see supra* Chapter 5 demonstrates that in the context of connected transboundary groundwater States either include groundwater as part of the overall hydrological cycle and treat it like surface water or tend to conclude informal arrangements that govern their relationships concerning the technical and scientific cooperation concerning these resources.

This is vitally important, not least of all because Israel does not feel it is in a position to accept the long-term commitments of a treaty at the time being, and is therefore not in a position to accomplish prompt ratification.[31]

The proposal for approaching the water conflict within the Palestinian-Israeli context, is therefore, to progressively establish a solid basis for long-term sustainable cooperative arrangement. The progressive approach starts with a non-binding flexible arrangement that can serve the short-term needs, which gradually builds up into a final legal binding arrangement based on international law. In between these two extreme solutions a combination of coordination and collaboration arrangement is proposed.[32]

The following analysis includes guidance on the structure of this approach including preconditions, measures and actions, shortcomings, impacts and funding requirements. The core difference between this research and earlier studies especially in this region, is the fact that the ultimate objective of the approach proposed here is to reach a final comprehensive, equitable and sustainable legal arrangement that governs the relation of the Palestinians and Israelis over their shared transboundary groundwater.

A very important contribution of this chapter is the "Framework Agreement concerning the Utilization, Development and Protection of the Mountain Aquifer", whose development was guided by the rules and principles of international law governing the utilization, development and protection of transboundary groundwater.[33] Additionally, this chapter proposes a framework for the assessment of legal entitlement from the shared transboundary groundwater, based upon the governing principle of equitable and reasonable utilization. The following discussion further elaborates on the contents and requirements ensuing from this approach.

6.2.3 Description and Preconditions

As a precondition for successful co-operation on the Mountain Aquifer, it is recommended that a more in-depth revision of the style of negotiations is undertaken in an attempt to give momentum for the currently quasi-frozen negotiation process. A mutual acknowledgement by the parties to the need for a common base in the interest of the resource optimisation and management and in

---

[31] *See supra* Section 6.1 and *supra* Chapter 4.

[32] Lessons could be learnt from Africa's practice concerning cooperation on their confined groundwater aquifers, or the European rnt experience concerning the Vechte Catchment, *see* generally *supra* Chapter 5 (Section 5.7).

[33] See relevant discussion *supra* Chapter 4. Reference is mainly made here to the 1997 UN Watercourses Convention and the 1989 Bellagio Draft Treaty. *See* Annex V for the proposed text of this Framework Agreement.

establishing appropriate mechanisms on the utilisation of the shared resources in an optimal manner for the present as well as for future generations is required. Additionally, the creation of the adequate dialogue and constructive communication is essential to bring the parties back to the table of negotiations.

The common objective among the parties and the appreciation of the necessity of co-operation with respect to their shared water resources must be genuine.[34] This would necessitate the confirmation by both parties through exchange of notes of competent authorities on their current position towards co-operation. This proposal requires the parties to reconsider their positions toward their future relationship relating to their shared groundwater resources. At the outset of this co-operation, the parties are advised to be cautious of the decisions they take. The most critical decision they may have to consider is the degree and intensity of their co-operation. This decision is dependent on many political, social and economic factors.[35]

The progressive approach is based on some important hypotheses. Firstly, there is very little schematic knowledge about transboundary groundwater especially on the side of the Palestinians. Secondly, both parties need guidance on the variety of existing options for international co-operation. Thirdly, the resolution of the conflict is envisaged to bear mutual benefit for the people and the shared resources. And finally, the ultimate common objective must envisage a binding legal arrangement on the basis of international law relating to transboundary groundwater. The design of a legal regime and an institutional structure responsive to the particular basin require a deep understanding of experience in this particular field.

The approach is expected to set the stage for thoughtful elaboration at the technical level, whereby scientists, economists and lawyers from both sides can establish an operational network of technical coordination and develop rules of procedures essential for protecting and harnessing the transboundary groundwater resources. It requires a substantial technical input by persons with specialized training and experience.

6.2.3.1 Identify the Problem

Among the actions that the Parties need to undertake is to identify the nature of the problem. This is perhaps the most difficult and most important step in conflict avoidance, particularly given the intense hostility between the parties involved. Once there is a decision for such kind of co-operation, the mobilisation of the available expertise in the various fields relevant to water resources management

---

[34] *Ibid.*

[35] UN/Natural Resources/Water Series No. 1, Management of International Water Resources: Institutional and Legal Aspects (United Nations: New York, 1975).

and the involvement of domestic governmental structures capable of effective international co-operation and collaboration becomes necessary.[36]

The need for a "Third Party's" or a "Mediator's" intervention arises, particularly in this case of power inequity.[37] The "Mediator" is recommended to be composed of one member nominated by each Party concerned and, in addition, a member not having the nationality of any of the Parties concerned chosen by the nominated members who shall serve as the chairman. [38] It is recommended that team's expertise is diverse and includes a lawyer, a hydrogeologist and an economist. The third party team could help the process of identifying the aspects of the problem, the actors, and act as advocate not just for reaching a compromise but also for ensuring that the presented facts and information are accurate and reliable. The emphasis within this stage on having reliable data and information articulates a universal need, without which rational management of aquifers at any level is not possible.

The next practical step is to identify the transboundary groundwater resources at issue, and agree on the geographic hydrogeologic and hydrographic concepts[39] they are willing to employ and the terminologies they are ready to adopt. [40] The task of identifying what groundwater aquifers are shared between them should be based upon scientific evidence and hydrogeological investigations[41]. By and large, it is agreed that the North-eastern and the Western Aquifer Basins are transboundary. As for the Eastern Aquifer Basin, there is a real need to jointly define the manner in which it has to be handled in the cooperation. A very

---

[36] *Ibid.*

[37] *Ibid,* international law considers Mediation as obligatory on States in case negotiations don't yield solutions. Mediation embraces the notion of co-basin states seeking the aid of an outsider that is of a third state or perhaps of an international body like the Untied Nations or one of its agencies. In essence, under this type of arrangement the parties to the dispute open themselves to the advice of persons charged with the responsibility of looking at the problem at issue objectively. The aid of third parties by way of good offices, mediation and conciliation in the resolution of conflicts not only has been advocated but also has been demonstrated in practice (e.g. the World Bank-Indus River, Eric Johnston-Jordan River)

[38] This step resembles Para (4) of Article 33 of the 1997 UN Convention on "Dispute Settlement".

[39] The choice of the conceptual hydrogeological management unit is a significant decision that has to be made. It was concluded in chapters 3 and 4 that the "Drainage Basin" concept in the best suited for the optimal management development and protection of transboundary groundwater resources. See *supra* Chapters 3 and 4.

[40] This step is basically identifying the "Scope and Use of Terms". Annex V of this thesis includes a water related "glossary of terms" for the use of the Palestinians. This can be further developed and agreed upon by the parties.

[41] These investigations are mainly available on the Israeli side, as during the years of occupation the Palestinians were denied access to important data related to the shared groundwater resources.

important issue in that regard is to mutually understand and respect the differences between the Parties and try to reconcile them. These differences are derived from the different stages of development, financial and institutional capabilities, and the need for more water, use priorities and environmental concerns.[42] This scoping exercise is not an easy task; in fact it might require the most crucial decisions to be made. Once this complicated task is successfully accomplished, the remaining tasks becomes simpler and the foundation is laid for agreeing on subsequent requirements and identifying commitments.

Of importance under this approach is the assessment of the strengths and weaknesses of each part with respect to technical personnel and equipment. The fact that Israel has developed faster than the Palestinians in the field of water resources management and regulation,[43] means that arrangements must be made for an exchange of personnel and training programmes between educational institutions in Israel and Palestine. The objective of such a step is to minimize and balance the asymmetries that exist among the Parties.[44]

6.2.3.2 Agree on Interests and not Positions

Having identified and agreed the problem and the contested groundwater resources the parties should proceed to agree on interests and not positions.[45] A vital step is to bring the two parties to agree on what their interests are and to identify options for mutual benefits and gains. Each party will negotiate only as long as its interests can be served best at the negotiating table. At this stage the parties are advised to conduct a joint assessment of how the international law could be employed to achieve an integrative, equitable, and mutually beneficial arrangement for the parties in conflict. Once this is done, the parties become more open to cooperate and collaborate in an environment of faith and good intentions. It is advisable at this stage that the Parties evaluate the legal assessments to, and value of, water, including legal, economic and social guidelines.[46] At this stage the

---

[42] The long years of occupation have negatively impacted all these areas of development for the Palestinians. These differences must be alleviated or modified through the proposed cooperation.

[43] *See supra* Chapter 2 pages 62 and 63.

[44] *See supra* Section 5.1.1.

[45] The emphasis on positions, as described in Chapter 2, has for many years been contributing to widening the gap between the Palestinians and the Israelis concerning their shared transboundary groundwaters. The recommendation here is to give up this approach, which failed in achieving feasible solution and follow the "common interest" approach which is more realistic.

[46] This is in line with the Legal Assessment Model (LAM) that was developed by the Knowledge and Research Project that was conducted by the International Water Law Research Institute at Dundee University entitled "*Transboundary Water Resources*

need will arise for specific activities to take place that are expected to pave the way for further cooperation. Such activities include:

A. Joint Inventory of Transboundary Groundwater

The main purpose of such an inventory is to identify and agree on the location, extent and the type of transboundary groundwater and to identify physical interrelationship between the various transboundary aquifers and between groundwater and the environment, mainly surface water. It is crucial to map all the monitoring activities pollution sources and contamination of transboundary groundwaters, and to identify the uses, problems and trends in the state of groundwater according to the observations up till now.

B. Joint inventory of Transboundary Groundwater Infrastructures

This activity includes the joint registering of all existing infrastructure related to transboundary groundwater utilization including data on current quotas, rates, taxes, restrictions, recharge and pollution abatement obligations and the like. It would as well be an expression of good intentions to agree on the accomplishment of commitments under Article 40 of the Interim Agreement.[47] This would require the Parties to identify and agree on the proposed well sites, expected yield, extraction and necessary installations and sources of funding and to embark on the implementation. From a technical and administrative point of view the Palestinians and Israelis might establish an institutional mechanism for the operation and maintenance of exiting infrastructure under some established scenarios.[48]

C. Creation of a Scientific Knowledge Base

Under this type of cooperation, it is desirable to start with the collection and exchange of physical, economic, and social data that is collected in a uniform manner to be analyzed by a joint body in order to establish sufficient knowledge and understanding on how to solve the specific problems. At the national level, the particular strengths and weaknesses of each with respect to technical personnel and equipment must be identified. This will enable the design of a program that captures the training and capacity building needs. Technical assistance can be

*Management: Using the Law to Develop Effective National Water Strategy: "Poverty Eradication through Enforceable Rights to Water"***.** This research seeks to provide guidance for States on how to assess their "equitable and reasonable utilization" through a practice-oriented and interdisciplinary approach in the context of shared unconfined groundwater aquifers between Israel and Palestine. The other two case studies are one that deals with upstream case (China) while the other handles the downstream (Mozambique). For more details *see* www.dundee.ac.uk/law/iwrli (*last visited* July 2003) [hereinafter referred to as the Dundee KAR Project].

47 Israeli Palestinian Interim Agreement, *supra* note 12.

48 The Joint Water Committee (JWC) is one important institutional mechanism that was established under the Interim Agreement. At present, the JWC can be developed and improved to serve the purposes of the desired cooperation.

sought from the States themselves if one has advanced technical capabilities and facilities than the others.

D. The Construction and Operation of Facilities for Providing Water Service

Most existing Palestinian water utilities were established before 1967.[49] These departments dealt with the planning, development, operation and maintenance of domestic supply networks, sewerage networks and other water resources development. Since the occupation in 1967, the role of these departments was limited to the operation and maintenance of the water and sewerage networks. They became subordinate to the Israeli civil administration officer.[50] Currently, these institutions do not possess competence in planning and design. These departments suffer from a multitude of constraints, which render their service inefficient, and limit the possibilities for improvements. Municipal water departments are integral parts of the organizational structure of the municipalities and have a common economy with the municipality administration. This subjects these departments to all the bureaucratic deficiencies in the existing municipal system, and consequently limits their efficiency.[51]

The Palestinian municipality networks are in desperate need of rehabilitation, as almost 50% of the water supply is lost due to leaky pipes in the West Bank.[52] Palestinians suffer from water shortages and an irregular water supply, especially in the summer months. Due to the water deficiency, it is estimated that Palestinian demand for water is at least twice as much as the present per capita consumption. [53] This situation has led to high loss percentages through pipe leakage. Unaccounted for water reaches a value of 50% in some of the West Bank Governorates.[54] Therefore, the inefficiency in those systems is conflict related and would have been avoided or mitigated had the Palestinians been authorized to conduct maintenance. After the establishment of the PA, and especially after the take-over by the Palestinian Water Authority, many projects have been

[49] WFED, Water Institutions, and Water Laws in Palestine: Gaza Strip and West Bank, March 1995 (found on file with the author).

[50] *Ibid,* by Military Order No. 92 of 15 August 1967 "Oder Concerning Jurisdiction Over Water Regulations" which provides:

" This vests all powers defined in any Jordanian law dealing with water in the hands of an Israeli officer appointed by the Area Commander. This Israeli official-in-charge- assumes full control over water resources. He has the right to authorize a given organization to operate or establish new water authority, he may control its method of operation and appoint its head...etc."

[51] *Ibid.*

[52] PWA database, 2002.

[53] *Found at* http://www.wws.princeton.edu/~wws401c/agriculture.html (*last visited* July 2003).

[54] These figures don't apply to the case of the Jerusalem Water Undertaking (JWU) in the Ramallah region, which has records of 15-25% of losses in the network.

implemented to construct new water networks or to rehabilitate existing ones. In order to reach the efficiency required by the modern water sector management, prodigious efforts have to be exerted by the Palestinians that should be complemented by a real commitment from the international community and Israel to continue supporting the build-up of the Palestinian water institutions.

6.2.3.3 Create Options for Mutual Benefits

By this stage the Parties have developed, hopefully, a common understanding on the availability of the resources, their respective vital human needs. There is a growing trend to recognize and protect the vital needs of the river, or "vital environmental needs" which includes in-stream flow requirements and conservation measures. This factor appears also in the 1997 UN Watercourses Convention (Article 10), which provides that any conflict of use should be resolved in accordance with the rule of equitable and reasonable use (Articles 5-7), giving "special regard to the requirements of vital human needs." The commentary defines these needs as "providing sufficient water to sustain human life, including both drinking water and water required for the production of food in order to prevent starvation." This wording does not adequately define the population that is considered to be dependent on the watercourse, which caused difficulties with its application in the case studies. The meeting of vital human needs as such as drinking water or food can be achieved other than from the local watercourse. In contrast ecological services do need to be delivered at the place of the watercourse.

There are three key global instruments related to human rights: the 1948 Universal Declaration of Human Rights,[55] the 1966 Covenant on Civil and Political Rights,[56] and the 1966 International Covenant on Economic, Social and Cultural Rights. [57] In order to provide greater interpretive clarity as to the intent, meaning of the latter Covenant, the Committee on Economic, Social and Cultural Rights adopted the General Comment on the Right to Water.[58] The 2002 General

---

[55] Universal Declaration on Human Rights, Dec.10, 1948, UN GA Res. 217A(III), U.N.Doc A/810 at 71 (1948), *found at* www.un.org/Overview/rights.html (*last visited* May 2004).

[56] International Covenant on Civil and Political Rights, UN GA Res. 2200A(XXI), 21 UN GAOR Supp. (No. 16) at 52, UN Doc. A/6316 (1966), 999 UNTS 171, entered into force in March 1976.

[57] International Covenant on Economic, Social and Cultural Rights, GA Res. 2200A (XXI), 21 UN GAOR Supp. (No. 16) at 49, UN Doc. A/6316 (1966), 993 UNTS 3, entered into force Jan. 3, 1976.

[58] UN Doc. E/C 12/2002/11, General Comment No. 15, The Right to Water, International Covenant on Economic, Social and Cultural Rights, Committee on Economic, Social and Cultural Rights, 29th Session, 26 November 2002, *found at* http://www.cedha.org.ar/docs/doc104-eng.htm (*last visited* May 2004)

Comment uses the term " personal and domestic uses" rather than "vital human needs". This term is wider in definition and content and emphasises that the right to water take precedence over all other water needs.[59] The General Comment confirms that although the adequacy of water required for the right to water may vary according to different conditions, some factors apply in all circumstances. The factor include the availability, good quality, physical, and economic accessibility, non-discrimination and finally information accessibility.[60]

In the area of the case study, no minimum water requirement has been formally defined and accepted by both parties in the study area. Peter Gleick suggests that based on the present urban water uses suggest that an appropriate level may be between 75 and 150 m3 per person/year. Hillel Shuval -a water quality scientist of international stature based at the Hebrew University in Jerusalem- provided an estimate for the minimum legitimate baseline water needs of the Israelis and Palestinians. This amount is required to ensure a reasonable minimum standard of living in a semi-arid area suffering from serious water shortages. He proposed that the Minimum Water Requirement--MWR for domestic, urban, and industrial use for Israelis and Palestinian alike be about 125 cubic meters/person/year (cm/person/year). The results of this study might recommend the reconsideration of water utilizations among the parties and therefore a decision in response will be crucial. As a second call for water, economic and social development needs and the extent to which the available resources can meet all the needs without compromising the sustainability and the long term safe yield of these resources. If the findings of the study reveal that under an equitable and reasonable utilization scenario, the existing resources are not adequate to compensate for the regional shortage, the parties are recommended to commonly plan for managing such situation.[61] They could, for example, cooperate in the areas of developing non-

---

[59] *Ibid,* The term "personal and domestic uses" includes: (i) drinking, which is defined by the UN Committee as "water for consumption through beverages and foodstuffs" (ii)Personal sanitation, which means disposal of human excreta, (iii)Food preparation, which includes food hygiene and preparation of food stuffs, whether water is incorporated into, or comes into contact with, food. And (iv)Personal and household hygiene, which means personal cleanliness and hygiene of the household environment.

[60] *Ibid,* comment no. 12.

[61] Many regional studies have been implemented under the multilateral talks, but their influence have not been controversial, due to the fact that the first step of assessing the legal entitlement in the fashion mentioned here was never done. Additionally, these projects were conducted under the umbrella of the Multilateral Working Group on Water, where donors have offered to pledge money for projects that are expected to contribute to the peace process. See *supra* note 13. The Zambezi Action Plan, which was discussed in chapter 4 provides a range of significant tasks and projects that could be implemented as part of the protection of the transboundary groundwater resources. See The Action Plan for the Environmentally Sound Management of the Common Zambezi River System Signed at

conventional water resources; i.e. desalinated water and the implementation of regional development projects.[62] Linkages could have been made to provide the Israelis with good will benefits while providing the Palestinians with technical assistance that could have raised the level of knowledge of the water resources in the Middle East in the Palestinian Authority to the level of other parties. There is no real reason to build another water data base in Israel (which no one uses in Israel) while there was a tremendous need for the Palestinian Water Authority to build its own water data base.

6.2.4 Joint Assessment of Legal Entitlement from the Mountain Aquifer

The study of the legal entitlements by each side to transboundary groundwater is of vital importance to the success of the proposed progressive approach. The agreement among the Parties to conduct such a study would indicate good intentions, which contributes to building confidence between them. It is recommended to follow a systematic approach in identifying the factors that affect and are affected by the existing transboundary groundwater recourses and their distribution. Furthermore relevant guidance is sought from the legal instruments that were analyzed in chapter 4. The recent research efforts conducted by the Knowledge and Research Project (KAR)[63] in this field are very relevant. One of the major outputs of the KAR Project was the development of a methodology to help States assess their legal entitlements and obligations with respect to their shared water resources. The methodology is supported by certain tools created for the purpose including: the "Glossary of Terms",[64] the "Relevant Factors Matrix",[65]

Harare, 28th May 1987, *found at* ocid.nacse.org/qml/research/tfdd/toTFDDdocs/177ENG.htm (*last visited* June 2003).

[62] These projects include the Mediterranean Sea Dead Sea and the Red Sea Dead Sea Conveyance projects. The later being lately identified as a Jordanians Israeli initiative with the possible involvement of the Palestinians. *See generally* http://www.american.edu/ted/deadsea.htm (*last visited* July 2003).

[63] The KAR Project, *supra* note 46. Annex IV builds on the methodology proposed by the KAR Project. It provides a legal entitlement assessment exercise in the case of Palestine. This exercise is illustrative and would highly benefit from an improved quality of the data used.

[64] Glossary of Terms includes all terms that are expected to have dual interpretations among hydrologists, economists, lawyers and other people working in the field of water.

[65] The "Matrix of Factors" is inspired by the list of factors that are set forth in Article 6 of the 1997 Watercourses Convention and Article V/a-k of the 1966 Helsinki Rules. The Matrix of Factors is a combination of both in addition to the research team's finding on how to best present these factors to be acceptable and

the "Legal Audit Scheme".[66] The Relevant Factors Matrix in particular, provide geographic,[67] hydrographic[68] hydrologic/hydrogeologic,[69] climatic,[70] population

---

[66] The purpose of the Legal Audit Scheme (LAS) is to provide a snapshot of the legal context within which the State operates. It provides an overview of the existing legal entitlements and obligations and is a tool to be used by a State in order to identify its rights and obligations as well as respective rights and obligations of other States sharing the same transboundary watercourse. The State can thus ascertain in a systematic manner the relevant provisions of international agreements in force for this State and other legal instruments, which may affect its existing or planned uses of the waters of an international watercourse. Additionally the LAS is designed to help ensure compliance by a State, with its existing commitments. It will also assist a State to monitor and verify performance by other States parties by assessing how effectively they implement and enforce their respective treaty obligations. Additionally, the LAS should identify whether a State's national law is in conflict with its international legal obligations. See generally the KAR Project *supra* note 46.

[67] Setting the geographic context include a statement on the location of the Watercourse State and whether it lies is downstream or upstream with respect to the specific groundwater aquifer and to what extent this factor is important and relevant to the whole assessment study. See generally *Ibid.*

[68] The hydrographic factor mainly concerns the extent of the international watercourse in the territory of each watercourse State. It is most usefully expressed as the proportion, in relation to the area of the whole basin that occurs in the State of interest. It can be measured from maps at scales appropriate to the size of the basin. In many cases this is a straightforward procedure, but difficulties can sometimes occur. The extent of the watercourse is interpreted to mean the extent of the aquifer, whether confined or not, in each territory. As the aquifer cannot be seen at the surface, its extent is not as obvious as is the drainage basin. Its assessment will depend on the availability of geological and hydrogeological maps or reports and studies of the hydrogeology of the area. In some cases accurate and large-scale maps are not available, and this will inevitably limit the accuracy of the result. The definition of the extent of the aquifer as used here is limited by the practicability of its exploitation. For instance, in some areas water bearing strata may be so deep underground that they could not practicably be used to supply water; such areas would not be included as part of the extent of the aquifer. Also in the case of transboundary groundwater it is wise to consider the overall surface area of the aquifers at issue together with any activities taking place on this area that might influence the quality and quantity of the recharged water (land use, industrial activities, and waste disposal..etc.). See generally the Dundee KAR project *supra* note 46.

[69] Hydrology is taken to include both surface water and groundwater. A number of different aspects need to be examined in order to assess this factor properly: mean water availability; variability of the resources; water quality; contribution of water to the watercourse by each watercourse State; potential climate change impacts; and possible interactions between groundwaters and surface waters. (i) mean annual availability: in most cases, mean water availability for groundwater is taken to be the long-term safe yield which could be extracted from the aquifer. The safe yield is the amount of water which can, on average, be extracted indefinitely without depleting the storage of the aquifer. This is equivalent to the average rate of recharge or replenishment of the aquifer from surface water. (ii) variability of

dependence,[71] economic and social factors[72] that must be jointly collected, compiled and analyzed in a consistent manner.[73] The impact of the existing and planned uses of one State on the other,[74] the efficiency and available alternatives

---

resources, which is relevant because the greater the variation over time, the less practicable it is to make effective use of the resource. Both seasonal and inter-annual variability are significant. For surface water, seasonal variability can usually be described by the average amount of variation between the wet season and the dry season. The inter-annual variability is the amount of variation of the annual flows over a long series of data, described, for instance, by the coefficient of variation of annual flows, as well as by the highest and lowest annual flows ever recorded. As noted above, river flows should be assessed as long time series (30 years or more) of monthly flows, and if this has been done, then there would already be sufficient information to describe the variability. Groundwater resources are often treated as invariant with time, and they are in any case usually much less variable than surface water. However, there are cases where the variation is significant, and this should therefore be assessed in the same way. Assessment of inter-annual variability is an important factor in allocating water as it may be that States require water for different purposes at different times. (iii) water quality: the quality of the water is important since this can determine whether or not water which is physically available can actually be used for a particular purpose.

[70] Assessment of the climatic factor does not nssarily require a detailed study of the climate of the area. The broad climate type should be identified, and the long-term mean monthly rainfall and potential evaporation rates determined. Observed data are available almost everywhere to enable this to be done, although data paucity may still be experienced (see para.1.3.1. above in relation to associated Case Studies). As for the hydrological factor, the seasonal and inter-annual variability of the precipitation should be identified. In addition, where long, reliable records are available, trends in climate patterns should be identified.

[71] Population in the basin including (i) Present and projected population, (ii) Populations in the study country and in the other States, (iii) Distribution of population, (iv) Growth and migration of population, (v) Livestock.

[72] This factor identifies the demands on or the uses of the watercourse, and the economic and other benefits that flow from the uses. The first step is to identify the uses themselves, both existing and potential, and then to consider the benefits derived from them. The existing and potential uses are the first two components, but certain types of use – meeting "vital human needs", social use, and ecological or environmental use – are considered as special ones and are also treated as separate components.

[73] Refer to the Relevant Factors Matrix that was developed within the KaR project *supra* note 46.

[74] Impacts are the effects, both positive and negative, which the uses of the watercourse have on other uses. It is important to determine the transboundary impacts, that is, the effects that uses in one State have on the other States in the basin, but consideration should also be given to impacts within national boundaries. The impacts of both existing and potential uses need to be examined, and there should be a clear separation between the two: the impacts of potential uses should be explicitly identified and separated from the impacts of the existing ones. Evidence of actual harm or beneficial effects should be clear and convincing if they are to be acceptable. It is essential that the projected impacts of potential uses be shown to be likely, rather than merely speculative.

including the potential for conjunctive use of the available waters are also factors that must be jointly studied and analyzed. For the evaluation of what is equitable and reasonable the Parties are recommended to adopt a unified method of evaluation[75] that is commonly developed and accepted. The success of such evaluation depends mainly on the availability of accurate and reliable information to be shared by the parties.[76] The study shall address the questions of how to operationalise the rule of customary international law of "equitable and reasonable utilization" in the case of transboundary groundwater and how to best inform and influence decision-making. This can be done through examining the type of data that is needed to incorporate legal, technical and social science existing requirements and needs of the future.

What is the format for data collection and presentation to allow exchange and easy access? How dynamic the process and what are the triggers for the reassessment? Is there a generic model that could be implemented in cases of transboundary groundwater aquifers?

The final conclusions of this study will offer guidance to the Parties on how to determine their legal entitlement from these shared resources. The study shall assess the quality and adequacy of the available data to identify gaps and needs for data improvement. The long term vision must be a unified comprehensive data base that incorporated all water data and related information pertinent to the development utilization and protection of the shared groundwater resources.[77] An assessment of the "vital human needs" and the degree of dependence of the population on the transboundary groundwater resources is the first step in the legal

---

[75] In this context, it is recommended to employ the Method of Evaluation that was developed under the KaR Project, *supra* note 46.

[76] Modern water management requires a large integrated knowledge base, data/information on the natural/ physical characteristics of the resource, on the social and economic uses needs and demands. The data has to be available in order to be shared and exchanged. A precondition for generating such data is availability of modern reliable monitoring networks and assessment systems. Monitoring needs to be embedded in a strategic 'holistic' approach. In addition to monitoring, institutional arrangements regarding data provision and exchange is nsary (e.g. water uses). In the Palestinian case none of the conditions applies. The generation of and access to data has been prohibited for the past years of occupation. The hydrological monitoring network dates back to the 1960's which is not in operation at present. As for Israel, in the last few decades it has been developing rapidly in the field of monitoring and assessment. Unfortunately most of the significant research and data relevant to the transboundary groundwater has been classified as confidential. The few reports that are published and exchanged with the Palestinians are those which are not considered a threat to Israel's "water security", (analysis by the author)

[77] The Regional Waternet project that was initiated under the multilateral talks can be reactivated and better developed to fulfil this objective. *See* relevant discussion *supra* chapter 4.

entitlement assessment exercise.[78] If it is practically difficult to carry out the assessment jointly, the determination of the factors and their relevance to the transboundary groundwaters can be done separately, the results of which can be discussed jointly to reach an agreement on to come out with a reasonable. Annex IV provides a framework for the assessment of the Palestinian legal entitlements from the transboundary groundwaters that is shared with Israel.

6.2.5 Collaboration and Joint Projects

This stage brings the efforts, studies and research under the technical co-operation phase into practical implementation. In fact, those efforts will be wasted if no action follows after them. Types of collaboration may include technical economic, financial administrative and political fields. The difference between this stage and the technical co-operation lies mainly in the need at this stage to implement projects and undertake programmes that might have an influence on the existing uses. Water officials from each Party have to be alerted on plans and the various options that can be made to meet the need of any stage of development of their shared groundwater system. To achieve the foregoing it is essential to establish of appropriate mechanisms for co-operation and for the fulfilment of crucial requisites. The said efforts include: (i) proactive support and long term commitment at the political level and (ii) the mobilisation of the available expertise in the various fields relevant to water resources management and (iii) the enactment of domestic governmental structures capable of effective international co-operation and collaboration.[79] Each of these projects must provide a specific contribution to this multidisciplinary initiative aiming to improve understanding of scientific, socio-economic, legal, institutional and environmental issues surrounding the management of transboundary aquifers. The resulting projects from the first stage are considered intermediary outputs and must be used as inputs for the next stage. These include conducting transboundary groundwater surveys,[80] and the development of monitoring and assessment guidelines.[81] The

---

[78] See *supra* note 56, also *See* S. Libiszewski, *and Water Disputes in the Jordan Basin Region and their Role in the Resolution of the Arab-Israeli Conflict*. ENCOP Occasional Paper No. 13.Center for Security Policy and Conflict Research/ Swiss Peace Foundation. Zurich/ Berne, August 1995. Internet version, *found at* http://www.fsk.ethz.ch/encop/13/en13.htm. (*last visited* May 2003).

[79] *Ibid.*

[80] Lessons could be learnt from the UNECE Task Force on Monitoring and Assessment, Inventory of transboundary groundwaters, (Lelystad, 1999). The survey could have the following scope and objectives: (a) to highlight the location, extension and type of transboundary groundwaters, (b) assess monitoring activities, pollution sources and contamination of transboundary groundwaters; (c) identify the uses, problems and trends in

integral basin area approach must be used as the basic concept for structuring the guidelines on monitoring and assessment. Also essential for the transboundary groundwater assessment is to develop and evaluate strategic policies for groundwater management in a harmonized way.[82] Projects that aim at the harmonization of national laws and regulations must be pursued. Lessons could be learnt from best practices in national regulations that responds to the challenges special issues and concerns in relation to transboundary groundwater. National regulations can be accordingly adapted and modified to provide a harmonized legal framework governing the utilization and development of these resources at the national level especially in the following areas: (a) where the matter is significant in the sphere of basin wide or system wide water policy and water management; (b) where there is discriminatory impact on non-national users and (c) where national users are placed at competitive disadvantage with respect to a similar class of users in the other State.[83] The removal of legal conflicts and the enactment of the essential regulation in the respective State codes are essential to water use optimization and to the promotion of regional development and cooperation generally.

---

the state of groundwater according to the observations and (d) propose the institutional and international aspects in their management in the Middle East region.

[81] Lessons could be learnt from the UNECE Task Force on Monitoring and Assessment, State of the Art on Monitoring and Assessment of Groundwaters, (Lelystad, 1999). The first step is to characterize the transboundary groundwater system in terms of flows, recharge discharge areas, interaction between surface and groundwater. Activities within the recharge areas at one side of the borders which might adversely affect the groundwater quality on the other side of the border must be monitored. The actual and possible future functions and uses of the aquifers must be assessed based on the quantity and quality features of the groundwater system concerned. The next step is assessing the vulnerability of the aquifer which identifies those zones of the aquifer system which are considered vulnerable to threats and therefore require an early warning system to avoid occurrence of pollution. This step will include the identification of types of threats, measures and the monitoring cycle. There will be a need to assess also the current monitoring and assessments practices and identify gaps, and based on that, design the groundwater monitoring systems.

[82] This means, for example, in order to assess trends in groundwater quality, the definition for trends should be comparable on both sides of the borders. Also, the methodologies for estimating annual recharge and safe yields of the shared aquifers must be compatible, see *ibid.*

[83] UN/ Natural Resources /Water Series No. 1, *supra* note 35.

6.2.6Treaty Formulation[84]

At this level of co-operation the Parties are hopefully in a position to establish common objective criteria in their efforts to conclude a water treaty. The Parties are recommended to recognize and accept their common responsibility in ensuring the reasonable and equitable development and management of groundwaters in the border region for the well being of their peoples.[85] In the treaty formulation process, attention must be paid to the fact that international water resources development is an undertaking that involves policy decisions and complex technical assignments. The current stage require the knowledgeable and persistent support of decision makers. It is recommended that there be an in-depth reconsideration of the negotiations approach in an attempt to give momentum for the negotiation process. As a precondition, the basis for treaty formulation is the Parties' common objective concerning the need to attain optimum utilization and conservation of transboundary groundwaters and to protect the underground environment. This type of understanding requires the Parties to develop and maintain reliable data and information concerning transboundary aquifers and their waters in order to use and protect these waters in a rational and informed manner. This next discussion will be based on the recommendations from chapter 3 to employ the UN Convention and the Bellagio Draft Treaty as a framework agreement governing the utilization and development of transboundary groundwater. Naturally, this framework will not be adopted in its entirety but rather adapted and modified so as to suit the local conditions.

6.2.6.1 The Glossary of Terms and Concepts

From the technical perspective, water resources management crosses many disciplines, which undoubtedly play a significant role in the efficient management and development of this valuable resource. Consequently, effective water administration calls for the active involvement of technical, legal, economic and political science experts in the final status negotiations and in the development of policies, strategies and regulation of the valuable resources. The problem of harmonizing terms and concepts in this regard is two fold: one is Palestinian-Palestinian, while the second is Israeli-Palestinian. The first invokes the need for harmonizing the understanding of the various terms and expressions used among the Palestinian experts, in order to ensure that experts from all disciplines are

---

[84] Annex V includes a "Proposed Framework Agreement between the Palestine Liberation Organization and the State of Israel Concerning The Utilization Development and Protection of Transboundary Groundwaters".

[85] See R. D. Hayton, and A. Utton, *Transboundary Groundwaters: The Bellagio Draft Treaty,* 29 NRJ (1989), Article II.

discussing the same topic and mean the same thing. This understanding will eventually lead to a solid platform for the negotiations, and for the enactment of the effective tools for the management of this resource. The second problem is Israeli-Palestinian, which requires the harmonization of terms and concepts between the Palestinians and the Israelis, is more complicated and is expected to be cumbersome to achieve. It would require prodigious efforts and willingness by both sides to reach a common understanding in that regard. A consensus on the meaning of terms such as the right to water, equitable and reasonable utilization, the definition of "international water", among many others, is crucial to the success of the final negotiations on water. This consensus will ensure compatibility in the understanding of the terms that are currently used by the politicians, the legal and technical people in varying meanings.

Annex V includes a proposed glossary of the essential technical and legal terms that might be used as a tool for negotiations and further development.

6.2.6.2 Define Hydrologic Management Unit

It is not always simple to define the geographic area to be covered by the legal arrangement. Although the previous analysis and recommendations from Chapter Three urge States to adopt the 'Drainage Basin Concept' in the case of transboundary groundwater, this is not always a realistic solution. State practice has demonstrated resistance on the side of States to adopt such a concept. Accordingly, the Parties are likely to resist the adoption of such a concept and favour the "watercourse system approach" as in the UN Convention which narrows the geographic scope of the resource in legal terms. Nevertheless, the treaty must ensure that land activities and other environment related activities that might adversely affect the quantity and quality of these resources are as well incorporated in this conceptualization. It is also possible that the Parties only agree that a portion of the basin in the border region are to be regulated. This kind of geographical focus involves deliberate decisions to work in one selected locale within the entire aquifer basin system either in an effort to achieve comprehensive development and management in at least the vulnerable sub regions, or in recognition of their inability to deal comprehensively with the entire aquifer basins at this time.[86] The treaty however should clearly indicate whether the

[86] This decision depends on the results from the monitoring and assessment in defining the vulnerable segments of the aquifer basin system. These results might indicate that some sub regions in the border areas are highly vulnerable and therefore are in need for joint efforts and regulation. The expert group who drafted the Bellagio Draft Treaty raised this issue in a very careful manner and considered that control at the border is to be asserted only in zones considered critical because withdrawals are exceeding recharge or contamination is threatening groundwater quality. See *ibid.* In the case of Israel and Palestine the Western Aquifer might be considered in the category of critical especially at the border region where

intention is a comprehensive administration along the entire border or otherwise a narrower and specified control is sought. In the situation of Israel and Palestine there are three aquifer basins to be considered within the proposed treaty the NEMAB, the WMAB and the EMAB. The first two aquifer basins are of the same category; i.e. aquifers that recharge and in one State and extend to another, and can be therefore treated in a similar manner. The EMAB is of a different category as it lies almost entirely in one State but is effluent to another international surface watercourse-the Jordan River. The Parties have to take this characteristic of the EMAB into account and respond accordingly in the provisions relating to its utilization development and protection.

6.2.6.4 Identify the Substantive Rules

Having agreed on terminologies, scope and geographic reach the next step is to formulate the provisions on the substantive rules and principles governing the utilization development and protection of the transboundary aquifer. These have been recommended to be derived from the UN Convention on the Law of Non-Navigational Uses of International Watercourses. The principle of equitable and reasonable utilization is set forth in Articles 5, 6 and 7 of the Convention. On the other hand Article 7 of the Convention sets forth the obligation not to cause significant harm to other Watercourse States. Where significant harm is caused, the Watercourse State causing this harm must take all appropriate measures to eliminate and mitigate such harm and where appropriate discuss the question of compensation.

6.2.6.5 Identify the Procedural Mechanisms

The rules of procedures within the proposed treaty may be borrowed from the UN Convention and the Bellagio Draft Treaty. Cooperation must be at the heart of these rules. Cooperation must include the exchange of data and information and the establishment of a joint data base relating to the transboundary groundwater aquifers. Notification and consultation should be required on any actual or planned significantly polluting discharge into transboundary groundwaters or recharge areas, or of other activity with the potential for significant leaching into transboundary groundwaters. Procedures for the protection of quality and quantity of transboundary groundwater include the declaration of transboundary groundwater Conservation Areas, Drought Alerts, Drought Emergencies and public health emergencies and the corresponding plans and Depletion Plans.

---

the Israelis have drilled hundreds of deep wells alongside the green line in the Israeli territories.

The criteria for the determination of Conservation Areas considers whether: (a) groundwater withdrawals exceed or are to exceed recharge even so as to endanger yield or water quality or are likely to diminish water, the quantity or quality of interrelated surface waters; (b) recharge has been or may become impaired; (c) the use of the included aquifer(s) as an important source of drinking source of water has been or may become impaired water; (d) the aquifer(s) have been or may become contaminated and; (e) recurring or persistent drought conditions which may necessitate management of all or some water supplies in the particular area.

For each declared Transboundary Groundwater Conservation Area, a Comprehensive Management Plan must be prepared for the rational development, use, protection and control of the waters in the Transboundary Groundwater Conservation Area. The Comprehensive Management Plan may (a) prescribe measures to prevent, eliminate or mitigate degradation of transboundary groundwater quality, (b) allocate the uses of groundwaters and interrelated surface waters taking into account any other allocation(s) previously made applicable within the Transboundary Groundwater Conservation Area; (c) prescribe measures including pumping limitations, criteria for well placement and number of new wells, retirement of existing wells, imposition of extraction fees, planned depletion regimes or reservations of groundwaters for future use; (d) arrange, where conditions are favourable, programs of transboundary aquifer recharge (e) articulate programs of conjunctive use where appropriate (f) prescribe the integration and coordination of water quality and quantity control programs and (g) include other measures and actions as may be deemed appropriate by the Commission.

The planning for drought is another critical issue that has to be addressed in the treaty which will be applicable to the border region for activation in the region, or in parts as agreed by the Parties. The Drought Management Plan must specify the hydrometeorological preconditions for the declaration of a Drought Alert and emergencies and the conservation measures to be observed by all water users within the border region. Additionally, the plan provides for the monitoring of the hydrometeorological conditions generally in the border region and compliance with prescribed conservation or other specific measures under any Drought Alert or Drought Emergency. Alternative resources have to be made available during drought alerts or emergencies. Accordingly the drought plan must designate and reserve certain transboundary aquifers or specific well sites for use in times of drought, schemes for the conjunctive management of groundwater and surface water, impose limitation on the normal allowable withdrawals and at variance with allocations made under a Comprehensive Management Plan for a Transboundary Groundwater Conservation Area or by prior agreements between the Parties maintaining to the extent practicable the established withdrawal ratios between the Parties and an equitable balance of all emergency obligations.

6.2.6.6 Institutional Mechanisms and Dispute Settlement

The foregoing proposed substantive and procedural rules governing the utilization and development of transboundary groundwaters between Israel and Palestine cannot be realistic and achievable unless there is an authorized joint institution responsible for the implementation. The Bellagio Draft Treaty introduces a Joint Commission that has wide authorities for oversight and supervision. To that end the existing Israeli-Palestinian Joint Water Committee that was established under the Interim Agreement between Israel and Palestine can assume the role of such institution. However, the mandate structure and objectives of this institution must be modified and adapted to comply with the responsibilities as given in the Bellagio Draft Treaty. In the context it is recommended however, not to limit the authorities of such institution to oversight and supervision rather expand it to include enforcement.[87]

---

[87] In defining responsibilities to the proposed institutions, lessons could be learnt from the .Bellagio Draft Treaty. Article III provides a list of the range of activities and functions such institutions can assume

" 1. The Commission is designated as the Parties agency to carry out the functions and responsibilities provided for by this Agreement. 2. The Commission shall be authorized a technical staff, which. In collaboration with the technical staffs of the Governments shall assist the Commission in the accomplishment of its functions and responsibilities. 3. The Commission is authorized to declare Transboundary Groundwater Conservation Areas, Drought Alerts, Drought Emergencies and Public Health Emergencies and to promulgate the corresponding plans and Depletion Plans, in accordance with the provisions of this Agreement. 4. The Commission shall have jurisdiction over such additional matters concerning the border region as are from time to time referred to it by the Governments jointly. 5. The Commission shall prepare and propose to the Governments a budget conforming insofar as practicable to the budget cycles and procedures of the Governments, covering the projected expenses and capital costs of the commission's joint operations plant and staff. The total amount of each budget shall be divided between the Governments in the proportions agreed upon by the Commission and approved by the Governments. 6. The budget for the separate operating costs of each national section shall be the responsibility of the respective Government. 7. The Governments may jointly refer a specific matter relating to transboundary groundwater to the Commission for investigation or action. Individually Governments may request the Commission advice relating to transboundary groundwaters on matters originating within the requesting Government portion of the border region. 8. The Commission shall cause each such referral and request to be taken up and investigated studied or acted upon as appropriate. The Commission shall render a report to the Governments on every referral and request taken up".

It is recommended as a first step for solving a dispute or accommodating for differences that the JWC expend its best efforts to resolve differences with respect to the facts and circumstances of a situation within the purview of the Agreement. In the event of failure to resolve such differences within a specified period (e.g. 6 months) at the technical level, the dispute should be submitted together with the entire record, to the Governments for its resolution by consultation. If after good faith consultations during another specified period (e.g. 12 months) Governments are unable to reach an accommodation of a difference or differences between them concerning the facts and circumstances of a situation, any Government should be entitled to appoint a commission of inquiry that should charged with a full and impartial study for the purpose of verification of the facts of the situation. This commission of inquiry should render a report to the Governments within the terms of its instructions and on the basis of independent and detailed examination of the data and information made available to it by the Governments and the JWC, and should be able to request such additional data and information as the commission of inquiry deems significant for its deliberations and findings. On the basis of the report of a commission of inquiry, the Parties should undertake promptly to enter into consultations for the purpose of reaching an agreed accommodation of the difference(s).

6.3 FINANCING OF INTERVENTIONS

The requirements of international co-operation cannot be accomplished unless there is a strong commitment from the multilateral institutions and governmental donor agencies to fund such activities. This commitment must derive from the governments' genuine intention to end the conflict between the Parties peacefully and amicably. It is true that current investments in the water sector in Palestine and Israel are huge; it is however believed that in the field of international co-operation these investments are improperly planned.[88] The reason for that is obvious since as of today there is no unified donor coordination policy towards the water conflict and the means for its resolution and accordingly projects are not planned procedures that ensures no duplications and overinvestment. There is a need for the harmonization of donor policies and strategies towards the water problems and more intensified efforts.

[88] Of the main donors funding the Multilateral Water Resources Working Group on Water are the USA, European Union, Japan, Canada, Norway, and France.

6.4 CONCLUSIONS

It is suggested here that if the Palestinians and Israelis begin with an informal arrangement that demonstrates tangible benefits and progress they could generate the required national willingness to enter into more formal arrangements. In the context of the case study, it is demonstrated that these problems derive mainly from political considerations, amplified by technical and terminology and conceptual problems related to communicating information; and uncertainties associated with technical analyses, impact assessments and grave management practices.

After many years of conflict, the Israelis and Palestinians have succeeded in achieving some level of cooperation, which is however, impeded by negotiations. The negotiations in question are characterized by their inequality in power among the disputants. Analysis of the style of negotiation reveal that the major barrier is that Israel, with superior power in every respect, does not recognize the Palestinian right to an independent existence, nor does it acknowledge the Palestinians' national aspirations. The arrangements that were established through the peace process in relation to water are politically sensitive, one reason that justifies its complete failure in addressing the needs of the population under the stressed political situation. Other reasons for this failure include the asymmetric conditions of the two Parties at the technical, political and financial levels. Above all, the fact that the issues of legal entitlements between the parties are not addressed appears to be at the heart of the problem. Israel continues to exercise full control over all the water resources including those transboundary groundwaters shared between the Palestinians.

To date, only a minimum of progress has been made during the four years of negotiations. No common agreement on the overarching principles for the future utilization of the international transboundary water resources has been yet achieved. The challenge viewed here is how the parties could negotiate an agreement irrespective of their unequal bargaining power relations relying on and the notion that there can be a role for ethical considerations and good intentions. The existing agreements are merely a temporary solution for solving only the immediate domestic needs of the Palestinians, in addition to being a mechanism that coordinates water-related activities within the two areas of jurisdiction for the transitional five years of the interim period, which has expired in September, 1998. It will require real prodigious efforts by the Palestinians negotiators and the international mediators to engage the Israelis in negotiations over water within the above-proposed approach to resume the final status negotiations. Of crucial importance is the mutual agreement on the benefits gained from establishing cooperation based on genuine intentions. Technical cooperation is suggested as the entry point for further intensive collaboration and cooperation. The next part will consolidate the main findings of this research in addition to introducing some vitally important recommendations that emanated from this work.

7

Main Findings and Recommendations

7.1 INTRODUCTION

This book examined the relevant rules and principles of international law that govern the utilization, development and protection of connected transboundary groundwater resources. The research undertaken identified what lessons could be learnt in approaching a solution for the Israeli-Palestinian conflict over the Mountain Aquifer. It was demonstrated that the complexity of issues, concerns and challenges associated with transboundary groundwater, coupled with the issue of States' territorial sovereignty, and the intervention of politics in the area of the case study, continue to play a key role in hindering or allowing the development of a clear legal framework that gains the acceptance of the Parties to the conflict.

In Part I, it was concluded that the prerequisite for the optimal and sustainable transboundary groundwater utilization, development and protection, is a deep understanding and knowledge of the various natural characteristics and behaviours of groundwater.[1] The illustrative review of the national groundwater regulations in some selected regions, suggests that the principles of groundwater regulations that apply on national groundwaters can be as well adapted at the international levels. However, the challenge is how to encourage States to adopt these principles regarding their shared transboundary groundwaters.[2] It was also demonstrated that the task of regulating groundwater is complex even under a "normal" political setting, a complexity that is further intensified where there exists a long history of conflict and unequal distribution of the resources.[3] This is true especially that the Parties to the conflict are not able to foresee the mutual benefits of cooperation concerning the shared resources. Part I also found that despite the fact that Palestine is under the effective control of belligerent occupation, international law continues to apply.[4] To date, however, international law appears not to have played an influential role in resolving the disputes over shared transboundary groundwaters in the region.[5]

Part II provided an overview of the modern development of international law in the field of transboundary groundwater. It was concluded that the fundamental rule governing the development and utilization of connected transboundary groundwater is the "equitable and reasonable utilization", which includes the factor of degree of harm among other factors.[6] The application of this rule requires rules of procedure that are tailor-made to groundwater.[7] The operationalization of this rule under the unique political situation of Israel and Palestine requires an innovative approach, which gives special attention to the existing challenges and establishes enough grounds for long term solutions.

---

[1] *See supra* Chapter 2, Summary and Conclusions.

[2] *See supra* Chapter 2, Section 2.4.

[3] Chapter 2 demonstrated the complexity of regulating transboundary groundwater

[4] *See* relevant discussion on Palestine's political status as an occupied territory as well as the discussion on Palestine's permanent sovereignty over natural its resources, *supra* Chapter 3 (sectionS 3.42-3.343).

[5] *Ibid,* also *see* the conclusions from chapter 3 (Conclusions). The fact that international law is believed to have played a minor role in solving the water conflict is further demonstrated *infra* Chapter 4 through analyzing the rules and principles governing the existing agreements between Israel and the PLO.

[6] According to the major international legal instruments including the 1966 Helsinki Rules and the 1997 UN Watercourses Convention, these factors include physical, hydrologic and socio-economic ones. For more details *see supra* Chapter 4 Section 4.4.

[7] The key features that continue to distinguish groundwater from surface water have been presented *supra* Chapter 2 Section 2.2 requiring a legal response that was described at the national level *supra* Chapter 2 Section 2.4. In Chapter 4, the comprehensive analysis of international legal instruments concerning international watercourses, demonstrated that the 1989 Bellagio Draft Treaty includes appropriately designed mechanisms and procedures that respond to the special needs of groundwater.

Although this research has demonstrated that States indirectly regulate transboundary groundwater under the international watercourse treaties, many of these aquifers remain formally unregulated.[8] States have favoured, without the formalities of treaties, to enter into cooperative technical non-binding arrangements that help the Parties build their knowledge and understanding concerning these resources.[9] Such practice has inspired the current research particularly in approaching the solution between Israel and Palestine. Accordingly, Part III of the research introduced a progressive multidisciplinary approach which envisions technical cooperative non-binding arrangements as the first step in resolving the water conflict, which ends up with a binding formal agreement. In between these informal and informal arrangements, lie a multitude of actions and measures that aim at building trust towards common interests and visions.

7.2 MAIN FINDINGS

7.2.1 Transboundary Groundwater Law- Current Issues

Chapter 1 demonstrated that modern trends in national groundwater regulation inspire the progressive development of transboundary groundwater law. This research reveals that what distinguishes international from national regulation of groundwater, are the legal and institutional management difficulties in the case of the latter.[10] This is true as decisions at the international level require two or more States to agree, with possibly two or more legal and political regimes involved. Having said that, however, the scientific and technical requirements of groundwater at national and international levels, were found to be analogous.[11]

The research also finds that despite the demonstrated efforts of governmental and non-governmental organizations and groups in addressing the topic of international watercourses, the law governing the utilization development and

---

[8] In Europe alone, this research found that there are 100 unregulated aquifers in Europe, 20 in Africa , and at least 4 in the Middle East (including the Mountain Aquifer shared between Israel and Palestine), *see supra* Chapter 1 page 3.

[9] *See supra* Chapter 5, Section 5.7.

[10] Refer to relevant discussion *supra* Chapter 2 (Section 2.4) and *see* generally *supra* Chapter 4.

[11] The fact that groundwater is finite, affects and is affected by the surrounding environment and the scarcity of groundwater data, are all key features that are common at national and international levels, *see supra* Chapter 2 (sections 2.2-2.3). The needed legal response is presented *supra* Chapter 3 (section 3.4), which was found to be relevant at the international level. These procedures include: the concept of "integrated water resources management", the "permitting and licensing system", "monitoring and assessment", and the "declaration of protection zones". See for comparison at the international level relevant discussion *supra* Chapter 4 Section 4.4 and relevant State Treaty Practice *supra* Chapter 5 Section 5.4.

management of transboundary groundwater remains uncertain. International instruments including the 1966 Helsinki Rules, the 1986 Seoul Rules, the 1989 Bellagio Draft Treaty and the 1997 UN Watercourses Convention, provide guidance for States in concluding agreements and cooperative arrangements over their international watercourses.[12] Of the mentioned instruments, the 1997 UN Watercourse Convention, although not yet in force, presents the only formal legal framework that codifies and progressively develops the law of international watercourses.[13] The Convention includes connected groundwaters within its scope, which implies that the rules and principles it embodies apply equally to surface and groundwater. This research finds that the equal treatment of surface and groundwater is justified as far as substantive rules are concerned. However, the Convention ignores the fact that transboundary the protection and management of groundwater require the development procedural rules that are unique and tailor-made. This is maybe justified given the framework nature of the Convention, which leaves it to the discretion of States to adapt, modify and adjust their exiting and future treaties to suit the condition of a particular watercourse, provided that they follow principles embodied in the 1997 UN Watercourses Convention.[14] Guidance on the development of such procedures can be sought in the 1989 Bellagio Draft Treaty, which embraces comprehensive mechanisms and procedures that respond directly to the needs of optimal and efficient regulation of transboundary groundwaters.[15]

Confined transboundary groundwater is another issue that remains unresolved under the 1997 UN Watercourses Convention.[16] Although this topic was not of particular relevance to the present research, a limited analysis to the status of development of international law concerning these resources has been undertaken. It was found that the "Resolution on Transboundary Confined Aquifers", which was submitted in conjunction with the ILC 1994 Draft set of Articles, was the compromise that the ILC could achieve on the matter. For many reasons, however, this Resolution was never adopted by the General Assembly. Guidance on this matter could as well be sought from the 1986 ILA Seoul Rules, the 1989

---

[12] *See supra* Chapters 3 and 4.
[13] *See supra* Chapter 4 (Sections 4.3-4.6)
[14] Article 3 of the 1997 UN Watercourses Convention provides:
"…….3. Watercourse States may enter into one or more agreements, hereinafter referred to as "watercourse agreements", which apply and adjust the provisions of the present Convention to the characteristics and uses of a particular international watercourse or part thereof…"
[15] *See* relevant discussion on the Bellagio Draft Treaty *supra* Chapter 3. *Also See infra* Section 6.2.2.
[16] See generally *supra* Chapter 4 (Section4.3), *also supra* chapter 5 (Section 5.3).

Bellagio Draft Treaty[17] and from the informal cooperative frameworks recently developing particularly in Africa.[18] Most recently, the issue of confined transboundary aquifers has returned to the agenda of the ILC. [19] In the latest recommendation, the need to formulate rules regulating confined transboundary has been emphasized. [20] An inventory of these resources worldwide and a breakdown of the different regional characteristics of these resources are accordingly very essential.

7.2.2 Rules and Principles Governing Transboundary Groundwater

The "equitable and reasonable utilization" principle was found to be the fundamental rule governing the utilization of transboundary groundwater.[21] This rule includes the factor of degree of harm among other factors.[22] The 1997 UN Watercourses Convention offers a compromised balance between the equitable and reasonable utilization and the no significant harm principle with respect to the utilization, development and management of international watercourses.[23] The operationalization of the equitable and reasonable utilization principle requires that hydrologic, economic social and legal factors are reflected and considered.[24]

In the context of procedural rules, the 1997 UN Watercourses Convention provides a useful -though not comprehensive- procedural framework, as far as transboundary groundwaters are concerned. Cooperation is at the heart of these

---

[17] *See* relevant discussion in Chapter 4 (section 4.2.3).This guidance provides that the same rules and principles applying to connected transboundary groundwater apply equally to confined transboundary groundwater.

[18] *See supra* Chapter 5 Section 5.7.

[19] UN General Assembly: International Law Commission Fifty Fifth Session, *Shared Natural Resources: First Report on Outlines*, First Report by the Special Rapporteur Chusei Yamada, UN Doc. A/CN.4/533, May 2003. *See* relevant discussion *supra* Chapter 4 (section 4.3.2).

[20] *Ibid.*

[21] The reviewed international instruments, including the 1966 Helsinki Rules, the 1997 UN Watercourses Convention and the 1989 Bellagio Draft Treaty, agree that this rule is governing the utilization of transboundary groundwater. *See supra* Chapter 4 section 4.4.

[22] *Ibid.*

[23] *Ibid.* The 1997 UN Watercourses Convention confirms that the "equitable and reasonable utilization" principle includes the factor of degree of harm. In this regard, some harm not reaching the threshold of "significant" is allowed to occur, if such occurrence serves the implementation of equity and reasonableness of the utilization.

[24] Guidance can be sought here from the Dundee KaR Project, which developed a "Legal Assessment Model" for the operationalization of the equitable and reasonable utilization principle, *see* relevant discussion *supra* Chapter 6 (section 6.2.4). *See also* Annex IV, which provides a framework for this assessment in the case of Palestine.

procedures, which, derives from the general principles of States' equality and sovereign rights.[25] The main pillars of cooperation are: prior notification, exchange of information, consultation and negotiation.[26] In addition to the fundamental procedural rules embodied within the UN Convention, it was found that the Draft Bellagio Treaty offers appropriate detailed procedures and mechanisms including: (i) the establishment of joint institutions for the implementation of the Agreement, (ii) the establishment and maintenance of a joint database, (iii) the identification of critical areas for joint administration, (iv) the declaration of conservation areas for concentrated management, regulations and control, (v) the development of comprehensive management and development plans, (vi) the provisions on public health emergencies, and (vii) the planning for drought.[27]

7.2.3 Transboundary Groundwater in State Practice

States have been appealing to formal and informal types of international cooperation, as far as their international watercourses are concerned.[28] It was found that, despite the improved awareness of groundwater behaviour and special characteristics, States remain reluctant to conclude international agreements that specifically regulate transboundary groundwater aquifers.[29] States' formal positions towards the regulation of transboundary groundwater have been reflected in the debate within the ILC and the Working Group of Whole.[30] Additionally, State practice has been examined through a treaty survey relating to international watercourses, in three selected regions: Africa, Europe and the Middle East.[31] States' informal cooperation concerning their international watercourses was also reviewed. The main findings concerning State practice in relation to transboundary groundwater provide that: (i) driven by the fear over their territorial sovereignty, States continue resist the conclusion of binding treaties over their transboundary groundwater aquifers (ii) State treaty practice provides a strong evidence that the inclusion of connected transboundary groundwater within the scope of an international watercourse treaty, is generally

---

[25] *See* relevant discussion *supra* Chapter 5 Section 5.5.

[26] *Ibid.*

[27] *Ibid.*

[28] *See* relevant discussion *supra* chapter 6.

[29] *See* relevant discussion *supra* Chapter 5. *Also t*he treaty survey included in Annexes I to III demonstrate that States either explicitly or explicitly include groundwater within the scope of the international watercourses treaty. Only few of these instruments treat groundwater as an element that requires a distinct treatment.

[30] *See* relevant discussion supra Chapter 4.

[31] See relevant discussion supra Chapter 5 and Annexes I to III.

accepted (iii) States explicitly exclude confined transboundary groundwaters from the scope of binding legal arrangements. Such exclusion, however, does not apply to informal cooperation between States with respect to confined transboundary aquifers.[32] The awareness of the importance of such resources has led to the development of programmes and schemes of significant net benefits of all States concerned. The exchange of hydrogeological and other data, the coordinated or joint construction and cooperation of projects has been the subject matter of numerous successful international arrangements.[33]

7.2.4 Transboundary Groundwater under Difficult Political Conditions - the Mountain Aquifer Case Study

This research reveals that politics plays a vital role in delaying the conclusion of a comprehensive formal agreement concerning the Mountain Aquifer. Accordingly, in addition to the factors that are mentioned in the 1997 UN Watercourses Convention,[34] the impact of the political factor, in this particular case, has to be well reflected and considered.

This research also finds that Palestine's sovereignty over their natural resources has not extinguished due to the Israeli occupation; more precisely it was not practiced for a period of time. Accordingly, Palestine is entitled to an equitable and reasonable share of the beneficial uses from all international water resources, including those from the Mountain Aquifer. It is concluded that despite the marked disparity between the stages of development and the size and strengths of Israel and Palestine, the challenge is how to convert this into a solution of the mutual benefits.[35] The Parties, however, have failed so far to foresee the benefits of a joint integrated action, as far as their shared groundwaters are concerned.[36] The inability on the side of the politicians to accept the consequences of international cooperation, is reflected in a lack of willingness to place confidence in joint institutions to manage the utilization and development of the Mountain Aquifer.[37] Finally, this research confirms that in the absence of genuine intentions

[32] *See supra* Chapter 5 Section 5.7.

[33] *Ibid.*

[34] These are the factors listed under Article 6, *see supra* Chapter 4.

[35] *See supra* Chapter 6 (section 6.2.3.2).

[36] To date, the Parties focus on positions and not on interests, *see supra* Chapter (3 section 3.3). Additionally existing agreements present a temporary arrangement that does not reflect an integrated mutually beneficial arrangement. *See* relevant discussion on the content of the Agreements between Israel and Palestine, *see* relevant discussion, *supra* Chapter 5.

[37] The Joint Water Committee (JWC), which was established under the 1995 Israeli-Palestinian Interim Agreement, is practically not functioning. The spirit of joint decisions

to cooperate, even legally binding treaties are not helpful to solve the current dispute.[38] International law is envisioned to offer enough guidance for the Parties on the steps needed to initiate and implement cooperation. Such guidance has been reflected in the proposed approach for solving the conflict that was presented in Chapter Five.

7.3 RECOMMENDATIONS

7.3.1 At the International Level

7.3.1.1 The Need for a Modified Transboundary Groundwater Framework Agreement

In the view of this book, a transformation in the manner transboundary groundwater is treated, is urgently needed. This research cited many efforts by non-governmental institutions and groups towards this transformation.[39] The serious attention by NGOs, expert groups and governmental organs of the UN highlighted the complexity and importance of this area of regulation. Further development and integration are still needed, in order to adopt multidisciplinary approaches for the progressive development of this particular area of the law. The additional challenge is the operationalization of the law, by offering mechanisms for its implementation and enforcement, under the specific watercourse socio-economic and political conditions.[40]

The Framework Treaty is recommended to build on the strengths of the 1997 UN Convention relating specifically to transboundary groundwater, and the complementary role that the Bellagio Draft Treaty was found to play in that regard. Additionally, this Treaty may benefit from the modern trends in national groundwater regulation.[41] Therefore, the strengths of the UN Convention on the substantive level,[42] combined with the strengths of the Bellagio Draft Treaty in the procedural mechanisms are expected to yield a Framework Treaty that responds to the special needs of transboundary groundwater. Additionally, major procedures and concepts can be adapted and modified from the national practice. These

---

and consensus is lacking and the Committee became a platform for VETO for both Parties, especially the Israelis. See *supra* Chapter 5 (section 5.6).

[38] *See supra* chapter 4 and 6.

[39] These efforts include the 1986 ILA Seoul Rules, the 1989 Draft Bellagio Treaty and the work of the UNECE on transboundary groundwater, and the ISARM programme, *see* relevant discussion *supra* Chapter 4.

[40] An example of such approaches is the one provided in Chapter 6 (section 6.2) "Progressive Approach". See also the Dundee KaR project *supra* note 24.

[41] *See supra* Chapter 2.

[42] See *supra* Section 7.2.2.

include the enactment of a Joint Permitting/Licensing System on part of the transboundary aquifer that is under joint regulation,[43] and the concept of integrated water resources management.[44]

7.3.1.2 The Need for Third Party

Under the framework of the proposed "progressive approach",[45] for solving the Israeli-Palestinian conflict over shared transboundary groundwater, the intervention of a "Third Party" was recommended.[46] In particular, this intervention was encouraged at the initial stages of "identification of the problem" and for setting the stage for collaboration and cooperation. Additionally, at end stages - namely "the treaty formulation"- this requirement is also valid. The role of a "Third Party" must be designed to produce wise outcomes efficiently and amicably. This role must ensure the development of an integrative arrangement, of equal benefit to the Israelis and Palestinians.

7.3.1.3 The Progressive Approach

In light of the political sensitivities, it was found inappropriate to propose a long term binding agreement over the Mountain Aquifer. The proposal within this research is to follow and implement a progressive approach,[47] which starts with the identification of the problems and agreeing on common objectives concerning the shared groundwater. Coordination, expands into cooperation and collaboration, and ends up with the formulation of a binding long term sustainable agreement.[48] The active participation of informed, top level political representatives, with solid political support within their respective countries is crucial and important. Once the intention for cooperation has been expressed genuinely at the political level, the rest of the process can be placed in the hands of the middle management and technical people. The mobilization of joint technical experts including engineers, lawyers, economists ...etc. is expected to enhance and improve international cooperation which constitutes the basis for a long term legal arrangements.[49]

---

[43] *Ibid.*

[44] *See supra* Chapter 2 (section 2.2).

[45] *Ibid.*

[46] *See supra* Chapter 4, and *supra* Chapter 6.

[47] *See supra* Chapter 6 (section 6.2).

[48] *Ibid, also see* V, which includes a proposal for a "Draft Framework Agreement Concerning the Mountain Aquifer".

[49] *Ibid.*

7.3.1.4 Institutionalization of Cooperation

Cooperation normally requires an institutional home. It is proposed that since the parties already have a joint international institution, the JWC, this co-operation could be achieved through the restructuring and reactivation of its work. Having said that, it is important to study of the reasons of its failure, and agree on the new mode of work that ensures successful cooperation.

7.3.2 At the National Level

7.3.2.1 Strengthen the Regulatory and Environmental Framework

Although the Water Law No 3 for 2002 presents an adequate framework for managing and conserving the water resources, there is a need to draft regulations, outline the new operating license, and train staff, among other activities. There is a need to review, coordinate, and update national water policies, legislation, and institutions to guide the preparation of water-resources assessments and management plans; and to promote the use of sustainable management practices to meet the growing needs for drinking water and access to sanitary services, especially in urban areas. There is an urgent need to activate the work of the NWC to ensure that the conditions of intersecotral planning and coordination are fulfilled.

7.3.2.2 Proceed with the Water Sector Reform

Palestine is currently undertaking reform in the water and sanitation sector through management contracts with performance targets. However in the last few years a number of Political, financial and technical factors have contributed to significant changes in the way this reform is proposed, designed and implemented. An important factor influencing the water sector reform in Palestine has much to do with a general reluctance on the part of the PA to move towards transactions. Uncertainty about their options and the impact these have both on their own roles and on their constituencies, entrenched interests. The other important factor relates to the donors' and private sector's uncertainty of the viability of such arrangements given the unstable political situations. The There is a need to greater analysis to understand how, within this environment, corporate incentives to serve the poor are affected, the public sector's oversight function is enhanced, and ultimately how poor communities are impacted. There is a need to understand the impacts of certain contract types, tariff structures and sector reforms in order to propose the most appropriate implementation models. There is as well need for a multi-stakeholder dialogue on the issue of Private Sector Participation (PSP) in developing countries. Though civil society organisations are demanding more of a say in decision-making, clearly some capacity building needs to occur to allow them to engage productively. Similarly issues around NGOs' own accountability, incentives and drivers are increasingly being challenged.

7.3.2.3 Strengthen the Management and Technical Capacities

At the national level the particular strengths and weaknesses with respect to technical personnel and equipment must be identified. This will enable the design of program that captures the training and capacity building needs. Technical assistance can be sought from or outside the region. Parallel to that some studies relating to enhancing and strengthening the knowledge on the shared water resources could be undertaken these include (i) an inventory of shared water resources (ii) inventory of wells and other water related infrastructure (iii) inventory of existing Monitoring Stations (iv) design and creation of a Joint Monitoring and Assessment systems, (vi) institutional mechanism.

7.3.2.4 Improve Knowledge on Transboundary Groundwater

The Palestinians need to improve their level of knowledge concerning the Mountain Aquifer. The current desk studies of the aquifer basins by the Palestinians need to be complemented by geophysical and hydrogeological investigations and modelling. Understanding the features of groundwater namely the interaction with the surface water and the environment, is essential for the choice of planning, management and regulatory hydrogeologic decision at the national and international levels. Such a step is prerequisite for the establishment of physical and legal frameworks essential to view the entire aquifer system.

7.3.2.5 Harmonize National Laws with Israel Pertaining to Transboundary Groundwater

National water legislation concerning the utilization development management and protection of transboundary groundwater should be harmonized. Only those general policies and specific provisions that inhibit and obstruct the desired development need to be harmonized. The removal of legal conflicts and the enactment of needed provisions in the respective State codes are essential to water use optimization and to the promotion of regional development generally. Preparatory studies for such harmonization would be

(i) the conduction a comparative studies of water laws of some countries.

(ii) a manual for the drafting of a water code based on the principles and consideration s outlined in the recommendations of the comparative study.

7.3.2.6 Create Drainage Basin Authorities/Boards

Palestine has to restructure its water system. There is a need to review the institutional strategies for the efficient and integrated management of all water resources within Palestine. PWA in its capacity as the regulator and administrator of the water resources cannot accomplish its tasks without the support of local decentralized management institutions at the drainage basin level. It is also a fact that successful implementation of IWRM can only be guaranteed through the establishment of appropriate institutional frameworks, such as independent drainage basin authorities or boards. The IWRM should be linked to social and economic development and should address land and water uses and the conservation of biodiversity within the context of the transboundary aquifers. This policy decision is very crucial to the future cooperation and coordination with Israel concerning the shared water resources. To that end a regional concentration of projects and integrated projects in erosion control and resource protection, assists in the management of watersheds and drainage basin areas.

Bibliography

1. PRIMARY SOURCES

1.1 Treaties

1.1.1 Treaty Collections-General

Caponera D.A. and D. Alheritiere, (1981), Principles for International Groundwater Law, in L. Teclaff, L. & A. Utton (eds.), International Groundwater Law (25) also found in Natural Resources Journal (18) 589-618.

Teclaff, L.A. and A.E. Utton, (1981) International Groundwater Law, Oceana Publication New York.

International Water Law Research Institute Treaty Collection, *found at* http://www.dundee.ac.uk/law/iwrli (*last visited* May 2003).

The Transboundary Freshwater Dispute Database (TFDD) at the Department of Geosciences at Oregon State University 1820-2002 *found at* http://www.transboundarywaters.orst.edu/ (*last visited* April 2003).

The United Nations Treaty Collection *found at* http://untreaty.un.org/English/treaty.asp (*last visited* April 2003).

The FAO Legislative Studies No. 50, 55 and 61. The latter *found at* http://www.fao.org/docrep/W7414B/W7414B00.htm (*last visited* May 2003).

The United Nations, Office of Legal Affairs, Codification Division, The Law of International Groundwater, December 1990.

1.1.2 Multilateral-General

Hague Convention on Land Warfare, 1 American Journal of International Law (done at the Hague, 29 July 1899, *also found at* http://www.yale.edu/lawweb/avalon/lawofwar/final99.htm (*last visited* June 2003).

Hague Convention on Land Warfare, (No. IV), done at the Hague 18 October 1907, UK Treaty Series No. 9 *also found at* http://library.byu.edu/~rdh/wwi/hague/hague5.html (*last visited* June 2003).

Article 22 of the Covenant of the League of Nations, 28 June 1919, *found at* http://www.mideastweb.org/leaguemand.htm (*last visited* June 2003).

Montevideo Convention on the Rights and Duties of States, Signed at Montevideo, 26 December 1933 Entered into Force, 26 December 1934, *found at* http://newtaiwan.virtualave.net/montevideo01.htm (*last visited* May 2003).

Charter of the United Nations, San Francisco, 26 June 1945,

found at http://www.un.org/aboutun/charter/ (*last visited* June 2003).

The Geneva Conventions:

(I) for the Amelioration of the Condition of the Wounded and Sick in Armed Forces in the Field,

(II) for the Amelioration of the Conditions of the Wounded Sick and Shipwrecked Members of Armed Forces at Sea; Convention

(III) Relative to the Treatment of Prisoners of Water; and Convention

(IV) Relative to the Protection of Civilian Persons in Time of Water; done at Geneva, 12 August 1949: Final Record of the Diplomatic Conference of Geneva of 1949, Vol I, Federal Department, Berne, at pp. 195-341; 75 United Nations Treaty Series 5-417. Also found at: http://www.icrc.org/ihl.nsf, (last visited June 2003).

Vienna Convention on the Law of Treaties, United Nations Conference on the Law of Treaties, UN Doc. A/CONF.39/27, May 23, 1969, reprinted in 8 ILM 1969, 679-713, *also found at* http://www.greenpeace.org/~intlaw/vien-tr.html (*last visited* June 2003).

Conference on Security and Co-operation in Europe, Final Act, Helsinki 1975.

Protocol 1: Additional to the Geneva Conventions, 1977 *found at* http://www.deoxy.org/wc/wc-proto.htm (*last visited* June 2003).

United Nations Framework Convention on Climate Change (1992), *found at* http://unfccc.int/resource/docs/convkp/conveng.pdf (*last visited* February 2003).

Vienna Convention on Succession of States in Respect of Treaties, UN Doc. A/CONF 80/31, August 22, 1978, 17 ILM 1978, 1488-1517.

Convention on the Protection and Use of Transboundary Watercourses and International Lakes, Helsinki, March 17, 1992, 31 ILM 1992, 1312-1329.

Convention on Environmental Impact Assessment in a Transboundary Context, Espoo, Finland, February 25, 1991; 30 ILM 1991, 800.

The United Nations Convention on the law of Non-Navigational Uses of International Watercourses 21 May 1997, U.N.Doc.A/RES/51/229, July 8, 1997, also *found at* http://www.un.org/law/ilc/texts/nonnav.htm#abstract (*last visited* June 2003).

1.1.3 Multilateral/Bilateral-By Region

1.1.3.1 The Middle East

Convention Regarding the Water Supply of Aden between Great Britain and the Sultan of Abdali, April 1910, *found at* http://ocid.nacse.org/qml/research/tfdd/toTFDDdocs/21ENG.htm (*last visited* May 2003).

League of Nations Mandate for Palestine, Nineteenth Session of the Council thirteenth meeting held at St James' Palace, London on July 24th, 1922, *found at* http://www.mfa.gov.il/mfa/go.asp?MFAH00pr0 (*last visited* June 2003).

Agreement between the Republic of Syria and the Hashemite Kingdom of Jordan concerning the Utilization of the Yarmuk Waters, signed at Damascus on 4 June 1953, (came into force on 8 July 1953), *found at* http://ocid.nacse.org/qml/research/tfdd/toTFDDdocs/82ENG.htm (*last visited* May 2003).

Johnston Negotiations on the Jordan River Basin, 1955 *found at* http://ocid.nacse.org/qml/research/tfdd/toTFDDdocs/92ENG.htm (*last visited* May 2003).

The Agreement between Italy and Egypt on the Ramba Well, 1925, UN Treaty Series No. 6 U.N. Doc. ST/LEG/SER.B/12(1963).

Israel-Palestine Liberation Organisation Declaration of Principles on Interim Self-government Arrangements, 13 September 1993, Washington DC., 32 ILM 1525 (1993), (came into force 13 October 1993), *found at* http://www.jmcc.org/peace/agreements.html (*last visited* May 2003).

Israel-Jordan: Common Agenda for the Bi-lateral Peace Negotiations, 14 September 1993, Washington, 32 ILM 1522 (1993).

Israel-Palestine Liberation Organisation: Agreement on the Gaza Strip and the Jericho Area, 4 May 1994, Cairo, 33 ILM. 622 (1994), also *found at* http://www.us-israel.org/jsource/Peace/gazajer.html (*last visited* June 2003).

including:

Annex I – Protocol Concerning Withdrawal of Israeli Military Forces and Security Arrangements;

Annex II – Protocol Concerning Civil Affairs;

Annex III – Protocol Concerning Legal Matters;

Annex IV – Protocol on Economic Relations between the Government of the State of Israel and the PLO, representing the Palestinian people.

Gaza – Jericho Agreement – letters.

Treaty of Peace between the State of Israel and the Hashemite Kingdom of Jordan, 26 October 1994, Arava/Araba Crossing Point, 34 ILM 1995, 43, *found at* http://ocid/nacse.org/qml/research/tfdd/toTFDDdoc/168ENG.htm (*last visited* May 2003),

including:

Annex I (a) - Israel-Jordan International Boundary Delimitation and Demarcation;

Annex II – Water Related Matters;

Annex IV – Environment.

Israeli-Palestinian Interim Agreement on the West Bank and the Gaza Strip, 28 September 28 1995, Washington D.C., 36 ILM 551 (1997), *found at* http://www.ariga.com/treaties/interim.html (*last visited* May 2003),

including:

Annex III – Protocol Concerning Civil Affairs;

Annex IV – Protocol Concerning Legal Affairs;

Annex VI – Protocol Concerning Israeli-Palestinian Cooperation Programs.

Israel-Jordan-Palestine Liberation Organisation Declaration on Principles for Co-operation on Water-Related Matters and New and Additional Water Resources,

13 February 1996, Oslo, 36 ILM 761 (1997), *found at* http://www.mfa.gov.il/ (*last visited* March 2003).

Palestinian Statements to the Middle East Peace Multilateral Negotiations First Round Vienna 1992, Second Round Geneva 1993, Third Round China 1993, Fourth Round Jordan 1994, Sixth Round Greece 1994, Seventh Round Jordan 1995 and Eighth Round Tunis 1996 (on file with the Author).

Agreement Concerning Water Distribution Of Al-Asi River Between Syria And Lebanon, signed on the 20 September 1994, (came into force December 11th 1994) (In Arabic), *found at* http://ocid.nacse.org/qml/research/tfdd/toTFDDdocs/273ARA.htm (*last visited* May 2003).

The 1998 Wye River Memorandum, *found at* http://www.palestinecenter.org/cpap/pubs/19990729ib.html (*last visited* June 2003).

Statement to the Middle East Peace Multilateral Negotiations Palestine Delegation, Position Paper Presented to the Working Group on Water (1st – Seventh Rounds). (on file with the author).

Israel - Palestine Liberation Organisation: Wye River Memorandum (Interim Agreement), 23 October 1998, 37 ILM 1251 (1998).

Israel - Palestine Liberation Organisation: The Sharm El Sheikh Memorandum, 4 September 1999, 38 ILM 1465 (1999).

1.1.3.2 Europe

Treaty Between Germany and Poland for the Settlement of Frontier Questions, Signed At Poznan, January 27, 1926, *found at* http://ocid.nacse.org/qml/research/tfdd/toTFDDdocs/35ENG.htm (*last visited* May 2003).

Agreement between the United Kingdom and Belgium Regarding Water Rights on the Boundary between Tanganyika and Ruanda-Urundi, London, November 22, 1934 (Ratifications Exchanged In London, May 10, 1938.) UN Treaty Series No.42 *also found at* http://ocid/nacse.org/qml/research/tfdd/toTFDDdoc/45ENG.htm (*last visited* May 2003).

The Treaty of Peace Between the Allies and Italy which Outlines Guarantees Between Italy and Yugoslavia concerning Springs in the Commune of Gorizia 1947, *found at* http://ocid.nacse.org/qml/research/tfdd/toTFDDdocs/394ENG.htm (*last visited* May 2003).

Agreement between the Austrian Federal Government and the Bavarian State Government concerning the Diversion of Water in the Rissbach, Durrach and Walchen Districts, concluded on 16 October 1950. (Came into force on 1 May 1951), *found at* http://ocid.nacse.org/qml/research/tfdd/toTFDDdocs/73ENG.htm (*last visited* May 2003).

Protocol to Establish Triparti Standing Committee on Polluted Waters between Luxemburg, France and Belgium, Brussels 8 April 1950, (came into force 8 April 1950.) *found at* http://ocid.nacse.org/qml/research/tfdd/toTFDDdocs/284ENG.pdf (*last visited* May 2003).

Agreement between the Federal People's Republic of Yugoslavia and the Republic of Austria concerning Water Economy Questions in respect of the Frontier Sector of the Mura and the Frontier Waters of the Mura, signed At Vienna on 16 December 1954 And Protocol Signed On November 1954(came into force on 9 February1956), *found at* http://ocid.nacse.org/qml/research/tfdd/toTFDDdocs/88ENG.pdf (*last visited* May 2003).

Treaty between the Hungarian People's Republic and the Republic of Austria Concerning the Regulation of Water Economy Questions in the Frontier Region, signed at Vienna, on 9 April 1956, (came into force On 31 July 1956), *found at* http://ocid.nacse.org/qml/research/tfdd/toTFDDdocs/94ENG.htm (last visited May 2003).

Agreement between the Government of the Federal People's Republic of Yugoslavia and the Government of the People's Republic of Albania Concerning Water Economy Questions, together with the Statute of the Yugoslav-Albanian Water Economic Commission and With the Protocol Concerning Fishing in Frontier Lakes and Rivers, signed on the 12 May 1956 (came into force on 6 August 1957), *found at* http://ocid.nacse.org/qml/research/tfdd/toTFDDdocs/97ENG.htm (*last visited* May 2003).

Agreement between the Government of the Czechoslovak Republic and the Government of the Polish People's Republic Concerning the Use of Water Resources in Frontier Waters, signed at Prague, on 21 March 1958, (came into force On 7 August 1958), *found at* http://ocid.nacse.org/qml/research/tfdd/toTFDDdocs/104ENG.htm (*last visited* May 2003).

Agreement Concerning Water-Economy Questions between the Government of the Federal People's Republic of Yugoslavia and the Government of the People's Republic of Bulgaria, signed at Sofia, on 4 April 1958 (came into force on 29 December 1959), *found at* http://ocid.nacse.org/qml/research/tfdd/toTFDDdocs/105ENG.htm (*last visited* May 2003).

Protocol Concerning International Commission to Protect the Moselle Against Pollution, between France, Federal Republic Of Germany and Luxemburg, signed at Paris 20 December 1961, (came into force on July 1 1962), *found at* http://ocid.nacse.org/qml/research/tfdd/toTFDDdocs/298ENG.htm (*last visited* May 2003).

Agreement on International Commission for the Protection of the Rhine against Pollution, Signed in Bern On 29 April 1963 between Switzerland, Federal Republic Of Germany, Luxemburg, Netherlands and France, (came into force on May 1965), *found at* http://ocid.nacse.org/qml/research/tfdd/toTFDDdocs/294ENG.pdf (*last visited* May 2003).

Agreement between the Government of the Polish People's Republic and the Government of the Union of Soviet Socialist Republics concerning the Use of Water Resources in Frontier Waters, signed At Warsaw, 17 JULY 1964; (came into force on 16 February

1965), *found at* http://www.internationalwaterlaw.org/RegionalDocs/Poland-USSR.htm (*last visited* May 2003).

Franco-Italian Convention Concerning the Supply of Water to the Commune of Menton, Done At Paris, On 28 September 1967, (came into force on 14 October 1972), *found at* http://ocid.nacse.org/qml/research/tfdd/toTFDDdocs/138ENG.htm (*last visited* May 2003).

Convention Concerning the Protection of Italo-Swiss Waters Against Pollution Between Italy and Switzerland, signed at Rome on 20 April 1972, *found at* http://ocid.nacse.org/qml/research/tfdd/toTFDDdocs/259ENG.htm (*last visited* May 2003).

Convention on the Treaty between the Hungarian People's Republic and the Czechoslovak People's Republic on 16 September 1977 Concerning the Construction and Operation of the Gabcíkovo-Nagymaros System of Locks, *found at* http://www.icj-j.org/icjwww/idocket/ihs/ihsjudgement/ihs_ijudgment_970925_frame.htm (*last visited* May 2003).

Convention on the Protection, Utilization and Recharging of the Geneva Aquifer. Canton of Geneva in Switzerland and the Department of Upper Savoy in France in 1977, *found at* www.dundee.ac.uk/law/iwrli/documents (*last visited* May 2003).

Treaty Between the Government of Romania and the Government of Hungary on the Regulation of Water Problems in Watercourses Forming or Crossing the Boundary, June 25 1986, *found at* http://ocid.nacse.org/qml/research/tfdd/toTFDDdocs/101ROM.htm (*last visited* May 2003).

Agreement on Co-operation on Management of Water Resources in the Danube Basin, between the Federal Republic of Germany the European Economic Community, signed at Regensburg on the 1st December 1987, *found at* http://ocid.nacse.org/qml/research/tfdd/toTFDDdocs/278ENG.htm (*last visited* May 2003).

Charter on Ground-Water Management adopted by the Economic Commission for Europe at its forty-fourth session (1989) by decision E (44), United Nations, New York, 1989E/ECE/1197, ECE/ENVWA/12, *also found at* www.internationalwaterlaw.org/Regional/Docs/Groundwater_Charter.htm (*last visited* May 2003).

Convention Between the Federal Republic of Germany and the Czech and Slovak Federal Republic and the European Economic Community on the International Commission for the Protection of the Elbe, done at Magdeburg on 8 October 1990, *found at* http://ocid.nacse.org/qml/research/tfdd/toTFDDdocs/183ENG.htm (*last visited* May 2003).

Agreement on Joint Activities in Addressing the Aral Sea and the Zone around the Sea Crisis, Improving the Environment, and Enduring the Social and Economic Development of the Aral Sea Region, between the Republic of Kazakhstan, Republic of Kyrgyzstan, Republic of Tajikistan, Turkmenistan, and Republic of Uzbekistan) signed on 26 March 1993, *found at* http://ocid.nacse.org/qml/research/tfdd/toTFDDdocs/166ENG.htm (*last visited* May 2003).

Agreements on the Protection of the River Meuse and Scheldt, done at Charleville Mezieres, April 26 1994, between France Netherlands and Belgium, *found at* http://ocid.nacse.org/qml/research/tfdd/toTFDDdocs/328ENG.htm (*last visited* May 2003).

Convention o Cooperation for the Protection and Sustainable Use of the Danube, signed in Sofia, on June 29th, 1994, *found at* http://ocid.nacse.org/qml/research/tfdd/toTFDDdocs/331ENG.htm (*last visited* May 2003).

Draft Agreement on Water Quality Management of Zapadnaya Dvina/Daugava River Basin. First Draft Version Presented at the Seminar in Novopolotsk 11-12 November 1997, between the Federation of Russia, the Republic of Latvia, and the Republic of Belarus (not known whether this Agreement is in force or not), *found at* http://ocid.nacse.org/qml/research/tfdd/toTFDDdocs/349ENG.htm (*last visited* May 2003).

Convention on Environmental Impact Assessment in a Transboundary Context, Espoo, September 10 1997, *found at* http://ocid.nacse.org/qml/research/tfdd/toTFDDdocs/339ENG.htm (*last visited* May 2003).

Convention on the Protection of the Rhine, Rotterdam, January 22, 1998 between the Governments of the Federal Republic of Germany, the French Republic, the Grand Duchy of Luxembourg, the Kingdom of the Netherlands, the Swiss Confederation and The European Union, *found at* http://ocid.nacse.org/qml/research/tfdd/toTFDDdocs/193ENG.htm (*last visited* May 2003).

Agreement between the Government of the Republic of Kazakhstan, (The Government of the Kyrgyz Republic and the Government of the Republic of Uzbekistan on Cooperation in the Area of Environment and Rational Nature Use, done in one original copy in Russian, in Bishkek on 17 March, 1998, *found at* http://ocid.nacse.org/qml/research/tfdd/toTFDDdocs/196ENG.htm (*last visited* May 2003).

Convention on Access to Information, Public Participation in Decision-Making and Access to Justice in Environmental Matters done at Aarhus, Denmark, on 25 June 1998 (Not yet in force Signatories: 40, Parties 7), *found at* http://www.mem.dk/aarhus-conference/issues/public-participation/ppartikler.htm (*last visited* May 2003).

Convention about the Cooperation for the Protection and Sustainable Use of the Waters of Portuguese-Spanish Hydrological Basins and Additional Protocol, Nov 30 1998, *found at* http://ocid.nacse.org/qml/research/tfdd/toTFDDdocs/251POR.htm (*last visited* May 2003).

UN/ECE Protocol on Water and Health to the 1992 Convention on the Protection and Use of Transboundary Watercourses and International Lakes 24 March 1999, adopted on 17 June 1999 at the Third Ministerial Conference on Environment and Health, *found at* http://www.who.dk/Document/Peh-ehp/ProtocolWater.pdf (*last visited* May 2003).

The European Union Water Framework Directive 2000/60/EC, Published in the Official Journal of the European Communities on 22 December 2000, (entered into force on that date), *found at* http://www.europa.eu.int/eur-lex (*last visited* May 2003).

1.1.3.3 United States and Mexico

Convention between the United States and Mexico Equitable Distribution of the Waters of the Rio Grande Signed at Washington, May 21, 1906, (came into force 16 January,

1907) Treaty Series 455, *found at* http://www.ibwc.state.gov/FORAFFAI/1906_convention.htm (*last visited* March 2003).

Treaty between the United States of America and Mexico relating to the Waters of the Colorado and Tijuana Rivers, and of the Rio Grande, Signed at Washington November 14, 1944, entered into force in November 8, 1945, *found at* http://www.ibwc.state.gov/FORAFFAI/body_1944_treaty.HTM (*last visited* March 2003).

Minute No. 242 of the International Boundary Water Commission (30 September 1973) of the Treaty between the United States of America and Mexico Relating to the Waters of the Colorado and Tijuana Rivers, and of the Rio Grande, 1944, 59 Stat.1219, Treaty Series No.994 United Nations Treaty Series (U.N.T.S) 313, United States Treaties and Other International Agreements, vol. 24, part 2 (1973) (Washington D.C. United States Government Printing Office, 1974, also found at: www.usembassy-mexico.gov (last visited March 2003).

Agreement on Co-Operation for the Protection and Improvement of the Environment in the Border Area Signed at La Paz, Baja California, on 14 August 1983, XXII (5) ILM 1983, *also found at* http://Www.Usembassy-Mexico.Gov/Elapaz.html (*last visited* March 2003).

1.1.3.4 Africa

Act regarding Navigation and Economic Co-operation between the States of the Niger Basin, Niamey 1963, (came into Force 1 February 1966). United Nations, Treaty Series, vol. 587, No. 8506, p. 9. Also the FAO Treaty Collection No. 2, *found at* http://www.fao.org/docrep/W7414B/w7414b04.htm#TopOfPage (*last visited* May 2003), *also found at* http://ocid.nacse.org/qml/research/tfdd/toTFDDdocs/120ENG.htm (*last visited* June 2003).

Agreement Concerning the River Niger Commission and the Navigation and Transport on the River Niger, done At Niamey, On 25 November 1964, between Benin, Cameroon, Chad, Ivory Coast, Guinea, Mali, Niger, Nigeria, Burkina Faso, (came into force 12 April 1966). United Nations, Treaty Series, vol. 587, No. 8507, p. 19. FAO Treaty Collection No. 4, *found at* http://www.fao.org/docrep/W7414B/w7414b06.htm#TopOfPage (*last visited* May 2003).

Convention Creating the Niger Basin Authority. 21 January 1980 between Benin, Cameroon, Chad, Ivory Coast, Guinea, Mali, Niger, Nigeria, Upper Volta, *found at* http://ocid.nacse.org/qml/research/tfdd/toTFDDdocs/374ENG.htm (*last visited* May 2003).

The Revised Convention Creating the Niger Basin Authority Done in D'jamena 1987 between Benin, Cameroon, Chad, Ivory Coast, Guinea, Mali, Niger, Nigeria, Upper Volta. FAO Treaty Collection No. 12 *found at* http://www.fao.org/docrep/W7414B/w7414b0e.htm#TopOfPage (*last visited* May 2003) (In French).

Convention and Statutes Relating to the Development of the Chad Basin. Signed at Fort Lamy, On 22 May 1964 between Cameroon, Chad, Niger, Nigeria, Journal Officiel de la République Fédérale du Cameroun, p. 1003 (came into force on 15 September

1964), FAO Treaty No. 3, *found at* http://www.fao.org/docrep/W7414B/w7414b05.htm#TopOfPage (*last visited* May 2003).

African Convention on the Conservation of Nature and Natural Resources (1968), (came into force on 16 June 1969), FAO Treaty No. 1, *found at* http://www.fao.org/docrep/W7414B/w7414b03.htm#TopOfPage (*last visited* May 2003).

Agreement for the Establishment of the Organization for Management and Development of the Kagera River Basin, concluded at Rusomo, Rwanda 24 August 1977. between Tanzania, Rwanda, Burundi, (came into force on 5 February 1978), United Nations Treaty Series No. 16695, Vol. 1089, pp 172-177, FAO Treaty No. 8 *found at* the http://www.fao.org/docrep/W7414B/w7414b0a.htm#TopOfPage (*last visited* May 2003), (in French), and the TFDD Treaty Collection *found at* http://ocid.nacse.org/qml/research/tfdd/toTFDDdocs/200ENG.pdf (*last visited* May 2003).

Convention Relating to the Status of the River Gambia signed at Kaolack, on 30 June 1978 between the Republic of the Gambia, the Republic of Senegal, and Guinea), FAO Treaty No. 9 *found at* http://www.fao.org/docrep/W7414B/w7414b0b.htm#TopOfPage (*last visited* May 2003).

The Convention Relating to the Creation of the Gambia River Basin Development Organization Signed at Kaolack on 30 June 1978, FAO Treaty Collection No. 10 *found at* http://www.fao.org/docrep/W7414B/w7414b0c.htm#TopOfPage (*last visited* May 2003). and the TFDD Treaty Collection http://ocid.nacse.org/qml/research/tfdd/toTFDDdocs/160ENG.htm (*last visited* May 2003).

Agreement Between the Federal Republic of Nigeria and the Republic of Niger Concerning the Equitable Sharing in the Development, Conservation and Use of Their Common Water Resources Done at Maiduguri, 18 July 1990, found at the FAO Treaty No. 34, *found at* http://www.fao.org/docrep/W7414B/w7414b10.htm#TopOfPage (*last visited* May 2003) and the TFDD Treaty Collection http://ocid.nacse.org/qml/research/tfdd/toTFDDdocs/182ENG.htm (*last visited* May 2003).

Tripartite Ministerial Meeting Of Ministers Responsible For Water Affairs Held On The 15th February, 1991 In Swaziland, *found at* the TFDD Treaty Collection http://ocid.nacse.org/qml/research/tfdd/toTFDDdocs/234ENG.htm (*last visited* May 2003).

Agreement on the Development and Utilisation of the Water Resources of the Komati River Basin Between the Government of the Republic of South Africa and the Government of Kangwane, 1992, *found at* the TFDD treaty collection http://ocid.nacse.org/qml/research/tfdd/toTFDDdocs/206ENG.htm (*last visited* May 2003).

Treaty on the Development and Utilisation of the Water Resources of the Komati River Basin between the Government of the Kingdom of Swaziland and the Government of the Republic of South Africa, signed on 13 March 1992, FAO Treaty Collection No. 38 *found at* http://www.fao.org/docrep/W7414B/w7414b14.htm#TopOfPage (*last visited* May 2003).

Treaty on the Establishment and Functioning of the Joint Water Commission Between the Government of the Kingdom of Swaziland and the Government of the Republic of South Africa, signed on 13 March 1992, *found at* the TFDD treaty collection http://ocid.nacse.org/qml/research/tfdd/toTFDDdocs/185ENG.htm (*last visited* May 2003).

Agreement between the Government of the Republic of Namibia and the Government of the Republic of South Africa on the Establishment of a Permanent Water Commission, signed at Noordoewer on 14 September 1992, *found at* the TFDD treaty collection http://ocid.nacse.org/qml/research/tfdd/toTFDDdocs/186ENG.htm (*last visited* May 2003), also FAO Treaty No. 32 *found at* http://www.fao.org/docrep/W7414B/w7414b0y.htm#TopOfPage (*last visited* May 2003).

Agreement between the Governments of the Republic of Angola, the Republic of Botswana, and the Republic of Namibia on the Establishment of a Permanent Okavango River Basin Water Commission (OKACOM), done at Windhoek on 16 September 1994, (came into force on 15th September 1994), *found at* the TFDD treaty collection http://ocid.nacse.org/qml/research/tfdd/toTFDDdocs/190ENG.htm (*last visited* May 2003), also FAO Treaty No. 20 http://www.fao.org/docrep/W7414B/w7414b0m.htm#TopOfPage (*last visited* May 2003).

Protocol on shared watercourse systems in the Southern African Development Community (SADC) Region, signed at Johannesburg, 28 August 1995, between the Republic of Angola, the Republic of Botswana, the Kingdom of Lesotho, the Republic of Malawi, the Republic of Mozambique, the Republic of Namibia, the Republic of South Africa, the Kingdom of Swaziland, the United Republic of Tanzania, the Republic of Zambia and the Republic of Zimbabwe, (not in force), FAO Treaty No. 21 *found at* http://www.fao.org/docrep/W7414B/w7414b0o.htm#TopOfPage (*last visited* May 2003), and TFDD Treaty Collection http://ocid.nacse.org/qml/research/tfdd/toTFDDdocs/205ENG.htm (last visited May 2003).

Revised Protocol on Shared Watercourses in the Southern African Development Community (SADC) Johannesburg 2000. entered into force on the 7 Sep. 2000 (Signatories: the Republic of Angola, the Republic of Botswana, the Kingdom of Lesotho, the Republic of Malawi, the Republic of Mauritius, the Republic of Mozambique, the Republic of Seychelles, the Republic of South Africa, the Kingdom of Swaziland, the Republic of Tanzania, the Republic of Zambia and the Republic of Zimbabwe), *found at* the TFDD treaty collection http://ocid.nacse.org/qml/research/tfdd/toTFDDdocs/ENG.htm (*last visited* May 2003).

1.2 Decisions of Courts and Tribunals

Lake Lannoux Arbitration (Fr. v Spain) (1957) 24 Int'l L. Rpt. 101

Trail Smelter Case, awards of 16 April 1938 and 11 March 1941, 3 RIAA 1905-1937; 1938-1982.

North Sea Continental Cases, International Court of Justice Reports 1969, 3-257.

Württemberg and Prussia v. Baden, (18 June 1927), 4 ADPILC 1927-1928, 128-133; 116

Entscheidungen des Reichsgerichts in Zivilsachen 1927, 18-45.

Case Concerning the Gabcikovo-Nagymaros Project (Hungary v. Slovakia) 1997 I.C.J. 7(25 September), *found at* http://www.icj-cij.org/icjwww/idocket/ihs/ihsjudgement/ihs_ijudgment_970925_frame.htm (*last visited* August 2002).

Factory at Chorzów 1928 P.C.I.J. (Ser. A) No. 17.

1.3 Declarations and Statements

Universal Declaration of Human Rights, G.A. Res. 217A(III), U.N. GAOR, U.N. Doc. A/810 (1948), *found at* https://www1.umn.edu/humanrts/instree/b1udhr.htm, (*last visited* June 2003).

Resolutions adopted by the Institute of International Law at its Session at Salzburg (4-13 September 1961), on the Utilization of Non-maritime International Waters, 11 September 1961, Annuaire de l'Institute de Droit International, Volume 49, Tome II, 381 (1961).

Declaration of the United Nations Conference on the Human Environment, June 16 1972, *found at* http://www.unep.org/Documents/Default.asp?DocumentID=97&ArticleID=1503 (*last visited* May 2003).

Declaration on the Right to Development, G.A. Res. 41/128, U.N. GAOR, 41st Sess., 97th plen. mtg., U.N. Doc. A/Res/41/128 (1986), reprinted in Raushcing, D., et al., Key Resolutions of the United Nations General Assembly 1946-1996, 241 (Cambridge University Press, Cambridge 1997)

The Dublin Statement on Water and Sustainable Development, 1992, *found at* http://www.gdrc.org/uem/water/dublin-statement.html (*last visited* April 2003).

Rio Declaration on the Environment and Development (3-14 June 1992) UN Doc. A/CONF.151/26 United Nations, *also found at* http://www.unep.org/unep/rio.htm (*last visited* April 2003).

Agenda 21: Chapter 18: Protection of the Quality of and Supply of Freshwater resources: Application of Integrated Approaches to the Development, Management and Use of Water Resources. -United Nations Conference on Environmental Development, 1992, *found at* http://www.unep.org/unep/agenda21.htm (*last visited* August 2002).

Vienna Declaration and Programme of Action, World Conference on Human Rights, Vienna, Austria, June 14-25, 1993, UN Doc. A/Conf.157/23, *found at* http://habitat.igc.org/undocs/vienna.html, (*last visited* June 2003).

Ministerial Declaration on Water and Sustainable Development, Paris, France, Mar. 21, 1998, *found at* http://www.iisd.ca/sd/frh2o.html , (*last visited* May 2003).

The Ministerial Declaration of The Hague on Water Security in the 21st Century, the Hague 22 March 2000, *found at* http://www.idi.or.jp/vision/declarat.htm (*last visited* May 2003).

Declaration of African Ministers responsible for Water Resources at the Bonn Conference on Water – A Key to Sustainable Development, Bonn, Germany, Dec. 5, 2001, *found at* www.thewaterpage.com/Documents/bonn_africa.doc, (*last visited* July 2003)

Johannesburg Declaration of the World Summit on Sustainable Development, Report of World Summit on Sustainable Development, Johannesburg, South Africa, Aug. 26 – Sep. 4, 2002, at 1, U.N. Doc. A/Conf.199/20 (2002), *found at* http://www.johannesburgsummit.org/html/documents/documents.html, (*last visited* June 2003).

Ministerial Declaration of the 3rd World Water Forum, Kyoto, Japan, Mar. 23, 2003, found at http://www.world.water-forum3.com/jp/mc/md_final.pdf, (*last visited* June 2003).

1.4 United Nations Resolutions and Reports

GA Res. 49/52, UN Doc. A/RES/49/52.

GA Res. 3281(xxix), UN GAOR, 29th Session, Supp. No. 31 (1974) *found at* http://www.dal.ca/~wwwlaw/kindred.intllaw/EcRtsandDuties.htm (*last visited* August 2003).

A/RES/46/59, 9 December 1991, found at http://www.un.org/documents/ga/res/46/a46r059.htm (*last visited* May 2003).

United Nations General Assembly Resolutions 3175 (XXVIII) of 17 December 1973, 3336 (XXIX) of 17 December 1974, 3516 (XXX) of 15 December 1975, 31/186 of 21 December 1976, 32/161 of 19 December 1977, 34/136 of 14 December 1979, 35/110 of 5 December 1980 and 36/173 of 17 December 1981, 37/135 of 1982, 38/144 of 1983, 51/190 of 16 December 1996 and 52/207, December 18, 1997.

Security Council S/RES/242, 22 November 1967, found at http://domino.un.org/UNISPAL.NSF/9a798adbf322aff38525617b006d88d7/59210ce6d04aef61852560c3005da209?OpenDocument&Highlight=2,242 (*last visited* March 2003).

United Nations General Assembly Resolution 43/177, *found at* http://www.danielpipes.org/article/185 (*last visited* March 2003).

United Nations General Assembly A/Res/181(Ii) (A+B), 29 November 1947, *found at* http://domino.un.org/UNISPAL.NSF/9a798adbf322aff38525617b006d88d7/7f0af2bd897689b785256c330061d253!OpenDocument (*last visited* March 2003).

Resolution on Confined Transboundary Groundwater of 1994 (Doc. A/51/275 of 6 August 1996, Doc. A/51/275/Add.1 of 12 September 1996, Doc. A/51/257/Corr.1 of 4 October 1996, Doc. A/51/275/Add.2 of 14 October 1996, Doc. A/51/275/Add.3 of 14 October 1996).

1.5 Reports of the ILC on the Non-Navigational Uses of International Water Courses

1.5.1 Studies Conducted by the Secretariat and Reports by the Secretary General

23rd session of the International Law Commission (1971), Note by the Secretariat, U.N. Doc. A/CN.4/244/Rev.1 (in Ybk, vII(2)).

25th session of the International Law Commission (1973), Supplementary report on the legal problems relating to the non-navigational uses of international watercourses requested by the General Assembly in resolution 2669 (XXV) - Advance report submitted by the Secretary-General pursuant to General Assembly resolution 2926 (XXVII), UN Doc. A/CN.4/270 (in Ybk, 1973, vII), ILC Report, A/9090/Rev.1 (A/28/10), 1973, chp. VI(A)(2), paras.146-150.

26th session of the International Law Commission (1974), Legal Problems relating to the utilization and use of international rivers - Report by the Secretary-General, UN Doc. A/5409 (in Ybk, 1974, vII(2)).

Legislative texts and treaty provisions concerning the utilization of international rivers for other purposes than navigation (UN Legislative series, vol.12), ST/LEG/SER.B/12 (United Nations Publication, Sales No. 63.V.4).

Legal problems relating to the non-navigational uses of international watercourses: supplementary report by the Secretary-General, UN Doc. A/CN.4/274 (in Ybk, 1974, vII(2)).

1.5.2 Reports of the Special Rapporteurs

First report of the Special Rapporteur, Mr. Richard D. Kearney (28th session of the ILC (1976)), UN Doc. A/CN.4/295 (in Ybk, 1976, vII(1)),

ILC Report, A/31/10, 1976, chp. V, paras.139-166.

First Report of the Special Rapporteur, Mr. Stephen Schwebel (31st session of the ILC (1979)),UN Doc. A/CN.4/320 and Corr.1 (in Ybk, 1979, vII(1)), ILC Report, A/34/10, 1979, chp. V(A), par. 109.

Second report of the Special Rapporteur, Mr. Stephen Schwebel (32nd session of the ILC (1980)), UN Doc. A/CN.4/332 and Corr.1 and Add.1 (in Ybk, 1980, vII(1)), ILC Report, A/35/10, 1980, chp. V(A)(1), par. 85.

Third report of the Special Rapporteur, Mr. Stephen Schwebel (34th session of the ILC (1982)), UN Doc. A/CN.4/348 and Corr.1 (in Ybk, 1982, vII(1)), ILC Report, A/37/10, 1982, chp. VII(A), par. 251.

First report of the Special Rapporteur, Mr. Jens Evensen (35th session of the ILC (1983)) UN Doc. A/CN.4/367 and Corr.1 (in Ybk, 1983, vII(1)), ILC Report, A/38/10, 1983, chp. VI(B), paras.215-260.

Second Report of the Special Rapporteur, Mr. Jens Evensen (36th session of the ILC (1984)), UN Doc. A/CN.4/381 and Corr.1 and 2 (French only) (in Ybk, 1984, vII (1), ILC Report, A/39/10, 1984, chp. VI(B), paras. 279-343

Preliminary Report of the Special Rapporteur, Mr. Stephen C. McCaffrey (37th session of the ILC (1985)), UN Doc. A/CN.4/393 (in Ybk, 1985, VII(1))

ILC Report, A/40/10, 1985, chp. VII(B), paras.279-90.

Second Report of the Special Rapporteur, Mr. Stephen C. McCaffrey (38th session of the ILC (1986)), UN Doc. A/CN.4/399 and Add.1-2 (in Ybk, 1986, vII(1)), ILC Report, A/41/10, 1986, chp. VII(B), paras.231-243.

Third Report of the Special Rapporteur, Mr. Stephen C. McCaffrey (39th session of the ILC (1987)), UN Doc. A/CN.4/406 and Corr.1, Add.1 and Add.1/Corr.1, Add.2 and Add.2/Corr.1 (in Ybk, 1987, vII(1)), ILC Report, A/42/10, 1987, chp. III(B), paras. 84-116.

Fourth Report of the Special Rapporteur, Mr. Stephen C. McCaffrey (40th session of the ILC (1988)), UN Doc. A/CN.4/412 and Add.1 and Add./Corr.1, and Add.2 and Add.2/Corr.1 to 3 (in Ybk, 1988, vII(1)) , ILC Report, A/43/10, 1988, chp. III(B), paras. 118-188.

Fifth Report of the Special Rapporteur, Mr. Stephen C. McCaffrey (41st session of the ILC (1989)), UN Doc. A/CN.4/421 and Corr.1-4, and Add.1-2 (in Ybk, 1989, vII(1))

ILC Report, A/44/10 (1989) chp. VII(B), paras.632-683.

Sixth Report of the Special Rapporteur, Mr. Stephen C. McCaffrey (42nd session of the ILC (1990)), ILC Report, A/CN.4/427 and Corr. 1, and Add.1 (in Ybk, 1990, vII(1)), ILC Report, A/45/10, 1990, chp. IV(B), paras.251-311.

Seventh Report of the Special Rapporteur, Mr. Stephen C. McCaffrey (43rd session of the ILC (1991)), UN Doc. A/CN.4/436 and Corr.1-3 (in Ybk, 1991, vII(1)), ILC Report, A/46/10, 1991, chp. III(B), paras.35-56.

First Report of the Special Rapporteur, Mr. Robert Rosenstock (45th session of the ILC (1993)) UN Doc. A/CN.4/451, ILC Report, A/48/10, 1993, chp. V(B), paras.342-416.

1.5.3 Comments by the Governments

28th session of the International Law Commission (1976),Replies of Governments to the Commission's questionnaire UN Doc. A/CN.4/294 and Add.1 (in Ybk, 1976, vII(1)).

30th session of the International Law Commission (1978), Replies of Governments to the Commission's questionnaire, UN Doc. A/CN.4/314 (In Ybk, 1978, vII(1)).

31st session of the International Law Commission (1979), Replies of Governments to the Commission's questionnaire, UN Doc. A/CN.4/324 (in Ybk, 1979, vII(1)).

32nd session of the International Law Commission (1980), Replies of Governments to the Commission's questionnaire, UN Doc. A/CN.4/329 and Add.1 (in Ybk, 1980, vII(1))

34th session of the International Law Commission (1982)

Replies of Governments to the Commission's questionnaire, UN Doc. A/CN.4/352 and Add.1 (in Ybk , 1982, vII(1)).

45th session of the International Law Commission (1993), Comments and observations received from Governments on the draft articles, UN Doc. A/CN.4/447 and Add.1-5.

1.5.4 Reports of the ILC

Report of the International Law Commission on the work of its twenty-third session, 26 April to 30 June 1971, (A/26/10), 1971, chp. V(A), paras.119-122,

Ybk, 1971, vII(1).

Report of the International Law Commission on the work of its twenty-fourth session, 2 May to 7 July 1972 , ILC Report, A/8710/Rev.1 (A/27/10), 1972, chp. V(A), par.77,Ybk, 1972, vII.

Report of the International Law Commission on the work of its twenty-fifth session, 7 May to 13 July 1973, ILC Report, A/9090/Rev.1 (A/28/10), 1973, chp. VI(A)(2), paras.144-150, Ybk, 1973, vII.

Report of the International Law Commission on the work of its twenty-seventh session, 5 May to 25 July 1975, ILC Report, A/10010/Rev.1 (A/30/10), 1975, chp. VI(A), par.138 ,Ybk, 1975, vII.

Report of the International Law Commission on the work of its twenty-eighth session, 3 May to 23 July 1976, ILC Report, A/31/10, 1976, chp. V, paras.106-166 , Ybk, 1976, vII(2).

Report of the International Law Commission on the work of its twenty-ninth session, 9 May to 29 July 1977, ILC Report, A/32/10, 1977, chp. V(B), par.79 - Ybk, 1977, vII(2).

Report of the International Law Commission on the work of its thirtieth session, 8 May to 28 July 1978, ILC Report, A/33/10, 1978, chp. VIII(A), paras. 157-160, Ybk, 1978, vII(2).

Report of the International Law Commission on the work of its thirty-first session, 14 May to 3 August 1979, ILC Report, A/34/10, 1979, chp. V, paras. 86-148 , Ybk, 1979, vII(2).

Report of the International Law Commission on the work of its thirty-second session, 5 May to 25 July 1980, ILC Report, A/35/10, 1980, chp. V, paras. 59-98,Ybk, 1980, vII(2).
Report of the International Law Commission on the work of its thirty-fourth session, 3 May to 23 July 1982, ILC Report, A/37/10, 1982, chp. VII(A), paras.250-251, Ybk, 1982, vII(1).
Report of the International Law Commission on the work of its thirty-fifth session, 3 May to 22 July 1983, ILC Report, A/38/10, 1983, chp. VI, paras.191-260, k, 1983, vII(2),Ybk, 1983, vI.
Report of the International Law Commission on the work of its thirty-sixth session, 7 May to 27 July 1984, ILC Report, A/39/10, chp. VI, paras.258-343, Ybk, 1984, vII(2).
Report of the International Law Commission on the work of its thirty-seventh session, 6 May to 26 July 1985, ILC Report, A/40/10, 1985, chp. VII, paras.268-290, Ybk, 1989, vII(2).
Report of the International Law Commission on the work of its thirty-eighth session, 5 May to 11 July 1986, ILC Report, A/41/10, 1986, chp. VII, paras.220-243, Ybk, 1986, vII(2).
Report of the International Law Commission on the work of its thirty-ninth session, 4 May to 27 July 1987, ILC Report, A/42/10, 1987, chp. III, paras.68-118, Ybk, 1987, vII(2).
Report of the International Law Commission on the work of its fortieth session, 9 May to 29 July 1988, ILC Report, A/43/10, 1988, chp.III, paras.103-91, Ybk, 1988, vII(2).
Report of the International Law Commission on the work of its forty-first session, 2 May to 21 July 1989 , ILC Report, A/44/10, 1989, chp. VII, paras.614-685, Ybk, 1989, vII(2).
Report of the International Law Commission on the work of its forty-second session, 1 May to 20 July 1990, ILC Report, A/45/10, 1990, chp. IV, paras.244-313, Ybk, 1990, vII(2).
Report of the International Law Commission on the work of its forty-third session, (1991 draft Articles) 29 April to 19 July 1991, ILC Report, A/46/10, 1991, chp. III, paras.29-59, Ybk, 1991, vII(2).
Report of the International Law Commission on the work of its forty-fourth session, 4 May to 24 July 1992, ILC Report, A/47/10, 1992, chp. V(A), paras.350-352, Ybk, 1992, vII(2).
Report of the International Law Commission on the work of its forty-fifth session, 3 May to 23 July 1993, ILC Report, A/48/10, 1993, chp. V, paras.336-416, Ybk, 1993, vII(2).
Report of the International Law Commission on the work of its forty-sixth session, 2 May to 22 July 1994,(draft articles on the topic and a draft resolution on transboundary confined groundwater), ILC Report, A/49/10, 1994, chp. III, paras. 210-221, Ybk, 1994, vII(2).
UN General Assembly International Law Commission Fifty Fifth Session, Shared Natural Resources: First Report on Outlines by the Special Rapporteur Chusei Yamada, UN Doc. A/CN.4/533, May 2003.

1.5.5 General Assembly Resolutions

Res. 2669 (XXV) of 8 December 1970
Res. 2780 (XXVI) of 3 December 1971
Res. 3071 (XXVIII) of 30 November 1973
Res. 3315 (XXIX) of 14 December 1974
Res. 31/97 of 15 December 1976
Res. 33/139 of 19 December 1978
Res. 34/141 of 17 December 1979
Res. 35/163 of 15 December 1980
Res. 46/55 of 9 December 1991
Res.47/33 of 25 November 1992
Res.48/31 of 9 December 1993

Res.49/51 of 9 December 1994
Res.49/52 of 9 December 1994
Res.51/206 of 17 December 1996

1.6 Report of the Sixth Committee

Report of the Sixth Committee convening as the working group of the whole, First Session, UN Doc.A/51/624, 1994.

Report of the Sixth Committee convening as the working group of the whole, Second Session, UN Doc.A/51/869 (1997).

Proposals for States within the Working Group, UN Doc. A/C.6/51/NUW/WG/CRP.

1.6.1 Other United Nations Reports

General Assembly Economic and Social Council, Document A/55/84 E/2000/16, 14 June 2000.

Groundwater storage and Artificial Recharge, Natural Resources/Water Series No. 2 (United Nations publication, Sales No. E.74.IIA.11), p.1.

United Nations General Assembly, Social and Economic Council, Documents A/40/381, June 1985, and A/55/84, June 2000.

UN Natural Resources/ Water Series, No. 1, The Management of International Water Resources: Institutional and Legal Aspects, Sales No. E.17.II.A.2 1975.

UNECE Task Force on Monitoring and Assessment, under the Convention on the Protection and Use of Transboundary Watercourses and International Lakes (Helsinki 1992), Inventory of Transboundary Groundwaters , Vol. 1, ISBN 9036952743. Lelystad, 1991.

UNECE Task Force on Monitoring and Assessment, under the Convention on the Protection and Use of Transboundary Watercourses and International Lakes (Helsinki 1992), Problem-Oriented approach and the Use of Indicators, Vol. 2, ISBN 9036952751, Lelystad, 1991.

UNECE Task Force on Monitoring and Assessment, under the Convention on the Protection and Use of Transboundary Watercourses and International Lakes (Helsinki 1992), Application of Models, Vol. 3, ISBN 90369527X, Lelystad, 1991.

UNECE Task Force on Monitoring and Assessment, under the Convention on the Protection and Use of Transboundary Watercourses and International Lakes (Helsinki 1992), State of the Art on Monitoring and Assessment of Groundwaters, Vol. 4, ISBN 9036952778, Lelystad, 1991.

WCED, Our common future: the Bruntland Report, (Oxford University Press from the World Commission on Environment and Development: New York 1987).

The Helsinki Rules on the Use of the Waters International Rivers, Report of the Forty-sixth Conference of he international Law Association held at Helsinki, 1966, at 484-532 (1967).

1.7 Other Reports

1.7.1 General

ILA, Helsinki Rules Commentary , 1967 in Garretson, Hayton and Olmstead; Oceana Publications **779.**

Wouters, P.K. (1997) Rivers of the World: The Fundamental Principles of International Watercourse Law, unpublished PhD thesis.

International Law Association (1986) Report of the Sixty second Conference, 238.

Resolutions Adopted by the Institute of International Law (IDI), (1961) at its Session at Salzburg (4-13).

International Association of the 52nd Conference (1967) **xi**, 447-533.

UNESCO and WMO (2000) Report on the Proposal for the Establishment of an International Groundwater Resources Assessment Centre, IHP/IC-XIV/Inf.26.

The World Conservation Union- IUCN (2000) Vision For Water And Nature: A World Strategy For Conservation and Sustainable Management of Water Resources in The 21st Century.

1.7.2 Other Reports-Middle East

Rofe and Raffety Consulting Engineers, Nablus and Jerusalem Districts: Water Resources Survey, Hashemite Kingdom of Jordan, Central Water Authority, Jordan (1965).

Minutes of Meetings of the Joint Israeli Palestinian Water Committee, Neve Ellan Covering period from November 1995-January 1999.

Applied Research Institute, The Environmental Profile, Palestine, Bethlehem 1996.

Palestinian Central Bureau of Statistics, Population Census 1997.

Statistical Abstract of Israel, No. 49 for 1998.

UNDP, Water Resources Action Programme Study on Water Resources, 1993.

Middle East Water Supply and Demand Development Study, GTZ 1998.

Daibes, F., Summary on Water Resources of Palestine, unpublished paper prepared for the Palestinian Water Authority, Ramallah 1995.

Daibes, F.M., KaR, Palestine Case Study Data Collection Reports and Synthesis, IWLRI, March 2003, available at http://www.dundee.ac.uk/law/water (last visited June 2003).

Israeli Ministry of Agriculture, Hydrological Services Annual Report, 1998.

Reference Framework for the Palestinian Position on the Water Issue, Official Reports by the Palestinian Water Authority (April 16, 2000).

Alster, J., Water Laws, Rules and Regulation: Israel, Water Supply, Economics and Pricing.

Trolldallen J.(ed.),Water Laws, Water Institutions and Water Supply Economics: in the Hashemiste Kingdom of Jordan, the Palestinian Authority and the State of Israel (CESAR: Oslo, 1997).

The Palestinian-Israeli Peace Agreement: a Documentary Record (Institute for Palestine Studies, Washington, D.C. 1994).

1.8 National Regulation

The Jordanian Law No. 40 of 1952 on Land and Water settlement (on file with the Author).

The Jordanian Law No. 31 of 1953 on Water Control (on file with the Author)

The Jordanian Law No. 29 of 1955 on the Municipalities (on file with the Author)

The Jordanian Law No. 12 of 1956 on the Water Installation of Hebron (on file with the Author)
The Jordanian Law No. 51 of 1959 on the Settlement of Water Resources (on file with the Author)
Law No. 37 of 1966 on the Organization of Natural Resources (on file with the author).
The Jordanian Law No. 9 of 1966 on the Organization of matters of drinking water in Jerusalem District (on file with the Author)
The Jordanian Law No. 26 for 1968
The Jordanian Water Law No. 18 of 1988 (on file with the author).
The African Water Act of 1998, found at
www.dundee.ac.uk/law/iwrli/Research_Document_National.html (last visited May 2003).
The 1996 Albanian Water Law, found at
www.dundee.ac.uk/law/iwrli/Research_Document_National.html (last visited May 2003).
The Lithuania Lithuanian Law on Water Resources of 1997, found at
www.dundee.ac.uk/law/iwrli/Research_Document_National.html (last visited May 2003).
Oman, Directive No. 45/75 of 1975, establishing the Water Resources Wealth Council
Oman, Directive No. 76/77 of 1977, on Water Resources Development
Oman, Directive No. 63/79 of 1979, concerning the Establishment of a Public Authority for Water Resources.
Oman, the Sultan Decree No. 100/89 of 1989 establishing the Ministry of Water Resources.
Oman, Ministerial Decision No. 2/90 of 1990 regulating the registry of existing wells and regulating the licensing of new wells.
The African Water Act of 1998, found at
www.dundee.ac.uk/law/iwrli/Research_Document_National.html (last visited May 2003).
Survey and Compilation of Normative Instruments: Applied and Applicable Legislation, West Bank and Gaza, CESAR Institute, Oslo, 1997.
The Palestinian By-Lay No. 2 for 1996 concerning the establishment of the Palestinian Water Authority, (on file with the author).
The Palestinian Water Law No.2/2003 (on file with the Author).
The Israeli Water Law 571/1959. (on file with the Author).
Israeli Military Order 92 concerning Jurisdiction Over Water Regulations, found in Israeli Military Orders in the Occupied Territories 1967-1992, (Jerusalem Media and Communication Centre (2nd ed.): Jerusalem, 1995).
Israeli Military Order 158 Concerning the Amendment to Supervision Over Water Law , 19 November 1967 (Amendment to the Jordanian Law No. 31 of 1953) , found in Israeli Military Orders in the Occupied Territories 1967-1992, (Jerusalem Media and Communication Centre (2nd ed.): Jerusalem, 1995).
Israeli Military Order 291 Concerning Settlement of Disputes Over Land and Water 19 December 1968 (Amendment to the Land and Water Regulation Law No. 40 of 1952), found in Israeli Military Orders in the Occupied Territories 1967-1992, (Jerusalem Media and Communication Centre (2nd ed.): Jerusalem, 1995).

2. SECONDARY SOURCES

2.1 Books

Allan, J.A. (ed.), Water, Peace and the Middle East: Negotiating Resources in the Jordan River Basin, (Tauris Academic Studies: New York, 1996).

Biswas, A. (ed.) International Waters of the Middle East: From Euphrates-Tigris to Nile, (Oxford University Press: New York, 1994).

Birnie, P. and A. Boyle, International Law And the Environment, (Oxford University Press New York, 2002).

Bogdanovic, S., International Law of Water Resources: Contribution of the International Law Association1954-2000, (Kluwer Law International : Dordrecht, 2001).

Bourne, C.B., International Water Law. Selected Writings of Professor Charles B. Bourne (ed. P.K.Wouters) (Kluwer Law International: Dordrecht, 1997).

Brownlie, I, Principles of Public International Law, (5th ed.) (Oxford University Press: New York, 1998).

Brierly, J.L., The Law of Nations (6th ed. / ed. by Waldock, H.), (Oxford, Clarendon Press, 1967).

Briggs, H.W., The Law of Nations (2nd ed.), (Stevens and Sons: London, 1953).

Bulloch, J. & Darwish, A., Water Wars. Coming Conflicts in The Middle East, (Victor Gollancz: London, 1994).

Caponera, D., Principles of Water Resources Administration: National and International (Balkema: Rotterdam, 1992).

Fetter, C.W. Applied Hydrogeology , (3rd ed.) (Prentice Hall: New Jersey, 1994).

Hadawi, S., Bitter Harvest: A Modern History of Palestine, (Interlink Pub Group: New York, 1998).

Kally, E. & Fischelson, G., Water and Peace. Water Resources and the Arab-Israeli Peace Process, (Praeger: Westport, Connecticut, 1993).

Karmon, Y. Israel: Regional Geography, (Wiley-Enterprise, 1971).

Lowi, M.R., Water and Power: The Politics of a Scarce Resource in the Jordan River Basin, (Cambridge University Press: Cambridge, 1993).

Malanczuk, P., Akehurst's Modern Introduction to International Law, (Oxford University Press: New York, 1999).

Manner, E.J., & Metsalampi, V. (ed.), The Work of the International Law Association on the Law of International Water Resources, (Finnish Branch of the International Law Association: Helsinki,, 1988) [ILA REDBOOK].

McCaffrey, S., The Law of International Watercourses (Oxford University Press: USA, 2001).

Naff, T. and Matson, R. Water in the Middle East: Conflict or Cooperation, (Middle East Research Institute: Pennsylvania, 1984).

Plant G., (ed.), Environmental Protection and the Law of War, (Belhaven Press: New York, 1992).

Price, M., Introducing Groundwater (Champen and Hall: London, 1996).

Roberts, A., Documents on the Law of War, (Oxford 1989).

Sherk, G.W., Dividing the Water: The Resolution of Interstate Conflicts in the United States (Kluwer Law International, the Netherlands, 2000).

Shaw, E, Hydrology in Practice, (Chapman and Hall: London, 1994).

Smith, B.C., State Responsibility and the Marine Environment: The Rules of Decision, (Clarendon Press: Oxford, 1988).
Smith, H.A., The Economic Uses of International Rivers, (King & Son Ltd.: London, 1931).
Smith, S.C., & Castle, E.N. (ed.), Economics and Public Policy in Water Resources Development, (Iowa State University Press: Ames, 1964).
Smith, Z.A. (ed.), Water and the Future of the Southwest, (University of New Mexico Press: New Mexico, 1989).
Tanzi, A. and Acrari, M, The United Nations Convention on the Law of International Watercourses, (Kluwer Law International: the Netherlands, 2001).
Teclaff, L.A., The River Basin in History and Law, (Martinus Nijhoff: The Hague, 1967).
--, Abstraction and Use of Water: A Comparison of Legal Regimes, (United Nations: New York, 1972).
--, International Environmental Law, (Praeger: New York, 1974).
Teclaff, L.A., (ed.), International Groundwater Law, (Oceana: New York,1981).
Teclaff, L.A., Transboundary Resources Law, (Westview Press: Boulder, 1987).
Ward, R.C., Principles of Hydrology, (McGraw-Hill Book: London, 1975).
Winpenny, J. Managing Water as an Economic Resource, (Routledge: New York, 1994).

2.2 Articles

Alemans, F. (2001) Digital Waterway Vechte, TNO report, NITG 97-283-B.
Allison, G.B., (1988) A review of some of the Physical, Chemical and Isotopic Techniques Available for Estimating Groundwater Recharge, NATO **222**, 49-67.
Barberis, J. (1991) The Development of International Transboundary Groundwater Law, Natural Resources Journal **31**, 167-186.
Bosnjakovic, B. (2000) The Regulation of International Watercourses Under the UN/ECE Regional Agreements, Water International **25**, 544-552.
Bourne, C.B. (1961) International Water Law in a World of Sovereign States: Perspectives for the Future, Conference Proceedings, Committee on the Economics of Water Resources Development of the Western Agricultural Research Council **(No. 10)** 121-125.
-- (1965) The Right to Utilize the Waters of International Rivers, CYIL **3,**187-264.
-- The Development of International Water Resources: The Drainage Basin Approach, 47 CBR 62-87, (1969).
-- (1971) International Law and Pollution of International Rivers and Lakes, UBCLR **6,**115-122.
-- (1971) Mediation, Conciliation and Adjudication in the Settlement of International Drainage Basin Disputes, CYIL **9**, 114-158.
-- (1972) Procedure in the Development of International Drainage Basins: The Duty to Consult and to Negotiate, CYIL **10,** 212-234.
-- (1972) Procedure in the Development of International Drainage Basins, UTLJ **22,** 172-206.
-- (1981) Legal Aspects of Transfrontier Pollution: Canada-United States Experience, NethILR **28,** 188-194.
-- (1992) The International Law Commission's Draft Articles on the Law of International Watercourses: Principles and Planned Measures, CoJIELP **3,** 65-92.

-- (1996) The International Law Association's Contribution to International Water Resources Law, NRJ **36**, 155-216.

-- (1997) The Primacy of the Principle of Equitable Utilization in the Watercourses Convention, CYIL.

-- (1997)The Case Concerning the Gabcikovo-Nagymaros Project: An Important Milestone in International Water Law, YBIEL **8,** 6-12.

Boyle, A.E. (1997) The Gabcikovo-Nagymaros Project: New Law in Old Bottles, YBIEL **8**, 13-20.

Burchi, S. (1999) National Regulations for Groundwater: Options, Issues and Best Practices, found in World Bank Technical Paper No. 456 p. 55 found in (S. Salman Ed., Washington, 1999).

Burke, J., Moenech, M and Sauveplane, C. (1999) Groundwater and Society: Problems in Variability and Points of Engagement found in (S. Salman Ed., World Bank Technical Paper No. 456 Washington, 1999).

Caponera D.A. and Alheritiere D. (1981) Principles for International Groundwater Law, in Teclaff, L. & Utton A. (eds) (1981) International Groundwater Law **25**.

Caponera, D. (1996) The Requirement for Drafting a modern Code and Water Legislation in the ESCWA Region, Amman, 1996, E/ESCWA/ENR/1996/WG.1/WP.1 17 November

Chenevert, D.J. (1992) Application of the Draft Articles on the Non-Navigational Uses of International Watercourses to the Water Disputes Involving the Nile River and the Jordan River, Emory Int'l L. Rev. **6**, 495- 506.

Eckstein, G., Application of International Water Law to Transboundary Groundwater Resources, and the Slovak-Hungarian Dispute Over Gabcikovo-Nagymaros, Suffolk Transnational Law Review **19,** 67 (1995).

Foster, S.S.D. and Carington, S.A. (1980) The Interpretation of Tritium in the Chalk Unsaturated Zone, J. Hydrol. **46,** 343-364.

Foster, S. (1999) Essential Concepts for Groundwater Regulators, fund in S. Salman (ed.) Groundwater: Legal and Policy Perspectives, (S. Salman Ed., World Bank Technical Paper No. 456 Washington, 1999).

Gleik, H.P., Water and Conflict, Occasional Paper Series of the Project on Environmental Change and Acute Conflict, Number 1 (1992).

Gross, D. and A. Soffer, International Groundwater and International Rules: the Case of the Middle East and the Agreement between Israel and Jordan and between Israel and the Palestinians about the Gaza Strip, Occasional Papers on the Middle East series no. 12, University of Haifa, (1996).

Hadawi, S., The Palestine Diary, Volume 1, New York 1914-1915.

Heller, M.A., and Sari N., No Trumpets No Drums: A Two State Solution of the Israeli Palestinian Conflict, 1991.

Hayton, R.D. (1982) The Groundwater Legal Regime as Instrument of Policy Objectives and Management Requirement, Institutional Alternatives for Mexico US Groundwater Management, NRJ **22**.

Hayton, R.D. (1982) The Law of International Aquifers, NRJ **22**, 71-93 .

Hayton, R.D., and A. E. Utton (1989) Transboundary Groundwaters: the Bellagio Draft Treaty, NRJ 29, *also found at* www.ana.gov.br/guarani/gestoo/tratados/The%Bellagio%20Draft%20Treaty%20lingl es.pdf (last visited June 2003).

Hays, (1991) The All American Canal Lining Project: A Catalyst for Regional and Comprehensive Groundwater Management on the United States-Mexico Border, N RJ **31**, 803.

Krishna, R., and S. Salman (1999) International Groundwater Law and the World Bank Policy for Projects on Transboundary Groundwater, found in World Bank Technical Paper No. 456 (S. Salman Ed., Washington, 1999).

Lerner, D., Issar, A. and Simmers, I. (1990) Groundwater Recharge: A Guide to Understanding and Estimating Natural Recharge, International Association of Hydrogeologists **8**.

Lowi, R.M. (1992) West Bank Water Resources and Resolution of Conflict in the Middle East, Occasional Paper Series of the Project on Environmental Change and Acute Conflict, Number 1.

Lipper, J. Equitable Utilisation, in The Law of International Drainage Basins **15**, 20.

McCaffrey, S. (1996) The Harmon Doctrine after 100 years: Buried not Praised, NRJ **36,** 540-590, and Bourne, .B. (1971) Mediation, Conciliation and Adjudication in the Settlement of International Drainage Disputes, CYIL **9**, 114-158.

-- (1983) The Work of the International Law Commission Relating to the Environment, ELQ **11**, 189-214.

-- (1973) Transboundary Pollution Injuries: Jurisdictional Considerations in Private Litigation Between Canada and the United States, CalWILJ **3**, 191-259.

-- (1975) The OECD Principles Concerning Transfrontier Pollution: Commentary, EPL **1**,2-7.

-- (1977) Pollution of Shared Natural Resources, ASILP 56-61.

-- (1983) International Environmental Law and the Work of the ILC, ASILP 414-418.

-- (1984) Codification and Progressive Development: Law and the World Environment, HarvILJ **7,** 8-15.

-- (1985) An Update on the Contributions of the International Law Commission to International Environmental Law, Environmental Law **15,** 667-678.

-- (1989) The Law of International Watercourses: Some Recent Developments and Unanswered Questions, DenvJILP **17**, 505-526.

-- (1990) The International Law Commission and its efforts to codify the International Law of Waterways, ASDI **67,** 32-55.

-- (1990)The Non-Navigational Uses of International Watercourses, ASILP 228-236.

-- (1990) Fresh Water Report, 1 YBIEL 105-109.

-- (1990)International Organizations and the Holistic Approach to Water Problems, NRJ **31,** 139-165.

- (1991)Fresh Water Report, YBIEL **2**, 117-123.

-- (1992) Fresh Water Report, YBIEL **3**, 233-240.

-- (1992) A Human Right to Water: Domestic and International Implications, GeoIELR **5**, 1-24.

-- (1992) The Law of International Watercourses: The ILC Completes its Draft Articles, EPL **22**, 66-67.

-- (1992) The International Law Commission's Draft Articles on the Law of International Watercourses: Background and Overview, CoJIELP **3**,17-29.

-- (1993) Fresh Water Report, YBIEL **4**, 121-132.

-- (1993)The Evolution of the Law of International Watercourses, AJPIL **45**, 87-111.

-- (1995) Current Developments. The International Law Commission Adopts Draft Articles on International Watercourses, AJIL **89**, 395-404.

Rosenthal A. and Kronfeld J. (1982) 234U-238U Disequilibria as an Aid to the Hydrological Study of the Judea Aquifer Group in Eastern Judea and Samaria, Journal of Hydrology **55**, 149-161.

Rodier, J.A., Aspects of Arid Zone Hydrology, Facets of Hydrology **11**, 205-246

Shuval, H.I. (2000) A Proposal For An Equitable Resolution To The Conflicts Between The Israelis And The Palestinians Over The Shared Water Resources Of The Mountain Aquifer, Arab Studies Quarterly **22**, 33.

Teclaff L.A. (1972) What You Have Always Wanted to Know About Riparian Rights But Were Afraid to Ask, NRJ **12**, 30-55.

-- (1976) Harmonizing Water Resources Development and Use with Environmental Protection in Municipal and International Law, 16 NRJ 807-86.

Teclaff, L.A. and E. Teclaff (1979) Transboundary Groundwater Pollution: Survey and Trends in Law, NRJ **19**, 628-663.

Teclaff, L.A. (1979) International Comparison of Trends in Water Resources Management, ELQ **7**, 881-915.

-- (1987) International Control of Cross-Media Pollution: An Ecosystem Approach, NRJ **27**, 21-53.

-- (1982) Principles for Transboundary Groundwater Pollution Control, NRJ **22**, 1065-1079.

-- (1996) Evolution of the River Basin Concept in National and International Law, NRJ **36**, 359-392.

Utton, A. (1982) The Development of International Groundwater Law, NRJ **22**, 95-118.

-- (1991) The Transfer of Water from an International Border Region: A Tale of Six Cities and the all American Canal, NCJInt'l L **16**.

-- (1991) Commercial Regulation 477: Crane, diminishing Water resources and International Law: A US Mexico Case Study, Cornell Int'l L **24**, 299.

-- (1982) The Development of International Groundwater Law, NRJ **22**, 95-118.

Vinogradov, S.V. (1992) The International Law Commission's Draft Articles on the Law of International Watercourses: Management and Remedies in Domestic Tribunals, CoJIELP **3**, 235-259.

-- (1990) International Environmental Security: The Concept and its Implementation, in Carter, A., and G. Danilenko (ed.), Perestroika and International Law **196**.

-- (1996) Transboundary Water Resources in the Former Soviet Union: Between Conflict and Cooperation, NRJ **36**, 393-416.

Waterbury, J., Policy and Strategic Considerations Affecting Cooperative Use of the Nile Water Basin, 5th Nile 2002 Conference, Ethiopia, February 1997.

Wolf, A. From Rights to Needs: Equitable Allocations in International Water Agreements, a paper submitted to the 5th Nile 2002 Conference, Ethiopia 1997.

Wouters, P.K. (1992) Allocation of the Non-Navigational Uses of International Watercourse : Efforts at Codification and the Experience of Canada and the United States, The Canadian Yearbook of International Law **XXX,** 43-88.

-- (2000) National and International Water Law: Achieving Equitable and Sustainable Use of Water Resources, Water International **25** (4), 499-512.

-- (2000) The Legal Response to International Water Conflicts: the UN Watercourses Convention and beyond, *found in* German Yearbook of International Law **42**, (Duncker and Humbolt, Berlin 2000).

3. OTHER

3.1 Web-based Articles

Crawford, J., The Creation Of The State Of Palestine Too Much Too Soon? http://www.ejil.org/journal/Vol1/No1/art21.html#P24_139 (Challis Professor of International Law, Dean, Faculty of Law, University of Sydney; Associé, Institut de Droit International).

Eckstein, G., Hydrologic Reality: International Water Law and Transboundary Ground-Water Resources October 1998, fund at http:// www.internationalwaterlaw.org/Articles/GlobalSouth.htm (*last visited* April 2003).

Fitzsimons, V., S. Gilroy2, and J. Cudden, Groundwater Protection Schemes and GIS – a Powerful aid in the Planning Process, found at http://www.proteus.ie/pdf/GWPS-GIS.pdf (*last visited* June 2003).

Foster, S. and Lawrence, A.R., Hazards Induced By Groundwater Recharge Under Rapid Urbanisation, B L Morris & British Geological Survey, Wallingford Ox10 8bb, UK found at http://lnweb18.worldbank.org/ESSD/essdext.nsf/18ByDocName/HazardsInducesbyGroundwaterRechargeUnderRapidUrbanisation/$FILE/HazardsInducedbyGroundwaterResourcesStatus.pdf (*last visited* March 2003).

Hammer, J., and A. Wolf, Patterns in International Water Resources Treaties, The Transboundary Fresh Water Dispute Database, found at http://terra.geo.edu/users/tfdd/documents/patterns.html (*last visited* January 2003).

Hosh, L., Criteria For The Evaluation Of Aquaculture Potentials, 1995 found at www.arij.org/pub/aquacul/reference.htm (*last visited* January 2003)

Khameisi, S., Water Legislation in Sultanate of Oman, E/ESCWA/ENR/1996/ WG.1/CP.8 (in Arabic).

Khroda, G., Strain, Social and Environmental Consequences, and Water Management in the Most Stressed Water Systems in Africa, found at http://www.idrc.ca/books/focus/804/chap7.html (*last visited* May 2003).

Koechler, H., The Palestine Problem In The Framework Of International Law

Sovereignty as the Crucial Issue of a Peaceful Settlement of the Palestinian-Israeli Conflict, Madrid 1999, found at http://www.library.cornell.edu/colldev/mideast/koechler.htm (*last visited* May 2003).

Libiszewski, S., The Environment of Conflict: Water Crises in the Jordan Basin Region, ENCOP Occasional Paper No. 13. (1995) (http://www.fsk.ethz.ch/encop/13/en13.htm).

Mar'i, M., Permanent Status Negotiations in the Palestinian - Israeli Context: A Case for International Intervention, found at http://firms.findlaw.com/MariLaw/memo9.htm (*last visited* May 2003).

Normand, R., The Final Approach to Final Status, July 7, 2000, found at http://www.merip.org/pins/pin25.html (*last visited* August 2002).

Puri, S (editor), B. Appelgren, G. Arnold, A. Aureli, Internationally Shared Aquifer Resources Their Significance and Sustainable Management, A Framework Document: found at http://www.fao.org/Legal/Prs-OL/isarm.pdf (*last visited* June 2003).

Tickner, J. and Myers, N. Precautionary Principle: Current Status and Implementation Centre for Sustainable Production, University of Massachusetts, found at http://web.greens.org/s-r/23/23-17.html (*last visited* May 2003)

Wolf, A., Hydro strategic Territory in the Jordan Basin: Water War, and the Arab Israeli Peace Negotiations, Paper presented at a conference: Water, a Trigger for Conflict/ a reason for Cooperation, 1996) http://w3.pnl.gov.2080/ces/academic/midleas2.htm (*last visited* May 2003).

3.2 Programmes and Projects

Regional Aquifer Systems in Arid Zones – Managing Non-Renewable Resources" Tripoli, 20-24 of November 1999.

International Workshop On Managing Shared Aquifers Resources In Africa: 2nd To 4th June 2002, Tripoli - Workshop Report And Recommendations.

Press Release No.2002-35, Africa's hidden groundwater resources http://www.unesco.org/bpi/eng/unescopress/2002/02-39e.shtml (*last visited* April 2003).

The formulation of a Regional Strategy for the Utilization of the Nubian Sandstone Aquifer System, http://isu2.cedare.org.eg/Nubian/project/regional_coop.htm (*last visited* April 2003).

World Bank Group on Groundwater GW-MATE, Briefing Note No. 1, Groundwater Resource Management: An Introduction to its Scope and Practice, found at http://lnweb18.worldbank.org/ESSD/essdext.nsf/18DocByUnid/F5DC74EF8FF66E5585 256B9000547E5C/$FILE/BN1.pdf (*last visited* May 2003).

Johnson, L., International Law, 2001 *found at* http://www.geocities.com/harkless25/ internationallawjohnson--garriques.doc (*last visited* March 2003).

http://www.unesco.org/ubb/Forum24/HTML/000003.html. (*last visited* March 2003).

Groundwater & Surface Water: Understanding the Interaction A Guide for Watershed Partnerships, found at http://www.ctic.purdue.edu/KYW/Brochures/GroundSurface.html. (*last visited* May 2003).

Pipes, D., Declaring Statehood: Israel and the PLO Orbis, 1989 www.danielpipes.org (last visited March 2003).

International Law Commission website http://www.un.org/ilc (*last visited* May 2003).

European Freshwater Programme, www.wwfreshwater.org/initiative/wfd.html (*last visited* April 2003).

Shared Aquifers and International Law Based on the notes contributed by Gabriel Eckstein Hydrogeologist, lawyer, Director of the International Water Project (web site www.internationalwaterlaw.org) (*last visited* May 2003).

www.europa.eu.int/eur-lex/en/lif/dat/2000/en_300L0060.html (*last visited* May 2003).

3.3 Important Websites

http://www.fao.org/docrep/W7414B/W7414B00.htm
http://www.fao.org/Legal/legstud/ls65-e.pdf
www.dundee.ac.uk/law/iwrli

www.wwfreshwater.org/initiative/wfd.html
http://www.fao.org/Legal/Prs-OL/isarm.pdf
http://domino.un.org/unispal.nsf
http://water1.geol.upenn.edu//
http://waternet.rug.ac.be/
http://www2.soas.ac.uk/Geography/WaterIssues/Home.html
http://www.al-bab.com/arab/env/water.htm
http://www.btselem.org/
http://exact-me.org/
http://www.idrc.ca/waterdemand/docs/english/index_eng.shtml
http://www.smuc.ac.uk/RiversWEB/sources/middle_east_water.htm
http://www.columbia.edu/cu/lweb/indiv/mideast/cuvlm/water.html
http://www.biu.ac.il/SOC/besa/meria/research-g/water.html
http://www.fao.org/waicent/faoinfo/agricult/agl/aglw/aquastat/main/index.stm
http://www.medrc.org.om/
http://www.arij.org/
http://gurukul.ucc.american.edu/ted/deadsea.htm
http://www.american.edu/projects/mandala/TED/ice/JORDAN.HTM
http://www.american.edu/projects/mandala/TED/ice/GAZA.HTM
http://www.nad-plo.org/permanent/water.html
http://www.cesr.org/PROGRAMS/waterpalestine.htm
http://www.mni.gov.il/english/units/water/water.shtml
http://www.iepic.org.il/pages/iepic/index_eng.html
http://www.emwis.org
http://www.transboundarywaters.orst.edu/
http://www.thewaterpage.com/index.htm

Appendix I

Treaty Survey in Africa

TREATY	Governing Rules and Principles	Implicit Reference to GW	Explicit Reference to GW
1. Act Regarding Navigation and Economic Co-operation between the States of the Niger Basin, Niamey 1963 (Came into Force 1 February 1966) *found at* the FAO Treaty Collection No. 2 http://www.fao.org/docrep/W7414B/w7414b04.htm#TopOfPage (*last visited* May 2003) and the TFDD Treaty Collection (ocid.nacse.org/qml/research/tfdd/toTFDDdocs/120ENG.htm (*last visited* May 2003)and United Nations, Treaty Series, vol. 587, No. 8506, p. 9.	**Objectives and Scope** Regulating the utilization of the River Niger, its tributaries and sub tributaries and Navigation **General Principles** *Sovereign Rights:* Article 2:" The utilisation of the River Niger, its tributaries and sub-tributaries, is open to each riparian State in respect of the portion of the River Niger basin lying in its territory and without prejudice to its sovereign rights in accordance with the principles defined in the present Act and in the manner that may be set forth in subsequent special agreements". **Substantive Rules** *"Duty to limit appreciable harm"* Article 4:" The riparian States undertake to establish close co-operation with regard to the study and the execution of any project likely to have an "appreciable effect" on certain features of the regime of the River, its tributaries and sub-tributaries, their conditions of navigability, agricultural and industrial exploitation, the sanitary conditions of their waters, and the biological characteristics of their fauna and flora". **Procedural Rules** Co-operation (Articles 4&5) mainly technical cooperation **Institutional Mechanism** The parties committed themselves to Establish Inter-governmental organization to implement the undertakings of this Act (preamble para 5, Articles 5 & 6) The functions of such an organization is mainly to encourage, promote and coordinate the studies and programmes concerning the exploitation of the resources of the River Niger basin. The composition, the functions	Groundwater was not explicitly excluded and there is a general reference the 'Resources of the River Basin'.	------

TREATY	Governing Rules and Principles	Implicit Reference to GW	Explicit Reference to GW
	and the procedures of such Inter-Governmental Organisation is to subject to a subsequent agreement. **Dispute Settlement Mechanisms** Amicable settlement of disputes then mediation conciliation and arbitration Article 7:" Any dispute that may arise between the riparian States regarding the interpretation or application of the present Act shall be amicably settled by direct agreement between them or through the Inter-Governmental Organisation referred to in Articles 5 and 6 above. Failing such settlement, the dispute shall be decided by arbitration, in particular by the Commission of Mediation, Conciliation and Arbitration of the Organisation of African Unity or by judicial settlement by the International Court of Justice".		
2. Agreement concerning the River Niger Commission and the Navigation and Transport on the River Niger. Done at Niamey, on 25 November 1964. Signatories: Benin, Cameroon, Chad, Ivory Coast, Guinea, Mali, Niger, Nigeria, Burkina Faso. (Came into force 12 April 1966) *found at* FAO Treaty Collection No. 4. *Found at* http://www.fao.org/docrep/W7414B/w7414b06.htm#TopOfPage (*last visited* May 2003) and in the United Nations, Treaty Series, vol. 587, No. 8507, p. 19	**Objectives and Scope** Economic Cooperation. Covers the same elements of the River Basin as the Niger Basin Act of 1963 (No 1 above). This Agreement forms an integral part of the Act of Niamey of 1963 and therefore is governed by the same rules and principles. The Commission is entrusted with the fostering, the promotion and the co-ordination of studies and programmes relating to the utilization and development of the resources of the River Niger Basin (Art. 16)	River Basin	

TREATY	Governing Rules and Principles	Implicit Reference to GW	Explicit Reference to GW
3 Convention Creating the Niger Basin Authority done on 21 January 1980. Signatories: Benin, Cameroon, Chad, Ivory Coast, Guinea, Mali, Niger, Nigeria, Upper Volta. http://ocid.nacse.org/qml/research/tfdd/toTFDDdocs/374ENG.htm (*last visited* May 2003), and in the FAO Treaty Collection No. 13 http://www.fao.org/docrep/W7414B/w7414b0f.htm#TopOfPage (*last visited* May 2003); and 4. The Revised Convention Creating the Niger Basin Authority Done In D'jamena 1987, *found at* the FAO Treaty Collection No. 12. Signatories: Benin, Cameroon, Chad, Ivory Coast, Guinea, Mali, Niger, Nigeria, Upper Volta. http://www.fao.org/docrep/W7414B/w7414b0e.htm#TopOfPage (In French). No information is available regarding the entry into force of this Convention	**Objectives and Scope** Transforming the Niger Commission that was established in 1964 and revised in 1968, 1973 and 1979 into an 'Authority' with wider and broader authorities' functions and responsibilities. **Institutional Mechanism** Granting more powers to the Niger Basin Authority which was originally a Commission. Defining the roles and functions of the institutions of this Authority (statistics and planning, infrastructure, water control and utilization, environment control and preservation, navigation control and regulation, **Dispute Settlement Mechanisms** Article. 15: "Any dispute that may arise among the member States over the interpretation and/or implementation of this Convention shall be amicably settled through direct negotiations In the event of failure to settle such disputes, the matter shall be referred to the Summit by a party to such disputes and the decision on the same shall be final".		Article 4: "The AUTHORITY shall be responsible for: the (d) The initiating and monitoring of an orderly and rational regional policy for the utilization of the surface and underground waters in the Basin".

TREATY	Governing Rules and Principles	Implicit Reference to GW	Explicit Reference to GW
5. Convention and Statutes Relating to the Development of the Chad Basin, signed at Fort Lamy, On 22 May 1964. Signatories: Cameroon, Chad, Niger, and Nigeria. Published in the Journal Officiel de la République Fédérale du Cameroun, 15 Septembre 1964, p. 1003), *also found at the* FAO Treaty Collection No. 3 *found at* http://www.fao.org/docrep/W7414B/w7414b05.htm#TopOfPage (last visited May 2003)	**Objectives and Scope** Regulating the utilization, economic development and protection of the waters of the Chad Basin **General Principles** Sovereign Rights of States sharing the Chad Basin **Substantive Rules** *The duty to limit Harm* "referred to in Article 5 "The Member States undertake a refrain from adopting, without referring to the Commission beforehand, any measures likely to exert a "marked influence" either upon the extent of water losses, or upon the form of the annual hydro graph and limnograph and certain other characteristics of the Lake, upon the conditions of their exploitation by other bordering States, upon the sanitary condition of the water resources or upon the biological characteristics of the fauna and the flora of the Basin. In particular, the Member States agree not to undertake in that part of the Basin falling within their jurisdiction any work in connection with the development of water resources or the soil likely to have a marked influence upon the system of the water courses and levels of the Basin without adequate notice and prior consultations with the Commission, provided always that the Member States shall retain the liberty of completing any plans and schemes in the course of execution or such plans and schemes as may be initiated over a period of 3 years to run from the signature of the present Convention". **Procedural Rules** Cooperation which necessitates prior notification and information Article 6 "With a view to obtaining the closest possible co-operation on the matters specified in article 5, the Member States agree to inform the		Article 4 "The exploitation of the Chad Basin and especially the utilization of surface and underground waters have the widest meaning and refer in particular to the needs of domestic and industrial and agricultural development and the collecting of its fauna and flora products".

TREATY	Governing Rules and Principles	Implicit Reference to GW	Explicit Reference to GW
	Commission of any studies and schemes which they propose to undertake, as from their initial stage." **Institutional Mechanism** Established The Chad Basin Commission. The terms of reference has been set in Art. 9 which mainly focus on technical and legal matters pertaining to the implementation of the current agreement. The Commission is also responsible for facilitating cooperation and assisting in settling disputes. The Commission's joint regulations and recommendations are not binding to the Member States unless they are approved by the Governments of the Member States. **Dispute Settlement Mechanisms** Two steps: -Chad Basin Commission -Commission of Mediation, Conciliation and Arbitration of the Organisation of African Unity for the purposes of decision.		
6. African Convention on the Conservation of Nature and Natural Resources (1968), (Came into force 16 June 1969), *found at* the FAO Treaty Collection No. 1 http://www.fao.org/docrep/W7414B/w7414b03.htm#TopOfPage (*last visited* May 2003), *also found at* TFDD treaty collection ocid.nacse.org/qml/research/tfdd/toTFDDdocs/341ENG.htm (*last visited* May 2003).	**Scope** Conservation of nature and natural resources including soil, water, flora and fauna. **Substantive Rules** Conservation and rational utilization of the natural resources including water Procedural Rules Cooperation including: Exchange of information data and reports and consultation Article V para 2 states:" 2. Where surface or underground water resources are shared by two or more of the Contracting States, the latter shall act in consultation, …"		Article V: Water "(1) The Contracting States shall establish policies for conservation, utilization and development of underground and surface water, and shall endeavour to guarantee for their populations a sufficient and continuous supply of suitable water….."

TREATY	Governing Rules and Principles	Implicit Reference to GW	Explicit Reference to GW
	Institutional Mechanism Encourages Member States to establish joint commissions. (see Article V (2) "….and if the need arises, set up inter-State Commissions to study and resolve problems arising from the joint use of these resources, and for the joint development and conservation thereof". **Dispute Settlement Mechanisms** (a) Negotiation (b) to be submitted to the Commission of Mediation, Conciliation and Arbitration of the Organization of African Unity. Institutional Mechanism		
7. Agreement for the Establishment of the Organization for Management and Development of the Kagera River Basin, concluded at Rusomo, Rwanda 24 August 1977. Signatories: Tanzania, Rwanda, Burundi, (Came into force 5th February 1978), *found at the* FAO Treaty Collection No. 8 http://www.fao.org/docrep/W7414B/w7414b0a.htm#TopOfPage (*last visited* June 2003) *also* at the TFDD Treaty Collection at http://ocid.nacse.org/qml/research/tfdd/toTFDDdocs/200ENG.pdf , also at the United Nations Treaty Series No. 16695, Vol. 1089, pp 172-177.	**Objectives and Scope** Setting up an organization for the management and development of the Kagera River Basin. **Substantive Rules** *"Limit substantial effects"* Art. 2: "A project work or programme shall be considered of an inter-State nature in terms of this article when :…(3) it is likely in the judgement of the organization to produce "substantial effects" whether these be beneficial or prejudicial in the territory of State or States different from the State of the project, work or programme to be undertaken." **Procedural Rules** Cooperation Comprehensive planning of the basin Institutional Mechanism Permanent Joint Technical Committee That has wide authorities nationally and internationally in the context of technical matters and finding means for their financing however, what	Reference to the general term 'River Basin' Considering the River Basin as a geographical unit which is interpreted here to include groundwater : Preamble para 4 states "Recognizing that the Kagera River Basin constitutes a geographical unit offering a valuable base for fruitful cooperation between Riparian Countries" Article 3 "The	

TREATY	Governing Rules and Principles	Implicit Reference to GW	Explicit Reference to GW
	they produce is considered as recommendations for the Government of the States to decide upon. **Dispute Settlement Mechanisms** (a) Consultation (b) resort to procedures as lay down in the Charter of the Organization of African Unity (United Nations Treaty Series vol. 479 p. 39.	Territorial jurisdiction of the organization is the area drained by the Kagera River and its tributaries and sub tributaries……. Provided that the Government may, by mutual agreement, assign to the Organization other geographical areas in order to facilitate or make possible the full proper study and comprehensive planning for the implementation of the projects, works and programmes established in the harmonization development of the Basin, or where service are to be provided to or from the Basin".	

TREATY	Governing Rules and Principles	Implicit Reference to GW	Explicit Reference to GW
8. Convention Relating to the Status of the River Gambia, signed At Kaolack, On 30 June 1978. (Parties: The Republic of the Gambia, The People's, The Republic of Senegal, Guinea became a party to the Convention by virtue of resolution 2/CEG/CG of 6 June 1981 of the Conference of Heads of State and Government of the Gambia River Basin Development Organization). *Found* at the FAO Treaty Collection No. 9 http://www.fao.org/docrep/W7414B/w7414b0b.htm#TopOfPage (*last visited* May 2003) 9. The Convention Relating to the Creation of the Gambia River Basin Development Organization, signed at Kaolack On 30 June 1978 *found at* (FAO Treaty Collection No. 10 http://www.fao.org/docrep/W7414B/w7414b0c.htm#TopOfPage (*last visited* May 2003), *also at* the TFDD Treaty Collection http://ocid.nacse.org/qml/research/tfdd/toTFDDdocs/160ENG.htm (*last*	**8. Objectives and Scope** Cooperation on the rational exploitation of the Resources of the Gambia River Basin **General Principles** Freedom, equality, justice and dignity Responsibility to harness the natural resources - Sovereignty and territorial integrity Peace Maintenance **Substantive Rules** *Limit Serious Modifications on the river regime* Article 4: "No project which is likely to bring about "serious modifications" on the characteristics of the river's regime, on its navigation conditions, the agricultural and industrial exploitation of the river, the sanitary state of the waters, the biological characteristics of its fauna and its flora, as well as its water level, will be implemented without the prior approval of the contracting States. Such approval should be received within six months. Beyond this deadline the State concerned may, as of right, proceed with the execution of its project". **Procedural Rules** Cooperation Notification Article 4 para 2: "The contracting States will be duly informed of any project in connection with the exploitation of the river". -Information **Institutional Mechanism** The Organisation for the River Gambia Development has the following permanent Organs The Conference of Heads of State and Government of the Organisation	The Resources of the Gambia River Basin	

TREATY	Governing Rules and Principles	Implicit Reference to GW	Explicit Reference to GW
visited June 2003) No information is available regarding the entry into force of this Convention	shall hold its ordinary session once a year. It may be convened into an extraordinary Session on the initiative of its Chairman or at the request of a member State. 2- The Council of Ministers shall consist of a single Minister representing each member State. The Ministers may be accompanied by members of their respective Governments. 3- The High Commission: Between sessions of the Council of Ministers, the High Commissioner shall represent the Organisation. He shall take all decisions within his capacity in accordance with the directives of the Council of Ministers and within the limits of the powers bestowed on him. 4- The Gambia Permanent Water Commission: The Permanent Water Commission shall be charged with the responsibility of defining the principles and modalities of the utilization of the Gambia River waters Between the States and between sectors using the water: industry, agriculture and transport. **Dispute Settlement Mechanisms** a)Discussion and Mediation b) Commission of Conciliation and Arbitration of the Organization of African Unity. c) Seek the assistance of the International Court of Justice at The Hague <u>Article 18:</u> "Any dispute that may arise between the contracting States regarding the interpretation or implementation of the present agreement shall be settled through conciliation or mediation. If no agreement is reached, the contracting States shall place the matter before the		

TREATY	Governing Rules and Principles	Implicit Reference to GW	Explicit Reference to GW
	Conciliation and Arbitration Committee of the OAU. As a last resort, they shall have recourse to the International Court of Justice at The Hague". **9. Objectives and Scope** Establishment of a Joint Organisation of Co-operation for the Development of the Gambia River Basin. The functions of this organization is to implement the Convention referred to in number 8. The organization shall have permanent bodies attached to it See <u>Article 3.</u> The decision of the organization permanent organs namely the "Conference of Head of States and Governments" shall be binding on all Member States.		
10. Agreement between the Federal Republic of Nigeria and the Republic of Niger Concerning the Equitable Sharing in the Development, Conservation and Use of their Common Water Resources, done At Maiduguri, 18 July 1990, *found at* FAO Treaty Collection No. 34 http://www.fao.org/docrep/W7414B/w7414b10.htm#TopOfPage (*last visited* May 2003), *also at* the TFDD Treaty Collection http://ocid.nacse.org/qml/research/tfdd/toTFDDdocs/182ENG.htm (*last visited* June 2003).	**Objectives and Scope** Development conservation and use of the waters of the Niger Basin <u>Article 1:</u> "These Articles shall govern the equitable development, conservation and use of the water resources in the river basins which are bi-sected by, or form, the common frontier between the Contracting Parties (hereinafter referred to as shared river basins). **Substantive Rules** *Equitable Sharing* in the development use and conservation of the water resources in the River Basin. <u>Article 2:</u> "Each Contracting Party is entitled, within its territory, to an equitable share in the development, conservation and use of the water resources in the shared river basins. For determining what is equitable sharing <u>Article 5</u> assigns a list of factors "1. In determining the equitable share to which each Contracting Party is entitled pursuant to Article 2, the following factors shall be taken into account:		Article 1(3) "Subject to the provisions of Article 9, a reference to the shared river basins shall include a reference to underground waters contributing to the flow of surface waters". Article 9: "Groundwater resources shall not be accounted for the purpose of equitable sharing determination unless: (a) such resources are part of shared river basins within the meaning of Article 1, paragraph (3); or

TREATY	Governing Rules and Principles	Implicit Reference to GW	Explicit Reference to GW
	(a) the climate of the region, and its influence on rainfall patterns; (b) rainfall patterns, and their influence on surface hydrology, and related hydro-geology; (c) surface hydrology and related hydro-geology; (d) existing uses of the waters; (e) reasonable planned water development requirements; (f) the economic and social development needs of the Contracting Parties; (g) the dependence of local populations on the waters in question for their own livelihood and welfare; (h) the availability of alternative sources of water to satisfy competing water demands; (i) the practicability of compensations either in cash or in kind one or the other Contracting Party as a means of adjusting competing water demands; (j) maintaining an acceptable environmental balance in and around a particular body of water; (k) the avoidance of unnecessary waste in the utilization of waters, with due regard for the technological and financial capabilities of each Contracting Party; (l) The proportion in which each Contracting Party contributes to the water balance of the basin. 2. Each factor is to be given the weight warranted by the circumstances peculiar to each individual river basin, or group of basins, and all factors so weighted are to be considered together and a determination arrived at on the basis of the whole." *"Limit Appreciable Impact"*: Article 3: "The Contracting Parties undertake to inform and consult with the Nigeria-Niger Joint		(b) Such resources lie in whole or only in part within the shared river basins and are bi-sected by the common frontier between the Contracting Parties."

TREATY	Governing Rules and Principles	Implicit Reference to GW	Explicit Reference to GW
	Commission for Co-operation in advance of undertaking any project, programme or plan for the implementation of agreed upon equitable sharing determinations, or likely to have an appreciable impact on any such determination." **Procedural Rules** Cooperation including: information and consultation Institutional Mechanism The Nigeria-Niger Joint Commission for co-operation **Dispute Settlement Mechanisms** Niger Nigeria Commission Commission of Mediation, Conciliation and Arbitration of the Organization of African Unity for binding determination Article 17:" 1. Any difference concerning the interpretation or implementation of this Agreement or of any equitable sharing determination made thereunder, shall be referred to the Nigeria-Niger Joint Commission for Co-operation. 2. Any such difference which cannot be settled by the Commission within six months after a reference to it under the last preceding paragraph shall be regarded as a dispute and shall, at the request of both Contracting Parties, be referred to the Commission of Mediation, Conciliation and Arbitration of the Organization of African Unity for binding determination".		

TREATY	Governing Rules and Principles	Implicit Reference to GW	Explicit Reference to GW
11. Tripartite Ministerial Meeting of Ministers Responsible for Water Affairs held on the 15th February, 1991 in Swaziland between Swaziland, Mozambique and Republic of South Africa, *found at the* TFDD Treaty Collection, *found at* http://ocid.nacse.org/qml/research/tfdd/toTFDDdocs/234ENG.htm (*last visited* June 2003). Sub region: South Africa	**Objectives and Scope** Conduct a joint study of the water resources, demands and development potential on the whole Inkomati River Basin	Whole River Basin	
12. Agreement on The Development and Utilisation of the Water Resources of the Komati River Basin between the Government of the Republic of South Africa and the Government of Kangwane, 1992, *found at the* TFDD Treaty Collection http://ocid.nacse.org/qml/research/tfdd/toTFDDdocs/206ENG.htm (*last visited* June 2003). No information on entry into force of the Agreement. Sub region: South Africa	**Objectives and Scope** Development and utilisation of the water resources of the Komati River Basin **General Principles** Good neighbourliness Common interests in the use of the water resources **Substantive Rules** *Equitable Utilization*: Article 2 Purpose of the Agreement:" The purpose of this Agreement is to provide for the equitable sharing and apportioning between KaNgwane and South Africa, of South Africa's benefits and obligations arising from the Treaty, a copy of which is attached as Annex 1 to this Agreement, and to provide for all matters related thereto". **Procedural Rules** Cooperation Consultation Exchange of Information	River Basin	Employing the term Water resources which includes interconnected groundwaters (ILA 1966 Helsinki Rules)

TREATY	Governing Rules and Principles	Implicit Reference to GW	Explicit Reference to GW
	Article 3 para 3:" (a) South Africa shall consult with KaNgwane through the PWC before concluding negotiations on any matter contemplated in the Treaty; and (b) KaNgwane shall furnish to South Africa through the PWC all information within its jurisdiction, required to be submitted by South Africa to the JWC or to Swaziland in terms of the Treaty. 4. Notwithstanding anything to the contrary contained in the Treaty, South Africa shall not without the prior consent of KaNgwane, obtained through the PWC, divert any water as stipulated from Article 3(2) of the Treaty. **Institutional mechanism** Permanent Water Commission (PWC) established under a separate Agreement between the Parties. **Dispute Settlement Mechanisms** (Article 11) (a) spirit of conciliation (b) arbitration		
13. Treaty on the Development and Utilisation of the Water Resources of the Komati River Basin Between the Government of the Kingdom of Swaziland and the Government of the Republic of South Africa, signed on 13 March 1992, *found at the* FAO Treaty Collection No. 38 http://www.fao.org/docrep/W7414B/w7414b14.htm#TopOfPage (*last visited* May 2003).	13.**Objectives and Scope** Regulate the development and utilization of the Komati River Basin **General Principles** (Preamble) Common Interests Good neighbourliness Peaceful Cooperation **Substantive Rules** *Reasonable and Equitable Utilisation* Article 3 (5) "The Parties recognise the right of the Republic of Mozambique to a reasonable and equitable share in the use of the waters of the Inkomati River Basin of which the Komati River Basin is an integral part". *Limiting substantial impact"* Article 13 "The Parties agree to take all reasonable measures to ensure that the design, construction, operation and		Employing the term Watercourse which includes interconnected groundwaters (ILA 1966 Helsinki Rules) "Komati River Basin" means the drainage area of the Komati River, and all its tributaries, upstream of its confluence with the Crocodile River near Komatipoort in the Republic of South Africa;

TREATY	Governing Rules and Principles	Implicit Reference to GW	Explicit Reference to GW
14. Treaty on the Establishment and Functioning of the Joint Water Commission between the Government of the Kingdom of Swaziland and the Government of the Republic of South Africa, signed on 13 March 1992, *found at the* TFDD Treat Collection http://ocid.nacse.org/qml/research/tfdd/toTFDDdocs/185ENG.htm (*last visited* June 2003). No information on entry to force of this Treaty. Sub region: South Africa	maintenance of the Project are compatible with the protection of the existing quality of the environment and, in particular, shall pay due regard to the maintenance of the welfare of persons and communities immediately affected by the Project." Article 15:" In the case of any substantial impairment of the implementation of the provisions of this Treaty caused by *Vis Major*, KOBWA, the JWC and the Parties shall at their various levels, jointly and severally, immediately take all necessary measures permitted by circumstances to mitigate such impairment where after the Parties shall agree on further action." **Procedural Rules** Cooperation Precautionary measures to protect the resources (control over the abstraction of water, curtailments to be imposed on the abstraction of water) Supply of information and plans relating to the utilization of common resources Participation **Institutional Mechanism** Joint water Institutions (advisory role): Article 1 para 2. "The object and functions of the Commission shall be to act as technical adviser to the Parties on all matters relating to the development and utilisation of water resources of common interest to the Parties and to perform such other functions pertaining to the development and utilisation of such resources as the Parties may from time to time agree to assign to the Commission." **Dispute Settlement Mechanisms** (Art. 16) (a) negotiations,(b) arbitration		

TREATY	Governing Rules and Principles	Implicit Reference to GW	Explicit Reference to GW
15. Agreement between the Government of the Republic of Namibia and the Government of the Republic of South Africa on the Establishment of a Permanent Water Commission, signed at Noordoewer, 14 September 1992, *found at* the TFDD treaty collection http://ocid.nacse.org/qml/research/tfdd/toTFDDdocs/186ENG.htm (*last visited* June 2003). No info on entry to force of this Agreement. Sub region: South Africa	**Objectives and Scope** Cooperation on the development of mutual projects in respect of water resources of common interest **General Principles** Neighbourliness Common Interests in the shared water resources **Procedural Rules** Cooperation which includes exchange of information Article 1(3) "In furtherance of the stated objective each Party shall, to the extent permitted by its own laws and procedures, provide such information and authorisations as the Commission may require for the performance of its functions." **Institutional Mechanism** Joint Permanent Commission (advisory role).		Employing the term Watercourse which includes interconnected groundwaters (ILA 1966 Helsinki Rules)
16. Agreement between the Governments of the Republic of Angola, the Republic of Botswana, and the Republic of Namibia on the Establishment of a Permanent Okavango River[1] Basin Water Commission (Okacom), done At Windhoek, 16 September 1994	**Objectives and Scope** Institutional Mechanism on matters relating to the conservation, development and utilization of water resources of common interest to the Contracting Parties Article 1: The objective of the Commission shall be to act as technical advisor to the Contracting Parties on matters relating to the conservation, development and utilization of water resources of common interest to the Contracting Parties and shall perform such other functions pertaining to		Employing the term "Watercourse" which includes connected groundwaters (ILA 1966 Helsinki Rules)

[1] The Okavango River rises in the Angolan highlands and flows over 1000 miles, passing through Namibia before entering Botswana and forming the Okavango Delta. The Okavango Delta is a 15,000-square-kilometer rich and varied habitat for thousands of mammals, birds, fish and other animals. It sustains tens of thousands of delta residents and a growing ecotourism industry

TREATY	Governing Rules and Principles	Implicit Reference to GW	Explicit Reference to GW
(came into force 15th September 1994, *found at* the TFDD Treaty Collection http://ocid.nacse.org/qml/research/tfdd/toTFDDdocs/190ENG.htm (*last visited* June 2003), *also at* the FAO Treaty Collection No. 20 http://www.fao.org/docrep/W7414B/w7414b0m.htm#TopOfPage (*last visited* May 2003).	the development and utilisation of such resources as the Contracting Parties may from time to time agree to assign to the Commission **General Principles** Good neighbourliness **Substantive Rules** *Equitable and reasonable utilization principle*. Preamble affirms that regional water resource development is to be promoted on the basis of the rules relating to the uses of the waters of international rivers approved in 1966 at Helsinki by the 52nd Conference of the International Law Association. **Procedural Rules** Technical Cooperation (Art.1) Exchange of information (Art.1) "In furtherance of the stated objective each Party shall, to the extent permitted by its own laws and procedures, provide such information and authorisations as the Commission may require for the performance of its functions." **Institutional Mechanism** The three basin states have formed the Permanent Okavango River Basin Commission (OKACOM) in 1994, with the purpose of promoting sustainable management of the river basin **Dispute Settlement Mechanisms** by the contracting Parties		
17. Establishment of a Joint Water Commission between South Africa and Mozambique, July1996.	**Objectives and Scope** Institutional mechanisms on technical matters relating to the conservation, development and utilization of water resources of common interest to the Contracting Parties		Employing the term Water resources which includes connected groundwaters (ILA 1966 Helsinki Rules)

TREATY	Governing Rules and Principles	Implicit Reference to GW	Explicit Reference to GW
	General Principle Common Interests in the shared water resources Good neighbourliness **Substantive Rules** No Harm **Procedural Rules** Technical Cooperation **Institutional Mechanism** Joint water Commission **Dispute settlement Mechanisms** (a) negotiations (b) arbitration		
18. Protocol on Shared Watercourse Systems in the Southern African Development Community (SADC) Region, signed At Johannesburg, 28 August 1995, by the Republic of Angola, the Republic of Botswana, the Kingdom of Lesotho, the Republic of Malawi, the Republic of Mozambique, the Republic of Namibia, the Republic of South Africa, the Kingdom of Swaziland, the United Republic of Tanzania, the Republic of Zambia and the Republic of Zimbabwe, *found at* the FAO Treaty Collection No. 21 http://www.fao.org/docrep/W7414B/	**General Principles** Equal rights of use of Watercourse States Sovereign rights of the States Principles of community of interests in the equitable utilization of those systems and related resources as rules of customary international law (Art. 2) **Substantive Rules** Equitable utilization: Article 2 para 6 Member States shall utilise a shared watercourse system in an equitable manner. In particular, a shared watercourse system shall be used and developed by member States with a view to attaining optimum utilisation thereof and obtaining benefits therefrom consistent with protection of the watercourse system Guiding Procedure to implement equitable utilization Article 2 para 7 "Utilisation of a shared watercourse system in an equitable manner within the meaning of paragraph 4 and 6 requires taking into account all relevant factors and circumstances including:	.	In Article 1 the Protocol defines the terms "Drainage Basin" as well as "Watercourse System" The "Drainage Basin" means " a geographical area determined by the watershed limits of a system of waters including underground waters flowing into a common terminus". And the Watercourse System means: "the inter-related hydrologic components of a drainage basin such as streams, rivers, lakes, canals and underground water which constitute a unitary whole

TREATY	Governing Rules and Principles	Implicit Reference to GW	Explicit Reference to GW
w7414b0o.htm#TopOfPage (*last visited* May 2003), *also at* the TFDD Treaty Collection http://ocid.nacse.org/qml/research/tfdd/toTFDDdocs/205ENG.htm (*last visited* June 2003).	geographical, hydrographical, hydrological, climatic and other factors of a natural character; (b) the social and economic needs of the member States concerned; (c) the effects of the use of a shared watercourse system in one watercourse state on another watercourse state; (d) existing and potential uses of the shared watercourse system; (e) guidelines and agreed standards to be adopted. **Procedural Rules** Cooperation including: -Exchange of Information Article 2 para 5 : "Member States within a shared watercourse system shall exchange available information and data regarding the hydrogeological, water quality, meteorological and ecological condition of such watercourse system. " - Prior Notification Article 2 paras 9, 10 and 11. **Institutional Mechanism** The establishment of 3 type of institutions one for monitoring at the level of the SADC regions, one at the level of each drainage basin and one at the national level within each member state. **Dispute Settlement Mechanisms** (a) Dispute between Member States -Amicable means - Tribunal for adjudication under Article 16 (1) of the Treaty of SADC. (b)Dispute between SADC and a member State - Request by the Council for an advisory opinion in accordance with Article 16 (4) of the Treaty of SADC and this opinion by the tribunal is binding and final.		by virtue of their physical relationship".

TREATY	Governing Rules and Principles	Implicit Reference to GW	Explicit Reference to GW
19. Revised Protocol on Shared Watercourses in the Southern African Development Community (SADC) Johannesburg 2000. entered into force on the 7 Sep. 2000 and was signed by the Republic of Angola, the Republic of Botswana, the Kingdom of Lesotho, the Republic of Malawi, the Republic of Mauritius, the Republic of Mozambique, the Republic of Seychelles, the Republic of South Africa, the Kingdom of Swaziland, the Republic of Tanzania, the Republic of Zambia and the Republic of Zimbabwe, *found* at the TFDD Treaty Collection http://ocid.nacse.org/qml/research/tfdd/toTFDDdocs/ENG.htm/ (*last visited* June 2003).	**General Principles** Sovereign Rights of States Respecting and applying the existing rules of general or customary international law relating to the utilization and management of the resources of shared watercourse systems. **Substantive Rules** *Equitable and Reasonable Utilization* Article 7. a) Watercourse States shall in their respective territories utilise a shared watercourse in an equitable and reasonable manner. In particular, a shared watercourse shall be used and developed by Watercourse States with a view to attain optimal and sustainable utilisation thereof and benefits therefrom, taking into account the interests of the Watercourse States concerned, consistent with adequate protection of the watercourse for the benefit of current and future generations. In determining equitable and reasonable utilization the Member States have to take all relevant factors and circumstances into account whereby the Protocol offer some guidance to what these factor might include. Art 3 para 8 i (i) geographical, hydrographical, hydrological, climatic, ecological and other factors of a natural character; (ii) the social, economic and environmental needs of the Watercourse States concerned; (iii) the population dependent on the shared watercourse in each Watercourse State; (iv) the effects of the use or uses of a shared watercourse in one Watercourse State on other Watercourse States; (v) existing and potential uses of the watercourse; (vi) conservation, protection, development and economy of use of the water resources of the shared watercourse and the costs of measures taken to that effect; and (vii) the availability of alternatives, of comparable value, to a particular planned or existing use. (b) The weight to be given to		Article 1: watercourse means a system of surface and ground waters consisting by virtue of their physical relationship a unitary whole normally flowing into a common terminus such as the sea, lake or aquifer;

TREATY	Governing Rules and Principles	Implicit Reference to GW	Explicit Reference to GW
	each factor is to be determined by its importance in comparison with that of other relevant factors. In determining what is an equitable and reasonable use, all relevant factors are to be considered together and a conclusion reached on the basis of the whole. Limiting Significant Harm Article 3(10) **Procedural Rules** Cooperation which includes: Duty to exchange information Article 3(6) Prior Notification Article 4(1) consultation and if necessary negotiation **Institutional Mechanism** Three types of institutions were established : a) SADC Water Sector Organs (i) the Committee of Water Ministers; (ii) the Committee of Water Senior Officials; (iii) the Water Sector Coordinating Unit; and (iv) the Water Resources Technical Committee and sub-Committees. b) Shared Watercourse Institutions c) The Committee of Water Ministers shall consist of Ministers responsible for water. d) The Committee of Water Senior Officials shall consist of the Permanent Secretaries or officials of equivalent rank responsible for water. e) The Water Sector Coordinating Unit which shall be the executing agency of the Water Sector shall be headed by a Coordinator appointed by the State Party responsible for coordinating the Water Sector, and he or she shall be assisted by such supporting staff of professional, administrative and secretarial personnel as the Coordinator may deem necessary. **Dispute Settlement Mechanisms** (a)Amicably (b) Tribunal		

Appendix II

Treaty Survey in Europe

TREATY	Summary	Implicit Reference to GW	Explicit Reference to GW
1. Agreement between Italy and Egypt concerning the Water Supply from the Ramba Well signed 1925.	The text of this Agreement could not be accessed. Description of the Agreement was cited from D. Caponera and D. Alherietre, …		
2. Treaty between Germany and Poland for the Settlement of Frontier Questions, signed At Poznan, signed January 27, 1926, *found at* the TFDD Treaty Collection http://ocid.nacse.org/qml/research/tfdd/toTFDDdocs/35ENG.htm (*last visited* June 2003).	**Objectives and Scope** Settling questions concerning frontier lines including the utilization of frontier waterways and frontier waters (Articles 30-33) General Principles Reciprocity in keeping frontier waters clean conditions and guarding against danger of flood **Substantive Rules** *Limiting Transboundary Impact:* Article 30: "Subject to reciprocity, the two contracting States shall each on its own side take all the measures provide for by the laws of the country with a view to maintaining the frontier waterways and frontier waters in a clean condition and also for the purpose of guarding against the danger of floods in the frontier waterways and frontier waters area. In so far as joint measures may be necessary for this purpose, an agreement shall be arrived at between the two contracting States". Article 31: "The flow of the water must not be impeded by installations set up on and in frontier waterways, for the purpose of utilizing the water. Places for watering cattle and geese must be situated on the banks so that they do not interfere with the flow of the water. The slopes and banks of such frontier waterways as come under the regulations concerning protection against flood must be kept free of trees and bushes."	Article 1 para 4: "The frontier line, as fixed and described in the documents, also determines the territorial boundary under the ground and in the air". Article 34: "The provisions of Article 30 to 33 shall apply by analogy to waterways in the frontier district which, without being frontier waterways, flow into such waterways or lead water from territory of one Party to that of the other….. The area within four kilometres of the frontier shall be regarded as the frontier district".	

TREATY	Summary	Implicit Reference to GW	Explicit Reference to GW
	Procedural Rules Cooperation which includes - Permitting system to carry out works on the frontier waters Article 29: "The strengthening of the banks, work on, or alterations in, the watercourse, and the linking up of inflow and outflow may, in the case of frontier waterways and frontier waters, only be carried out with the permission of the competent authorities of the country in which the work is to be executed. " Notification: Article 29 "…The said authorities shall before granting such permission notify the corresponding authorities of the other Party. The carrying out of new work, the effects of which might extend other installations by which the course of a frontier waterway might be altered, or the water level of a frontier waterway or of frontier waters might be influenced, or the discharge of flood water might be impeded , shall only be allowed on permission being given by the Committees provided for in Article 6." Agreed to conclude agreements for information exchange **Institutional Mechanism** A Boundary Commission whose decisions are binding Article 39 provides that :"The decisions of the Boundary Commission in connection with the present Treaty are enumerated in a special list and are hereby recognised by both Parties as binding. These decisions shall be executed by the Committees provided for in Article 6 in so far as the latter still regard this as necessary."		

TREATY	Summary	Implicit Reference to GW	Explicit Reference to GW
3. Agreement between the United Kingdom and Belgium regarding Water Rights on the Boundary between Tanganyika and Ruanda-Urundi-London, November 22, 1934 (Ratifications Exchanged In London, May 10, 1938) *found at* the United Nations Treaty Series, No.42 (1938), *found at* the TFDD Treaty Collection http://ocid/nacse.org/qml/research/tfdd/toTFDDdoc/45ENG.htm (*last visited* June 2003).	**Objectives and Scope:** Water Rights on the Boundary pollution prevention and control **General Principles** Territorial integrity Article 1(1):"Water diverted from a part of a river or stream wholly within the Tanganyika Territory or Ruanda-Urundi shall be returned without substantial reduction to its natural bed at some point before such river or stream flows into the other territory, or at some point before such river or stream forms the common boundary between the two territories" **Substantive Rules** *Limiting Transboundary Impact* Article 1 (3) : . " No operations of mining or industrial nature shall be permitted by either of the contracting Governments in Tanganyika or Ruanda-Urundi respectively which may pollute or cause the deposit of any poisonous, noxious or polluting substance in the waters of any river or stream forming part of the boundary between the Tanganyika Territory and Ruanda-Urundi or any tributary river or stream thereof, or in any river or stream flowing from one territory into the other. 4. Each contracting Government shall have the right to divert, for operations of a mining or industrial nature, at any point along any river or stream where such river or stream forms a common boundary between the two territories, up to a maximum of half the volume of water flowing at such point measured during the season of low water, provided that such water after use shall without substantial reduction be returned to its natural bed." **Procedural Rules** - Notification: Article 1 (6) "In the event of either contracting Government desiring to utilize the waters of any river or stream on	Streams	

TREATY	Summary	Implicit Reference to GW	Explicit Reference to GW
	the aforesaid boundary or to permit any person to utilize such water for irrigation purposes, such Contracting Government shall give to the other contracting, Government notice of such desire 6 months before commencing operations for the utilization of such waters, in order to permit of the consideration of any objections which the other contracting Government may wish to raise." - Monitoring and Assessment **Institutional Mechanism** None **Dispute Settlement Mechanisms** Settlement of disputes by an agreed upon arbitrator: Article 1(10). In the event of any dispute arising between the contracting Governments in respect of any matter covered by this agreement, the contracting Governments shall refer such matter to such arbitrator or court of arbitration as may be mutually agreed upon.		
4. The Treaty of Peace between the Allies and Italy which Outlines Guarantees between Italy and Yugoslavia Concerning Springs in the Commune of Gorizia 1947, *found at* the TFDD Treaty Collection http://ocid.nacse.org/qml/research/tfdd/toTFDDdocs/394ENG.htm (*last visited* May 2003).			

TREATY	Summary	Implicit Reference to GW	Explicit Reference to GW
5. Agreement between the Austrian Federal Government and the Bagarian State Government Concerning the Diversion of Water in the Rissbach, Durrach and Walchen Districts, reached on 29 June 1948, and Concluded On 16 October 1950. (Came Into Force On 1 May 1951) *found at* the TFDD Treaty Collection http://ocid.nacse.org/qml/research/tfdd/toTFDDdocs/73ENG.htm (*last visited* May 2003).	**Objectives and scope** Giving up (Austria) her right to use without compensation	Streams	
6. Protocol to Establish Tripartite Standing Committee on Polluted Waters, Brussels 8 April 1950, between Luxemburg, France and Belgium (Came Into Force 8 April 1950), *found at* The TFDD Treaty Collection http://ocid.nacse.org/qml/research/tfdd/toTFDDdocs/284ENG.pdf (*last visited* May 2003).	**Objectives and Scope:** Institutional mechanism to ensure protection of frontier waters against pollution	Waters	

TREATY	Summary	Implicit Reference to GW	Explicit Reference to GW
7. Agreement between the Federal People's Republic of Yugoslavia and the Republic of Austria concerning Water Economy Questions in Respect of the Frontier Sector of the Mura and the Frontier Waters of the Mura (Mura Agreement), signed at Vienna on 16 December 1954 and Protocol Signed On November 1954. (Came Into Force On 9 Feb 1956), *found at* the TFDD Treaty Collection http://ocid.nacse.org/qml/research/tfdd/toTFDDdocs/88ENG.pdf (*last visited* May 2003).	**Objectives and Scope:** Economic Questions on frontier waters. Applies to measures and works on the frontier waters that might influence the quantity and quality of the waters of the basin. A Permanent Commission to deal with Economy questions **Substantive Rules** *Limiting Transboundary Impact* Article 1: "Water economy questions, measures and works on the frontier sector of the Mura, and substantial effects brought about on that frontier sector by diversion of water from the Mura Basin or by pollution shall be dealt with by the Yugoslav-Austrian Commission for the Mura…" **Procedural Rules** Cooperation Notification Article 5, and Article 7: "The local authorities shall advise each other by the quickest possible means, of any danger from high water or ice and of any other impeding danger which comes to their notice in connexion with the Mura, the same shall apply to the frontier waters of the Mura where such danger come to the notice of the local authorities". **Institutional Mechanism** Mixed Institution Article 1(2): " The terms of reference of the Commission shall accordingly comprise in particular regulation, the erection of high water embankments, protection against flooding and ice, the utilization of water power, changes in the regime of the river, the reclamation of riparian areas, water supply, pollution by effluents, ferries and bridges." **Dispute Settlement Mechanisms** Negotiation Court of Arbitration	Waters of the Mura Basin	

TREATY	Summary	Implicit Reference to GW	Explicit Reference to GW
8. Treaty between the Hungarian People's Republic and the Republic of Austria concerning the Regulation of Water Economy Questions in the Frontier Region, signed at Vienna, on 9 April 1956. (Came Into Force On 31 July 1956), *found at* the TFDD Treaty Collection http://ocid.nacse.org/qml/research/tfdd/toTFDDdocs/94ENG.htm (*last visited* May 2003).	**Objectives and Scope** cooperation on economy question in frontier Article 1 **Substantive Rules** *Limiting Transboundary Impact* Article 2: "1. Each Contracting Party undertakes to refrain from unilaterally - without the consent of the other Contracting Party - carrying out any measures or works on frontier waters (article 1, sub-paragraph 1) which would adversely affect water conditions in the territory of the other Contracting Party. Consent may be refused only if the grounds for such refusal are duly set forth." **Procedural Rules** Cooperation which includes: *Exchange of information:* Article 5(3) "The Commission shall ensure that the Contracting Parties provide each other with any data required for planning or that the co-operation necessary for planning takes place in some other suitable manner." - *Notification* Article 11: "The authorities of the Contracting Parties, particularly the hydrographic service and local authorities, shall notify each other as quickly as possible of any danger of flood or ice or other danger arising in connexion with frontier waters which comes to their attention." -*Negotiation* (not compulsory) Article 13(3) "The two Governments may also enter into direct negotiations on matters lying within the competence of the Commission." **Institutional Mechanism** Joint Commission whose decisions are recommendations to the governments **Dispute Settlement Mechanisms** Mandated to the Joint Commission	Article 1 Frontier waters i.e. Frontier waters, i.e. : *(a)* Those sectors of bodies of water which form the frontier between Hungary and Austria or along which the frontier runs (frontier waters in the restricted sense); *(b)* Bodies of water which intersect the frontier and those situated in the frontier region, i.e., in the territory of either country within six kilometres of the frontier. 2. Hydraulic structures and installations situated in the frontier region (sub-paragraph 1). 3. Those sectors of watercourses and those hydraulic structures and installations which are enumerated in annex 1[2] of this Treaty or are designated at some future time by the Hungarian-Austrian Water Commission (article 12)."	

TREATY	Summary	Implicit Reference to GW	Explicit Reference to GW
9.Agreement between the Government of the Federal People's Republic of Yugoslavia and the Government of the People's Republic of Albania concerning Water Economy Questions, together with the Statute of the Yugoslav-Albanian Water Economic Commission and with the Protocol concerning Fishing in Frontier Lakes And Rivers, signed on 2 May 1956. Parties: Albania, Yugoslavia1 (Came Into Force on 6 August 1957), *found at* the TFDD Treaty Collection http://ocid.nacse.org/qml/research/tfdd/toTFDDdocs/97ENG.htm (*last visited* May 2003).	**Objectives and Scope** regulating water economy questions and establishing an institutional mechanisms on frontier lakes and rivers. Article 1 : " The provisions of this Agreement shall apply to all water economy questions, measures and works on watercourses which form the State frontier and watercourses, lakes and water systems which are intersected by the State frontier (especially Lake Ohrid, the Crni Drim, the Beli Drim, Lake Skadar and the Bojana), and which are of interest to both Contracting Parties, and in particular to: *(a)* The utilization of water power; *(b)* The regulation and canalization of watercourses and lakes and the maintenance of their beds; *(c)* The discharge of water, drainage and similar measures; *(d)* Protection against flooding; *(e)* Storage and retention works, *(f)* Water supply and pipe-laying; *(g)* Navigation; *(h)* Ground water; *(i)* Protection against soil erosion; *(j)* The utilization of water in agriculture; *(k)* Hydrological studies, the preparation of projects and the execution of work; *(l)* Fishing; *(m)* The apportionment of the cost of survey, planning and construction works, and of operation and maintenance; *(n)* The exchange of data and plans and of information on the above questions; and *(o)* The exchange of data on water levels.		See column two Article 1 item h and Article 1 para 3: "(3) The expression "water system " shall mean, in this Agreement, all watercourses (surface or underground, natural or artificial), installations, measures and works which may affect watercourses from the standpoint of water economy, and installations forming or intersected by the State frontier."

TREATY	Summary	Implicit Reference to GW	Explicit Reference to GW
	Substantive Rules *Limiting Transboundary Impact* Article 1(1) "The Contracting Parties undertake, pursuant to the provisions of this Agreement, to examine and to resolve by agreement all questions of water economy, including measures and works which may affect the quantity and quality of the water and which are of interest to both or either of the Contracting Parties, having due regard to the maintenance of a common policy in water economy relations and recognizing the rights and obligations arising out of such policy." And Article 6 "The Contracting Parties undertake, each in its own territory, to preserve and maintain and, where necessary, to augment or renew, such permanent benchmarks and datum marks along the frontier as are necessary for the purpose of works on the waters. Both Contracting Parties may use these marks. If it is necessary to cross the State frontier in order to use the said marks, the provisions of article 7 of this Agreement shall apply." **Procedural Rules** Cooperation which includes: *Notification* Article 8 "The local authorities of the Contracting Parties shall advise each other, by the quickest possible means, of any danger from high water and of any other impending danger which may arise on watercourses which form the State frontier and watercourses, lakes and water systems intersected by the State frontier." Exchange of Information: Article 1(n-o) Annex I Article1(8): "To submit proposals for the exchange of practical experience in the matter of water economy, for the exchange of hydrological and		

TREATY	Summary	Implicit Reference to GW	Explicit Reference to GW
	hydro-meteorological data, and for the operation of the information service established to transmit particulars of the water level and so forth;. . . ." **Institutional Mechanism** Joint Commission to implement the Agreement. (Yugoslav-Albanian Water Economy Commission) whose decisions are recommendations to the Governments. **Dispute Settlement Mechanisms** (a) Agreement (b) Commission (c) submit dispute to the Government. <u>Article 10:</u> "Any dispute between the Contracting Parties relating to the application and interpretation of this Agreement shall, unless the two parties to the dispute agree upon some other mode of settlement, be submitted at the request of either Contracting Party to a commission composed of two representatives of each Party. If this commission fails to reach agreement, the dispute shall be settled directly by the Governments of the two Contracting Parties."		
10. Agreement between The Government Of The Czechoslovak Republic and the Government of the Polish People's Republic Concerning the Use of Water Resources in Frontier Waters, signed at Prague, on 21 March 1958, (came into force on 7 August 1958), *found at* the TFDD Treaty Collection	**Objectives and Scope** Use of water resources in frontier waters **General Principles** Abatement of pollution **Substantive Rules** *Limiting Transboundary Impact* <u>Article 3</u>: (1) Neither Contracting Party may, without the consent of the other Contracting Party, carry out any works in frontier waters which may affect the latter Party's water economy.(2) The Contracting Parties shall come to agreement on the amount of water to be taken from frontier waters for domestic, industrial, power generation and agricultural requirements		Article 2: "For the purposes of this Agreement, the term "frontier waters" means : (b) Surface and ground waters flowing from the territory of one State to the territory of the other, at those places where they are intersected by the State frontier".

TREATY	Summary	Implicit Reference to GW	Explicit Reference to GW
http://ocid.nacse.org/qml/research/tfdd/toTFDDdocs/104ENG.htm (*last visited* May 2003).	and on the discharge of waste water. (3) The Contracting Parties shall come to an agreement in each particular case on what runoff ratios are to be preserved in frontier waters. (4) The Contracting Parties have agreed to abate the pollution of frontier waters and to keep them clean to such extent as is specifically determined in each particular case in accordance with the economic and technical possibilities and requirements of the Contracting Parties. When installations discharging polluted water into frontier waters are constructed or reconstructed, treatment of the waste water shall be required. **Procedural Rules** Cooperation which includes: - Permitting system for the use frontier waters (authorization held by the Contacting Parties) -Exchange of available information and observations, legal material relevant to the frontier waters <u>Article 8(b)</u> " Provide each other, to the extent agreed upon, with information on observations that have been made, on the results of hydrographic measurements on relevant hydrological research;. . ." and Article 8(c) "Provide each other with reports on high water, drifting ice and other hazards and notify each other of water-level forecasts based on water gauges specified by agreement between them;.." and Article 8(d): "Exchange the texts of important legal provisions relating to the use of water resources. **Institutional Mechanism** Appointing plenipotentiary from each Contracting Party whose decisions are not binding		

TREATY	Summary	Implicit Reference to GW	Explicit Reference to GW
11.Agreement concerning Water-Economy Questions between the Government of the Federal People's Republic of Yugoslavia and the Government of the People's Republic of Bulgaria, signed at Sofia, on 4 April 1958,(came into force on 29 December 1959), *found at* the TFDD Treaty Collection http://ocid.nacse.org/qml/research/tfdd/toTFDDdocs/105ENG.htm (*last visited* May 2003).	**Objectives and Scope** Water economy questions <u>Article 1</u> 2. The provisions of this Agreement shall, in so far as the Contracting Parties are interested in accordance with paragraph 1 of this article, apply water-economy questions, measures and works on rivers, tributaries and river basins followed or intersected by the State frontier, and in particular to: (a) The regulation and canalization of rivers and tributaries and the maintenance of river beds;(b) Drainage; (c) Storage and retention works; (d) Water supply and pipe-laying; (e) Protection of the waters against pollution; (f) The study and utilization of ground-water; (g) The utilization of water power; (h) Protection against soil erosion in forested and agricultural areas forestation, soil conservation, the erection of retaining-walls and silting control; (i) The utilization of waters for irrigation; (j) Hydrological, geological and other surveys and the planning and construction operations and use of water-economy installations and works; (k) The apportionment of the cost of survey, planning and construction operations and of the maintenance of installations; (3) The exchange of data and plans and of information on the above questions." **Substantive Rules** *Limiting Transboundary Impact* <u>Article 1:</u> "The Contracting Parties undertake, pursuant to the provisions of the Agreement, to examine and resolve all questions of water economy, including measures and works which may affect the quantity and quality of the waters and which are of interest to both or either of the Contracting Parties." And <u>Article 6</u> "The Contracting Parties undertake, each in its own territory, to preserve and maintain and, where necessary, to augment		See Article 1 (2) in column 2 of this table.

TREATY	Summary	Implicit Reference to GW	Explicit Reference to GW
	or adjust, such permanent benchmarks and other datum marks along the State frontier as are necessary for the purpose of regulation and other water-economy works. Each Contracting Party may use the other's datum marks. If it is necessary to cross the State frontier in order to use the said marks, the provisions of article 7 of this Agreement shall apply." **Procedural Rules** Cooperation which includes *Exchange of information* including hydrological meteorological and geological data, water levels..etc. *Notification* "The frontier and local authorities of the Contracting Parties shall advise each other, by the most rapid possible means, of any danger from high water or drifting ice and of any other danger which may arise on rivers and tributaries followed or intersected by the State frontier." **Institutional Mechanism** Joint Commission for implementation of the Agreement **Dispute settlement Mechanisms** (a) Commission (b) Governments of contracting Parties Article 10 "Any dispute between the Contracting Parties relating to the application and interpretation of this Agreement shall, unless the two Parties agree upon some other mode of settlement, be submitted at the request of either Contracting Party to a commission composed of two representatives of each Party. If this commission fails to reach agreement, the dispute shall be submitted to the Governments of the two Contracting Parties for decision."		

TREATY	Summary	Implicit Reference to GW	Explicit Reference to GW
12. Protocol concerning International Commission to Protect the Moselle against Pollution, between France, Federal Republic Of Germany Luxemburg, signed At Paris 20 December 1961, (came into force July 1 1962), *found at* the TFDD Treaty Collection http://ocid.nacse.org/qml/research/tfdd/toTFDDdocs/298ENG.htm (*last visited* May 2003).	**Objectives and Scope** Institutional mechanism to prevent and control the pollution of the waters of the Moselle, cooperation **Procedural Rules** Cooperation mainly in the form of establishing a Joint Commission	Waters of the Moselle	
13. Agreement on International Commission for the Protection of the Rhine against Pollution, between Switzerland, Federal Republic Of Germany, Luxemburg, Netherlands And France, signed In Bern On 29 April 1963 (Came Into Force May 1965), *found at* the TFDD Treaty Collection http://ocid.nacse.org/qml/research/tfdd/toTFDDdocs/294ENG.pdf (*last visited* May 2003).	**Objectives and Scope** Cooperation Protection from pollution and the establishment of a joint commission **Substantive Rules** Limiting Transboundary Impact **Procedural Rules** Cooperation **Institutional Mechanism** Established the International Commission for the Protection of the Rhine to implement the Agreement	Waters of the Rhine	

TREATY	Summary	Implicit Reference to GW	Explicit Reference to GW
14. Agreement between the Government of the Polish People's Republic and the Government of the Union of Soviet Socialist Republics Concerning the Use of Water Resources in Frontier Waters signed at Warsaw, 17 July 1964; (came into force, 16 February 1965), *found at* the TFDD Treaty Collection http://www.internationalwaterlaw.org/RegionalDocs/Poland-USSR.htm (*last visited* May 2003).	**Objectives and Scope** regulate matters relating to the use of water resources in frontier waters in a spirit of friendship and fraternal co-operation. **General Principles** Friendship Protection of frontier waters against pollution and depletion **Substantive Rules** *Limiting Transboundary Impact* <u>Article 9 (1)</u> : 1) Neither Contracting Party shall, save by agreement with the other Contracting Party, carry out any works in frontier waters which may affect the use of water resources by the latter Contracting Party". **Procedural Rules** Cooperation including - Regular exchange of data (hydrological, hydrometeorological and hydrogeological) Article 8. - Unified method for observation and assessment - Joint monitoring of pollution and joint procedures for controlling - Coordinated actions to deal with emergencies and natural phenomena **Institutional Mechanism** Appointing plenipotentiaries and deputies from both sides to follow up the implementation of the Agreement. **Dispute Settlement Mechanisms** <u>Article 13(3)</u> **"**the Plenipotentiaries or their Deputies may establish mixed Polish-Soviet working groups to consider particular problems and may call in experts to take part in the discussions."		Article 2: "For the purposes of this Agreement, the term "frontier waters" means: (1) the surface waters referred to in article 12, paragraph 1, of the Treaty between the Government of the Polish People's Republic and the Government of the Union of Soviet Socialist Republics concerning the regime of the Polish-Soviet State frontier and co-operation and mutual assistance in frontier matters, signed in Moscow on 15 February 1961, hereinafter referred to as the Treaty of 15 February 1961; (2) other surface waters intersected by the State frontier, (3) groundwaters intersected by the State frontier." <u>Article 3:</u> The purpose of this Agreement is to ensure co-operation, between the Contracting Parties in economic, scientific and technical activities relating to the use of water resources in frontier waters, including in particular:

TREATY	Summary	Implicit Reference to GW	Explicit Reference to GW
			(1) the regulation of frontier waters and the construction of hydraulic structures with a view to the integrated utilization of water resources; ……..(7) the protection of surface and ground waters against depletion and pollution; …. (9) the study of frontier waters for the purpose of determining their quantity and quality; (10) hydro-meteorological measurements and observation and the exchange of information in this regard;…..”
15.Franco-Italian Convention Concerning The Supply of Water to the Commune of Menton, done At Paris, On 28 September 1967, (came into force on 14 October 1972) *found at* the TFDD Treaty Collection http://ocid.nacse.org/qml/research/tfdd/toTFDDdocs/138ENG.htm (*last visited* May 2003)	**Objectives and Scope** Water Allocation and Supply **Procedural Rules** Cooperation Consultation **Dispute Settlement Mechanisms** Disputes that cannot be settled through diplomatic channels shall be submitted to an arbitrator.		Article 3. “The withdrawal of water provided for in article I shall be affected by pumping from the alluvial aquifer of the Roya at a level of approximately 30 meters above sea level, upstream from the confluence with the Bevera and downstream from the last hydroelectric power station on the Roya”.

.

TREATY	Summary	Implicit Reference to GW	Explicit Reference to GW
16. Convention concerning the Protection of Italo-Swiss Waters against Pollution between Italy and Switzerland, signed At Rome On 20 April 1972, *found at* the TFDD Treaty Collection http://ocid.nacse.org/qml/research/tfdd/toTFDDdocs/259ENG.htm (*last visited* May 2003).	**Objectives and Scope** Cooperation for the Protection of shared water against pollution and establishing a joint commission for the purpose of implementation **Substantive Rules** *Limiting Transboundary Impact* see <u>Article 1</u> in column 3 **Institutional Mechanism** Joint Commission that establishes mechanisms and procedures for the implementation of the Convention.		Article 1 states: " The Contracting Governments agree to collaborate closely with a view to protecting against pollution the following Italo-Swiss surface and groundwater including surface and groundwater of their affluent……"
17. Convention on the Protection, Utilization and Recharging of the Geneva Aquifer. Canton Of Geneva in Switzerland and the Department of Upper Savoy in France in 1977, *found at http://www.dundee.ac.uk/law/iwlri/Documents/Treaties/Basin%20Specific%20and%20Bilateral%20Treaties/Europe/arrangement.doc (last visited* May 2003).	**Objectives and Scope** Artificial recharge has been carried out from the Arve River into the Geneva aquifer in order to maintain the groundwater level and enable water resources management. **Procedural Rules** Cooperation joint monitoring exchange of information and data Institutional Mechanism The Arrangement establishes a management Commission of the Geneva Aquifer composed of three Swiss and three French members designated by the contracting parties and of which two each are to be specialists in water matters (Article 1). The Commission is purely consultative. The Commission convenes periodically, in practice twice a year, alternatively in France and in Switzerland. Each national delegation, which may be accompanied by any number of consultants and advisers, covers its own		

TREATY	Summary	Implicit Reference to GW	Explicit Reference to GW
	operational costs (Article 3). The conclusions reached by the Commission are recorded jointly and submitted to the consideration of the contracting parties which make sovereign decisions on the implementation thereof within their respective territory (Article 5.2).		
18. Treaty between the Government of Romania and the Government of Hungary on the Regulation of Water Problems in Watercourses Forming or Crossing the Boundary, June 25 1986, *found at* the TFDD Treaty Collection http://ocid.nacse.org/qml/research/tfdd/toTFDDdocs/101ROM.htm (*last visited* May 2003) (in Romanian)	**Objectives and Scope** Improve and develop cooperation in the solution of questions regarding transboundary and boundary forming watercourses and waterworks. Annex 1 contains the list of the sections of watercourses, waterworks falling under the present Agreement		The Agreement includes also provisions for the protection against damages which can be caused by water (flood, groundwater, etc.).
19. Agreement on Co-operation on Management of Water Resources in the Danube Basin, between the Federal Republic of Germany and the European Economic Community, Regensburg, Regensburg, 1 December 1987, *found at* the TFDD Treaty Collection http://ocid.nacse.org/qml/resea	**Objectives and Scope** Cooperation on the protection and management of the Danube **Substantive Rules** *Limiting Transboundary Impact* reference to *NO Substantial Influence* as in Article 2 "1. The Contracting Parties shall notify each other in good time of major projects on the territory of the Federal Republic of Germany or the Republic of Austria or where such projects might substantially influence the proper management of water resources on the territory of the other State". *No Substantial Adverse Effect* Article 3 "1. The Contracting Parties shall take the necessary measures within their respective legal		Article 2 "The maintenance and achievement of proper management of water resources within the meaning of this Agreement shall cover projects relating to:(a) protection of the aquatic environment including the groundwater, in particular the prevention of pollution, and the discharge of waste water and heat;…(c) the utilization of the

TREATY	Summary	Implicit Reference to GW	Explicit Reference to GW
rch/tfdd/toTFDDdocs/278ENG.htm (*last visited* May 2003).	systems to ensure that projects on stretches of water forming the frontier shall not have a substantial adverse effect on the condition of water resources on the territory of the Federal Republic of Germany or the Republic of Austria. They shall hold consultations with the aim of reaching mutual agreement, insofar as one party invokes these effects within a period of three months of notification by adducing serious grounds. 2. In the case of projects on all over waters with might have a substantial adverse effect on the condition of water resources on the territory of the other State, the Contracting Parties shall, at the request of the party concerned, discuss the possibilities of preventing such effects before the projects are carried out." **Procedural Rules**: Cooperation Article 1 "(a) exchange of experience,(b) exchange of information on water management regulations and measures, (c) exchange of experts, (d) exchange of publications, regulations and guidelines, (e) participation in scientific and specialist meetings, (f) consideration of projects on the territory of the Federal Republic of Germany or the Republic of Austria which might substantially influence the proper management of water resources on the territory of the other State. g) consultations in the Standing Committee on Management of Water Resources". *Notification* Article 2: "the Contracting Parties shall notify each other in good time of major projects on the territory of the Federal Republic of Germany or the Republic of Austria or where such projects might substantially influence the proper management of water resources on the territory of the other State".		aquatic environment including the groundwaters, in particular the use of water power and the diversion and abstraction of water;….." …"

TREATY	Summary	Implicit Reference to GW	Explicit Reference to GW
	Institutional Mechanism A Standing Committee is established to by Article 7 "1. A Standing Committee on Management of Water Resources shall be set up. Its duty shall be to contribute to the solution of questions arising from the application of this Agreement through joint consultations. For this purpose it may address to the Contracting Parties recommendations drawn up by agreement…." **Dispute settlement Mechanisms** Article 9 Diplomatic means Arbitral tribunal at the request of each party whose decision shall be binding		
Article 1: "The contracting parties shall cooperate in the International Commission for the Protection of the Elbe, hereinafter referred to as the Commission, to prevent the pollution of the Elbe and **its drainage area**". Article 2(1) " In particular, the Commission shall:….(m) prepare the basis for any regulation between the contracting parties regarding the protection of the Elbe and its drainage area".			

TREATY	Summary	Implicit Reference to GW	Explicit Reference to GW
21. Agreement on Joint Activities in addressing the Aral Sea and the Zone around the Sea Crisis, improving the Environment, and Enduring the Social and Economic Development of the Aral Sea Region , 26 March 1993, between the Republic Of Kazakhstan, Republic Of Kyrgyzstan, Republic Of Tajikistan, Turkmenistan, and Republic of Uzbekistan, *found at* the TFDD Treaty Collection http://ocid.nacse.org/qml/research/tfdd/toTFDDdocs/166ENG.htm (*last visited* May 2003).	**Objectives and Scope** Cooperation on the development and protection of the Aral Sea Basin **General Principles** Sovereign rights of States in matters of usage and protection of water resources in the basin, Mutual interests in the common resources of the basin **Substantive Rules** *Limiting Transboundary Impact* Preamble Article 1 (i-xi). **Procedural Rules** Cooperation including Exchange of Information, common information system for monitoring the environment **Institutional Mechanism** Establishing an Interstate Council for the Aral Sea basin crisis	Article I para 2 "States-participants recognize as common objectives:…maintaining the required water quality in the rivers, reservoirs, and springs, due to an, in future, preventing the release into these bodies of industrial and urban waste waters, and polluted and mineralized collector and drainage waters;…."	
22. The Helsinki Convention on the Protection and the Use of Transboundary Watercourses and International Lakes, Helsinki 1992, (came into force on October 6 1996 with 26 Signatories and 22 Parties), *found at the TFDD* Treaty Collection.	**Objectives and Scope** Cooperation on the Protection and Utilization of international watercourses **General Principles** Equality Reciprocity **Substantive Rules** *Equitable and Reasonable Utilization* Article 2 (c) "To ensure that transboundary waters are used in a reasonable and equitable way, taking into particular account their transboundary character, in the		Article 1 (1): Transboundary Waters means" any surface or groundwaters which mark, cross or are located on the boundaries between two or more States; wherever transboundary waters flow directly into the sea, these transboundary waters end at a straight line across their respective mouths between points on the low

TREATY	Summary	Implicit Reference to GW	Explicit Reference to GW
http://ocid.nacse.org/qml/research/tfdd/toTFDDdocs/338ENG.htm (*last visited* May 2003).	case of activities which cause or are likely to cause transboundary impact;..." *Limiting Transboundary Impact* Article 3 "To prevent, control and reduce transboundary impact, the Parties shall develop, adopt, implement and, as far as possible, render compatible relevant legal, administrative, economic, financial and technical measures, in order to ensure,….." **Procedural Rules** Cooperation (Article 9) Consultation (Art. 10) Joint Monitoring and Assessment (Art. 11) Exchange of Information (Art. 13) Warning and Alarm System (Art 14) **Institutional Mechanisms** Meetings of the Parties as regulated in Article 17 (1&2) **Dispute Settlement Mechanisms** Article 22 (a) negotiation or by any other means of dispute settlement acceptable to the parties to the dispute, if appropriate with assistance by the International Commission. (b) if not settled submit dispute to a compulsory decision either to the ICJ or arbitration according to Article IV of the Convention.		water line of their banks".
23. Agreements on the Protection of the River Meuse and Scheldt, France Netherlands and Belgium, done at Charleville Mezieres, April 26 1994, *found at* the	**Objectives and Scope** Cooperation on the protection of the Scheldt and the Meuse **General Principles** Common interests Protection against pollution and improvement of quality **Substantive Rules**	Article 1 (b) "The Meuse River Basin: the Meuse as well as all the waterways and canals which directly and indirectly run into it where situated on the	

TREATY	Summary	Implicit Reference to GW	Explicit Reference to GW
TFDD Treaty Collection http://ocid.nacse.org/qml/research/tfdd/toTFDDdocs/328ENG.htm (*last visited* May 2003), *also* in 34 ILM 851 (1995).	*Limiting Transboundary Impact* *Equitable and reasonable utilization* (Article 2 in the spirit of the 1992 Helsinki Convention see above) **Procedural Rules** Cooperation (Article 5) including Exchange of information Coordination of monitoring programmes Joint planning for the protection of the waters **Institutional Mechanism** (International Commission) to facilitate the implementation of the Agreement	territory of the Contracting Parties (c) The Meuse Drainage Area: the area, the waters of which run into the Meuse or its tributaries."	
24. Convention on Cooperation for the Protection and Sustainable Use of the Danube, signed In Sofia, On June 29th, 1994(came into force 22 October 1998), *found at* the TFDD Treaty Collection http://ocid.nacse.org/qml/research/tfdd/toTFDDdocs/331ENG.htm (*last visited* May 2003).	**Objectives and Scope** Cooperation on the management and protection of the waters of the Danube **General Principles** Equality Reciprocity **Substantive Rules** *Limit Transboundary Impact* Article 2 para 2 "The Contracting Parties pursuant to the provisions of this Convention shall cooperate on fundamental water management issues and take all appropriate legal, administrative and technical measures, to at least maintain and improve the current environmental and water quality conditions of the Danube River and of the waters in its catchment area and to prevent and reduce as far as possible adverse impacts and changes occurring or likely to be caused." See also articles 5 and 6. **Procedural Rules** Cooperation: The forms of cooperation under this Convention as a		Article 1: Definitions (g) "Water balance" means the relationship characterizing the natural water household of an entire river basin as to its components (precipitation, evaporation, surface **and underground run-off).** In addition a component of current man-made effects originating from water use and influencing water quantity is included." Article 2: OBJECTIVES AND PRINCIPLES OF COOPERATION (1) The Contracting Parties shall strive at achieving the goals of a sustainable and equitable water

TREATY	Summary	Implicit Reference to GW	Explicit Reference to GW
	rule are the following: Article 4 " FORMS OF COOPERATION (a) consultations and joint activities in the framework of the International Commission pursuant to the provisions of this Convention; (b) exchange of information on bi- and multilateral agreements, legal regulations and on measures in the field of water management; exchange of legal documents and directives and of other publications; other forms for the exchange of information and experiences." Consultation (Art. 11) - Exchange of Information, se (Art. 12, 13, 14)security of information and information to the public -Research and Development (Art. 15) - Communication, Warning Systems, emergency plans (Art. 16) **Institutional Mechanism** (Part III Article 18) (International Commission) whose role is to implement the Agreement and provide proposals and recommendations to the Contracting Parties. Adaptation of existing Agreements where necessary to eliminate any contradiction with the basic principles of this Convention. **Dispute Settlement Mechanisms** (Article 24) (a) negotiation or by any other means of dispute settlement acceptable to the parties to the dispute, if appropriate with assistance by the International Commission. (b) if not settled submit dispute to a compulsory decision either to the ICJ or arbitration according to Article V of the Convention.		management, including the conservation, improvement and the rational use of surface waters and **ground water** in the catchment area as far as possible. Moreover the Contracting Parties shall make all efforts to control the hazards originating from accidents involving substances hazardous to water, floods and ice-hazards of the Danube River. Moreover they shall endeavour to contribute to reducing the pollution loads of the Black Sea from sources in the catchment area. ARTICLE 6: SPECIFIC WATER RESOURCES PROTECTION MEASURES The Contracting Parties shall take appropriate measures aiming at the prevention or reduction of transboundary impacts and at a sustainable and equitable use of water resources as well as at the conservation of ecological resources, especially: (a) enumerate groundwater resources subject to a long-term

TREATY	Summary	Implicit Reference to GW	Explicit Reference to GW
			protection as well as protection zones valuable for existing or future drinking water supply purposes." (b) prevent the pollution of ground-water resources, especially those in a long-term perspective reserved for drinking water supply, in particular caused by nitrates, plant protection agents and pesticides as well as other hazardous substances;"
25. Draft Agreement on Water Quality Management of Zapadnaya Dvina/Daugava River Basin. presented at the Seminar in Novopolotsk 11-12 November 1997. The Federation of Russia, the Republic of Latvia, and the Republic of Belarus, *found at* the TFDD Treaty Collection http://ocid.nacse.org/qml/research/tfdd/toTFDDdocs/349ENG.htm (*last visited* May 2003). Could not find out whether this Agreement is in force or not	**Objectives and Scope** Management and Protection of the Zapadnaya River Basin **Procedural Rules** Cooperation including - Exchange of data - Joint Monitoring -Geographical database for the implementation of the Agreement **Institutional Mechanism** Ad hoc Commission which has an advisory role **Dispute Settlement Mechanisms** (a) negotiations (b) mediation by a contracting party of the 1992 Convention on the Protection of the Marine Environment of the Baltic Sea (HELCOM).		Preamble para 3: "MINDFUL of the interrelationship between surface and groundwater between upstream and downstream users and conditions, and between inland and coastal waters as well as marine waters of the Baltic Sea;"

TREATY	Summary	Implicit Reference to GW	Explicit Reference to GW
26. Convention on Environmental Impact Assessment in a Transboundary Context, ESPOO, September 10 1997, *found at* http://ocid.nacse.org/qml/research/tfdd/toTFDDdocs/339ENG.htm (*last visited* May 2003).	**Objectives and Scope** Procedures for evaluating the likely impact of a proposed activity on the environment in a transboundary context **Substantive Rules** *Limiting Transboundary Impact* Article 2: "The Parties shall, either individually or jointly, take all appropriate and effective measures to prevent, reduce and control significant adverse transboundary environmental impact from proposed activities. 2. Each Party shall take the necessary legal, administrative or other measures to implement the provisions of this Convention, including, with respect to proposed activities listed in Appendix I that are likely to cause significant adverse transboundary impact, the establishment of an environmental impact assessment procedure that permits public participation and preparation of the environmental impact assessment documentation described in Appendix II. 3. The Party of origin shall ensure that in accordance with the provisions of this Convention an environmental impact assessment is undertaken prior to a decision to authorize or undertake a proposed activity listed in Appendix I that is likely to cause a significant adverse transboundary impact." **Procedural Rules** Cooperation including: - Notification (Art. 3) - Exchange of information - Consultation (Art. 5) - Research Programmes (Art 90) **Institutional Mechanisms** Meeting of the Parties (Art 11)		

TREATY	Summary	Implicit Reference to GW	Explicit Reference to GW
	The Executive Secretary of the Economic Commission for Europe shall be responsible for the meetings of the Parties (Art 13): **Dispute Settlement Mechanisms** Stepwise Negotiation, amicable solutions International Court of Justice Arbitration Article 15 "1. If a dispute arises between two or more Parties about the interpretation or application of this Convention, they shall seek a solution by negotiation or by any other method of dispute settlement acceptable to the parties to the dispute. 2. When signing, ratifying, accepting, approving or acceding to this Convention, or at any time thereafter, a Party may declare in writing to the Depositary that for a dispute not resolved in accordance with paragraph 1 of this Article, it accepts one or both of the following means of dispute settlement as compulsory in relation to any Party accepting the same obligation: (a) Submission of the dispute to the International Court of Justice; (b) Arbitration in accordance with the procedure set out in Appendix VII".		
27. Convention on the Protection of the Rhine, Rotterdam, January 22, 1998. The Governments Of The Federal Republic Of Germany The French Republic The Grand Duchy Of Luxembourg	**Objectives and Scope** The Protection of the Rhine ecosystem Article 2: The scope of this Convention comprises 1. the Rhine 2. the ground-water interacting with the Rhine, 3. the aquatic and terrestrial ecosystems interacting with the Rhine or whose interaction with the Rhine could be re-established, 4. the Rhine catchment area, as far as its pollution adversely affects the Rhine 5. the Rhine catchment area, as far as it is of importance for issues of flood prevention and defence along the Rhine		Article 2 The scope of this Convention comprises ….(2) the ground-water interacting with the Rhine,…." Article 3: "**Targets**: With this Convention the Contracting Parties set the following goals:1. sustainable development of the Rhine ecosystem, in particular by

TREATY	Summary	Implicit Reference to GW	Explicit Reference to GW
The Kingdom Of The Netherlands The Swiss Confederation And The European Union, *found at* the TFDD Treaty Collection http://ocid.nacse.org/qml/research/tfdd/toTFDDdocs/193ENG.htm (*last visited* May 2003).	**General Principles** Article 4 "a) principle of prevention b) principle of precaution c) principle of mainly fighting environmental deteriorations at the source d) polluter-pays principle e) principle of not increasing adverse effects f) principle of compensation for considerable technical interventions g) principle of sustainable development h) application and further development of the Best Available Technique and of the Best Environmental Practice i) principle of not transferring environmental pollution into other environmental media. **Substantive Rules** *Limiting Transboundary Impact* Article 5 "4. They take the necessary autonomous measures on their territory, ensuring at all events that a) the discharge of wastewater liable to affect the water quality is subject to a prior discharge consent or is subject to generally obliging regulations fixing emission limits; b) the discharge of hazardous substances is gradually being reduced with the aim of not discharging such substances at all; c) the compliance with the discharge consents or general binding regulations and the discharge is subject to surveillance; d) the discharge consents or general binding regulations are regularly examined and adapted as far as allowed by considerable improvements of the development of the B.A.T. or required by the state of the receiving body of water; e) regulations will, as far as possible, reduce the danger of pollution due to incidents or accidents and that emergency measures are taken; f) technical interventions liable to considerably affect the Rhine ecosystem are subject to a prior consent with the necessary conditions or to generally binding regulations."		a) maintaining and improving the Rhine water quality as well as the quality of the suspended matter, the sediments and the ground water, in particular by avoiding, reducing or eliminating as far as possible pollution due to hazardous substances and nutrients from point sources (e.g. from industry and municipalities), from diffuse sources (e.g. from agriculture and traffic) equally passing by the ground water and from shipping. ensuring and improving the security of industrial plants and by preventing incidents and accidents; ………e) ensuring an ecologically sound and rational management of water resources"

TREATY	Summary	Implicit Reference to GW	Explicit Reference to GW
	Procedural Rules Cooperation - mutual information on planned measures. - International Monitoring programs and analysis j -Early warning system, institutional mechanism and above procedures to prevent harm. **Institutional Mechanism** Joint Institution (Commission) **Dispute Settlement Mechanisms** through arbitration whose procedures are found in Annex A		
28.Agreement between the Government of the Republic of Kazakhstan, the Government of the Kyrgyz Republic and the Government of the Republic of Uzbekistan on Cooperation in the Area of Environment and Rational Nature Use, done in One Original Copy in Russian, in Bishkek on 17 March, 1998, *found at* the TFDD Treaty Collection http://ocid.nacse.org/qml/research/tfdd/toTFDDdocs/196ENG.htm (*last visited* May 2003).	**Objectives and Scope** Cooperation on the Protection of the Environment and Rational natural resources use **General Principles** Friendship Cooperation Mutual benefit equality of rights **Substantive Rules** Limit Transboundary Impact Article 2(a-q). "a) to harmonize the ecological laws, interstate normative and legal regulations in the area of environmental protection and use of natural resources providing the mechanism for compensation of the damages and losses incurred by one Party to the other Party;.."	Subsoil Transboundary water resources	

TREATY	Summary	Implicit Reference to GW	Explicit Reference to GW
	Procedural Rules Cooperation (Article 2 a-q) *Notification* (Art. 3) " The Parties shall notify without delay about dangerous environmental situations on their territories, which may have ill effects on the adjacent countries, and to inform about the measures to liquidate these consequences, as well as about the results of investigations, aimed at estimating their influence on the environment." **Dispute Settlement Mechanisms** Negotiation Article 7 "All disputes arising under execution and interpretation of the Provisions of the Agreement shall be settled by means of negotiations".		
29. Convention on Access to Information, Public Participation in Decision-Making and Access to Justice in Environmental Matters, done at Aarhus, Denmark, on 25 June 1998 Not yet in force Signatories: 40, Parties 7, May 2003, *found at* http://www.mem.dk/aarhus-conference/issues/public-participation/ppartikler.htm (*last visited* May 2003).	**Objectives and Scope** Access to environmental information "Environmental information" means any information in written, visual, aural, electronic or any other material form on: (a) The state of elements of the environment, such as air and atmosphere, water, soil, land, landscape and natural sites, biological diversity and its components, including genetically modified organisms, and the interaction among these elements; (b) Factors, such as substances, energy, noise and radiation, and activities or measures, including administrative measures, environmental agreements, policies, legislation, plans and programmes, affecting or likely to affect the elements of the environment within the scope of subparagraph (a) above, and cost-benefit and other economic analyses and assumptions used in environmental decision-making;(c) The state of human health and safety, conditions of human life, cultural sites and built structures, inasmuch as they are or may be affected by the state of the elements		PUBLIC PARTICIPATION IN DECISIONS ON SPECIFIC ACTIVITIES 1. Each Party: (a) Shall apply the provisions of this article with respect to decisions on whether to permit proposed activities listed in annex I; Annex I Para 10: "Groundwater abstraction or artificial groundwater recharge schemes where the annual volume of water abstracted or recharged is equivalent to or exceeds 10 million cubic meters."

TREATY	Summary	Implicit Reference to GW	Explicit Reference to GW
	of the environment or, through these elements, by the factors, activities or measures referred to in subparagraph **Substantive Rules** Justice in access to information **Procedural Rules** Cooperation in the form of information exchange, public participation in decision making **Institutional Mechanisms** Meeting of the Parties (Art. 10) **Dispute Settlement Mechanisms** Negotiation, Arbitration, Article 16:" 1. If a dispute arises between two or more Parties about the interpretation or application of this Convention, they shall seek a solution by negotiation or by any other means of dispute settlement acceptable to the parties to the dispute.2. When signing, ratifying, accepting, approving or acceding to this Convention, or at any time thereafter, a Party may declare in writing to the Depositary that, for a dispute not resolved in accordance with paragraph 1 above, it accepts one or both of the following means of dispute settlement as compulsory in relation to any Party accepting the same obligation: (a) Submission of the dispute to the International Court of Justice;(b) Arbitration in accordance with the procedure set out in annex II. 3. If the parties to the dispute have accepted both means of dispute settlement referred to in paragraph 2 above, the dispute may be submitted only to the International Court of Justice, unless the parties agree otherwise".		

TREATY	Summary	Implicit Reference to GW	Explicit Reference to GW
30. Convention about the Cooperation for the Protection and Sustainable Use of the Waters of Portuguese-Spanish Hydrological Basins and Additional Protocol, Nov 30 1998, *found at* the TFDD Treaty Collection http://ocid.nacse.org/qml/research/tfdd/toTFDDdocs/251POR.htm (*last visited* May 2003).	**Objectives and Scope** Institutional Mechanism Convention forms joint commission concerning shared river basins. Focus on environmental protection, although document also discusses changes to hydrology of river courses and other general issues of potential transboundary impact.		Article 2: "to protect surface and groundwater and the terrestrial and aquatic systems on which these ecosystems directly depend".
31. UN/ECE Protocol on Water and Health to the 1992 Convention on the Protection and Use of Transboundary Watercourses and International Lakes 24 March 1999, adopted on 17 June 1999 at the Third Ministerial Conference on Environment and Health, *found at* http://www.who.dk/Document/Peh-ehp/ProtocolWater.pdf (*last visited* May 2003)	**Objectives and Scope:** Protection and use of transboundary watercourses and international lakes (Article 3) The provisions of this Protocol apply to: (a) Surface freshwater; (b) Groundwater; (c) Estuaries; (d) Coastal waters which are used for recreation or for the production of fish by aquaculture or for the production or harvesting of shellfish; (e) Enclosed waters generally available for bathing; (f) Water in the course of abstraction, transport, treatment or supply; (g) Waste water throughout the course of collection, transport, treatment and discharge or reuse. General Principles Sovereign right to exploit own resources pursuant to their own environmental and developmental policies.		Preamble: "Aware that surface waters and groundwater are renewable resources with a limited capacity to recover from adverse impacts from human activities on their quantity and quality, that any failure to respect those limits may result in adverse effects, in both the short and long terms, on the health and well-being of those who rely on those resources and their quality, and that in consequence sustainable management of the hydrological cycle is essential for both meeting human needs and protecting the

TREATY	Summary	Implicit Reference to GW	Explicit Reference to GW
	Substantive Rules Equitable access to water, adequate in terms both of quantity and of quality Limiting Transboundary Impact **Procedural Rules** Cooperation Articles 11 and 12 Notification Article 12 "(f) The prompt and clear notification by the competent authorities of one Party to the corresponding authorities of other Parties which may be affected of…." The exchange of Information "(g) The exchange of information on effective means of disseminating to the public information about water-related disease." And Article 13 (a) Exchange information and share knowledge about the transboundary waters and the problems and risks which they present with the other Parties bordering the same waters;" *Consultation* Article 13 (d) Consult each other, at the request of any one of them, on the significance of any adverse effect on human health which may constitute a water-related disease." **Institutional Mechanisms** Meeting of the Parties The Executive Secretary of the Economic Commission for Europe and the Regional Director of the Regional Office for Europe of the World Health Organization shall carry out the following secretariat functions for this Protocol.(Art 17)		environment" Article 2 (3) "Groundwater" means all water which is below the surface of the ground in the saturation zone and in direct contact with the ground or subsoil; (5) "Transboundary waters" means any surface or ground waters which mark, cross or are located on boundaries between two or more States; wherever transboundary waters flow directly into the sea, these transboundary waters end at a straight line across their respective mouths between points on the low-water line of their banks;

TREATY	Summary	Implicit Reference to GW	Explicit Reference to GW
	Dispute Settlement Mechanisms Article 20 Negotiation Any other accepted means of dispute settlement International Court of Justice		
32. The European Union Water Framework Directive 2000/60/Ec, Published in the Official Journal Of The European Communities on 22 December 2000, (entered into force on that date). The fifteen Members States of the Community are: Austria, Belgium, Denmark, Finland, France, Germany, Greece, Ireland, Italy, Luxembourg, The Netherlands, Portugal, Spain, Sweden, and the United Kingdom: http://www.europa.eu.int/eur-lex (visited on 29 March 2002). European Freshwater Programme, www.wwfreshwater.org/initiative/wfd.html (visited on 1st March 2002).	**Scope and Objectives** to establish a framework for the protection of inland surface waters, transitional waters, coastal waters and groundwater **Substantive Rules** *Limiting Transboundary Impact* Article 3: "5. Where a river basin district extends beyond the territory of the Community, the Member State or Member States concerned shall endeavour to establish appropriate coordination with the relevant non-Member States, with the aim of achieving the objectives of this Directive throughout the river basin district. Member States shall ensure the application of the rules of this Directive within their territory." **Procedural Rules** Coordination Exchange of Information Protected Areas		(preamble para 20 "The quantitative status of a body of groundwater may have an impact on the ecological quality of surface waters and terrestrial ecosystems associated with that groundwater body". Preamble paragraph 23: "Common principles are needed in order to coordinate Member States' efforts to improve the protection of Community waters in terms of quantity and quality, to promote sustainable water use, to contribute to the control of transboundary water problems, to protect aquatic ecosystems, and terrestrial ecosystems and wetlands directly depending on them, and to safeguard and develop the potential uses of Community waters".

TREATY	Summary	Implicit Reference to GW	Explicit Reference to GW
			Preamble paragraph 28: "Surface waters and groundwaters are in principle renewable natural resources; in particular, the task of ensuring good status of groundwater requires early action and stable long-term planning of protective measures, owing to the natural time lag in its formation and renewal. Such time lag for improvement should be taken into account in timetables when establishing measures for the achievement of good status of groundwater and reversing any significant and sustained upward trend in the concentration of any pollutant in groundwater". Preamble paragraph 33: "The objective of achieving good water status should be pursued for each river basin, so that measures in respect of surface water and groundwaters belonging to the same ecological, hydrological and hydrogeological system are coordinated".

TREATY	Summary	Implicit Reference to GW	Explicit Reference to GW
			Paragraphs 34 of the Preamble states that: " For the purposes of environmental protection there is a need for a greater integration of qualitative and quantitative aspects of both surface waters and groundwaters, taking into account the natural flow conditions of water within the hydrological cycle". Article1: "Purpose: The purpose of this Directive is to establish a framework for the protection of inland surface waters, transitional waters, coastal waters and groundwater……" Article 2: (2) "Groundwater" means all water which is below the surface of the ground in the saturation zone and in direct contact with the ground or subsoil" Article 2 (11)Aquifer" means a subsurface layer or layers of rock or other geological strata of sufficient porosity and permeability to allow either a significant flow of groundwater or

TREATY	Summary	Implicit Reference to GW	Explicit Reference to GW
			the abstraction of significant quantities of groundwater. Article 2 (12). "Body of groundwater" means a distinct volume of groundwater within an aquifer or aquifers.. Article 2 (15) "River basin district" means the area of land and sea, made up of one or more neighbouring river basins together with their associated groundwaters and coastal waters, which is identified under Article 3(1) as the main unit for management of river basins. <u>Article 4 Environmental Objectives</u> :" (b) for groundwater (i) Member States shall implement the measures necessary to prevent or limit the input of pollutants into groundwater and to prevent the deterioration of the status of all bodies of groundwater, subject to the application of paragraphs 6 and 7 and without prejudice to paragraph 8 of this Article and subject to the application of Article 11(3)(j);

TREATY	Summary	Implicit Reference to GW	Explicit Reference to GW
			(ii) Member States shall protect, enhance and restore all bodies of groundwater, ensure a balance between abstraction and recharge of groundwater, with the aim of achieving good groundwater status at the latest 15 years after the date of entry into force of this Directive, in accordance with the provisions laid down in Annex V, subject to the application of extensions determined in accordance with paragraph 4 and to the application of paragraphs 5, 6 and 7 without prejudice to paragraph 8 of this Article and subject to the application of Article 11(3)(j); (iii) Member States shall implement the measures necessary to reverse any significant and sustained upward trend in the concentration of any pollutant resulting from the impact of human activity in order progressively to reduce pollution of groundwater. Measures to achieve trend reversal shall be

TREATY	Summary	Implicit Reference to GW	Explicit Reference to GW
			implemented in accordance with paragraphs 2, 4 and 5 of Article 17, taking into account the applicable standards set out in relevant Community legislation, subject to the application of paragraphs 6 and 7 and without prejudice to paragraph 8; Article 15 b (ii) states that "Member States shall protect, enhance and restore all bodies of groundwater, ensure a balance between abstraction and recharge of groundwater, with the aim of achieving good groundwater status at the latest 15 years after the date of entry into force of this Directive"

Appendix III

Treaty Survey in the Middle East

TREATY	Summary	Implicit Reference to GW	Explicit Reference to GW
1. Convention Regarding the Water Supply of Aden between Great Britain and the Sultan of Abdali, April 1910 between General E. DeBrath, C.B., C.I.E., Political Resident, Aden, on behalf of the British Government on the one part and SIR Ahmed Fadthl Mohsin, K.C.B.I., Sultan Of The ABDALI, his heirs and successors on the other part. Found at http://ocid.nacse.org/qml/research/tfdd/toTFDDdocs/21ENG.htm (last visited May 2003).	**Objectives and Scope** Water Supply from a Groundwater Well **Substantive Rules** No Harm the Parties agreed not to allow the depletion or pollution of the groundwater as stipulated in <u>Article I :</u> "(II) not to do or allow to be done anything that will reduce or contaminate the supply of water yielded by the wells sunk on the above site, i.e., the working of wells by machinery and the throwing of dirt within a distance of 400 feet of the above site. And in " And as stipulated from Article IV that the failure to protect the water would reduce the terms of payment towards the use of water: " the Sultan of Lahej undertakes that, in case the water is diminished owing to his failing to protect the works against damage by the Subehis or his tribesmen the Abdalis, then he is entitled to no more than Rs.15 per 100,000 gallons. Out of the aforesaid amount. Rs. 1,000 to be paid to the land-owners in consideration of the utilisation of the water for Aden and as rout of the land on which the wells are sunk. In the event of Government ceasing from drawing the water the Treaty shall become null and void." (c) with reference to the agreement and the water supply herein projected the British Government undertake on their part that nothing shelf be done to diminish or affect the existing rights and absolute independence of the Sultan of Lahej that they shall not interfere in any way in the administration of his State."		
2. Agreement between the Republic of Syria and the Hashemite Kingdom of Jordan concerning the Utilization of the Yarmuk	**Objectives and Scope** Cooperation on the utilization of the Yarmuk River for water supply and electric power generation and the establishment of a commission for the implementation of the Agreement	Article 8:" a) Syria shall retain the right to the use of the waters of all springs welling up within its territory in the basin of the Yarmuk and its tributaries, with	

TREATY	Summary	Implicit Reference to GW	Explicit Reference to GW
Waters. Signed at Damascus, On 4 June 1953 (came into force on 8 July 1953) found at the TFDD Treaty Collection http://ocid.nacse.org/qml/research/tfdd/toTFDDdocs/82ENG.htm (last visited May 2003).	**General Principles** - Joint use of the waters for mutual benefits of the countries Article 11: "Both Governments recognize the importance of the Yarmuk scheme and the expediency of its joint use for the benefit of the two countries, and to this end agree, to assign it a high priority among their projected schemes." - Precedence of States' rights to the use of the water over the private rights Article 12:" The rights of the two States to the use of the water of the Yarmuk basin shall take precedence over all private rights and each State shall be responsible in accordance with its own laws, and independently of the other State, for the settlement of all claims arising in this connexion within its territory in consequence of this Agreement." **Procedural Rules** Cooperation **Institutional Mechanism** Joint Commission	the exception of the waters welling up above the dam below the 250-metre level and shall retain the right to use water from the river and its tributaries below the dam for the irrigation of Syrian land in the lower Yarmuk basin and eastward of Lake Tabariyya or for other Syrian schemes.	
3. Johnston Negotiations on the Jordan River Basin, 1955, found at the TFDD Treaty Collection http://ocid.nacse.org/qml/research/tfdd/toTFDDdocs/92ENG.htm (last visited May 2003).	**Objectives and Scope** Allocation of the water of the Jordan Rivers and its tributaries between Jordan(including the West Bank at that time) , Israel, Lebanon and Syria General Principles 3: Division of Water: "In evolving the present plan, the basic principle was adopted of assuring to the Arab states enough water to meet the needs of all their lands that could feasibly be irrigated; and the division of water in the present plan accomplishes this objective." "A technical factor affecting the division of the water however is the need to assure that the quality of the water in the system as a whole is maintained at a usable standard of salinity and yet permit as much as	The term spring is used. The purpose of mentioning springs is in the context of preventing saline springs from flowing into Tabariyya Lake is mentions The Saline springs: 3 Division on Water "The Plan, therefore, contemplates the possibility that some of these springs around the edge of the Lake might be	

TREATY	Summary	Implicit Reference to GW	Explicit Reference to GW
	possible of all the waters to be used". **Procedural Rules** Cooperation **Institutional Mechanism** The Plan proposes the creation of an impartial Engineering Board together with a Watermaster for the purpose of supervising operation of the water system and compliance of the parties. The Engineering Board would consist of three eminent engineers who would be selected from a list prepared by the Secretary General of the United Nations.	collected and prevented from flowing into the Lake as a means of keeping the salinity of the Lake within usable limits."	
4. Agreement concerning Water Distribution of Al-Asi River between Syria and Lebanon, signed on the 20 September 1994 and came into force December 11th 1994 (In Arabic), found at the TFDD Treaty Collection http://ocid.nacse.org/qml/research/tfdd/toTFDDdocs/273ARA.htm (last visited June 2003).	**Objectives and Scope** Allocation of the waters of Al Asi River including runoff water, springs, falls, groundwater, rainfall water **Procedural Rules** Cooperation Declared protection areas 500 meters from each side of the border and 1500 from the centre of springs (Hermel, Ras Elmal, Zerqa, permanent springs of Al Asi river and any other spring identified by the Joint Water Committee) Article 2 Rationing of water use during drought periods: Article 4: " When the flow in the river measured at the Hermel Bridge is less than 400 mcm/yr the year is considered a drought year, in this case the Lebanese share is reduced by the same proportion in the reduction of the flow of the river…" Prevention of well drilling in the recharge area of Al Asi River after the entry of this Agreement into force. As for the existing wells they continue to operate and their data concerning flows and abstractions are registered and signed by both Parties. (Article 8).		

TREATY	Summary	Implicit Reference to GW	Explicit Reference to GW
	Institutional Mechanism Article 5 Joint Water Commission to monitor the flow in the river, wells springs and installations on the river and to supervise the implementation of the allocation schemes **Dispute Settlement Mechanisms** Article 7 Establishment of an arbitral committee for settling any dispute on the implementation of the provision of the Agreement. This committee is derived from the Joint Water Commission		
5. Treaty of Peace between the State of Israel and the Hashemite Kingdom of Jordan, 26 October 1994, Arava/Araba Crossing Point, 34 ILM 1995, 43, found at http://www.mfa.gov.il and http://ocid/nacse.org/qml/research/tfdd/toTFDDdoc/168ENG.htm (last visited May 2003).	**Objectives and Scope** Cooperation to Achieve settlement on all water issue including allocation, Storage, Water quality protection Article 6: "1. The Parties agree mutually to recognise the rightful allocations of both of them in Jordan River and Yarmouk River waters and Araba/Arava ground water in accordance with the agreed acceptable principles, quantities and quality as set out in Annex II, which shall be fully respected and complied with." **Substantive Rules** Rightful Allocations of both sides from the River No Harm Article 6 (2). "2. The Parties, recognizing the necessity to find a practical, just and agreed solution to their water problems and with the view that the subject of water can form the basis for the advancement of co-operation between them, jointly undertake to ensure that the management and development of their water resources do not, in any way, harm the water resources of the other Party." Article 6 Annex II Article III "Israel and Jordan each undertake to protect, within		See Article 6 (column 2 f this table)

TREATY	Summary	Implicit Reference to GW	Explicit Reference to GW
	their own jurisdiction, the shared waters of the Jordan and Yarmouk Rivers, and Arava/Araba groundwater, against any pollution, contamination, harm or unauthorized withdrawals of each other's allocations. 2. For this purpose, Israel and Jordan will jointly monitor the quality of water along their boundary, by use of jointly established monitoring stations to be operated under the guidance of the Joint Water Committee. 3. Israel and Jordan will each prohibit the disposal of municipal and industrial wastewater into the courses of the Yarmouk and the Jordan Rivers before they are treated to standards allowing their unrestricted agricultural use. Implementation of this prohibition shall be completed within three years from the entry into force of the Treaty. 5. Saline springs currently diverted to the Jordan River are earmarked for desalination within four years. Both countries shall cooperate to ensure that the resulting brine will not be disposed of in the Jordan River or in any of its tributaries. 6. Israel and Jordan will protect water systems each in its own territory, supplying water to the other, against any pollution, contamination, harm or unauthorised withdrawal of each other's allocations. 4. The quality of water supplied from one country to the other at any given location shall be equivalent to the quality of the water used from the same location by the supplying country. Article IV. "- Groundwater in Emek Ha/Earava/Wadi Araba 1. In accordance with the provisions of this Treaty, some wells drilled and used by Israel along with their associated systems fall on the Jordanian side of the borders. These wells and systems are under Jordan's sovereignty. Israel shall retain the use of these wells and systems in the quantity and quality detailed in an Appendix to this Annex that shall be jointly prepared by 31st December, 1994. Neither country		

TREATY	Summary	Implicit Reference to GW	Explicit Reference to GW
	shall take, nor cause to be taken, any measure that may appreciably reduce the yields or quality of these wells and systems." **Procedural Rules** Notification Article 6 Annex II Article V: "1. Artificial changes in or of the course of the Jordan and Yarmouk Rivers can only be made by mutual agreement. 2. Each country undertakes to notify the other, six months ahead of time, of any intended projects which are likely to change the flow of either of the above rivers along their common boundary, or the quality of such flow. The subject will be discussed in the Joint Water Committee with the aim of preventing harm and mitigating adverse impacts such projects may cause." Cooperation Article 6 Annex II Article VI 1. Israel and Jordan undertake to exchange relevant data on water resources through the Joint Water Committee. 2. Israel and Jordan shall co-operate in developing plans for purposes of increasing water supplies and improving water use efficiency, within the context of bilateral, regional or international cooperation. **Institutional Mechanism** Establishing the Joint Water Committee for the implementation of the Agreement. **Dispute Settlement Mechanisms** None		
6. Israel-Palestine Liberation Organisation Declaration of Principles on Interim Self-government	**Objectives and Scope** The aim of the Israeli-Palestinian negotiations within the Middle East peace process is, among other things, was to establish a Palestinian Interim Self-Government Authority, the elected Council (the `Council'), for the Palestinian	Annex III 1.: ". Cooperation in the field of water, including a Water Development Program prepared by experts from both	

TREATY	Summary	Implicit Reference to GW	Explicit Reference to GW
Arrangements, 13 September 1993, Washington DC., 32 ILM 1525 (1993), (came into force 13 October 1993), found at http://www.jmcc.org/peace/agreements.html, (last visited May 2003).	people in the West Bank and the Gaza Strip, for a transitional period not exceeding five years, leading to a permanent settlement based on Security Council Resolutions 242 and 338. In the field of water the Declaration included commitment from both sides to cooperate in the field of water and the protection of the environment including the initiation of studies and research to strengthen this cooperation. **Dispute Settlement Mechanisms** Article XV,1: "Disputes… (arising over this or any subsequent agreement) shall be resolved through the Joint Liaison Committee." **Article XV,2: "Disputes which can not be settled by negotiations may be resolved by a mechanism of conciliation to be agreed upon by the parties."** **Article XV,3: "The parties may agree to submit to arbitration disputes relating to the interim period, which cannot be settled through conciliation. To this end, upon the agreement of both parties, the parties will establish an Arbitration Committee."** Article X **"**…a Joint Israeli-Palestinian Liaison Committee will be established in order to deal with issues requiring coordination other issues of common interest and disputes."	sides, which will also specify the mode of cooperation in the management of water resources in the West Bank and Gaza Strip, and will include proposals for studies and plans on water rights of each party, as well as on the equitable utilization of joint water resources for implementation in and beyond the interim period." Annex IV included a Protocol concerning the Regional Development Programs part of which is the water development programs.	
7. Israel-Palestine Liberation Organisation: Agreement on the Gaza Strip and the Jericho Area, 4 May 1994, Cairo, 33 ILM. 622 (1994), also found http://www.us-	**Objectives and Scope** Peace arrangement in the Gaza Strip and Jericho Areas. **General Principles** Mutual recognition Mutual dignity and coexistence **Substantive Rules** Preventing Harm		

TREATY	Summary	Implicit Reference to GW	Explicit Reference to GW
israel.org/jsource/Peace/gazajer.html (last visited June 2003).	Annex II, Article 31h "the subcommittee shall deal with)..all mutual prevention of harm to the water resources." **Procedural Rules** Cooperation Harmonization of technical standards for the use and handling of hazardous substances: Annex II PROTOCOL CONCERNING CIVIL AFFAIRS Paragraph 35 "Israel and the Palestinian Authority shall respectively adopt, apply and ensure compliance with internationally-recognized standards concerning acceptable levels of land, air, water and sea pollution, and acceptable levels of treatment and disposal of solid and liquid wastes; for the use and handling of hazardous substances, including pesticides, insecticides and herbicides, and standards for the prevention and abatement of noise, odour, pests and other nuisances which may affect each side and the Settlements and the Military Installation Area. Notification through Early Warning Systems and Emergency Measures: "Israel and the Palestinian Authority shall respectively operate an emergency warning system in order to respond to events or accidents which may generate environmental pollution, damage or hazards. A mechanism for mutual notification and coordination in cases of such events or accidents shall be established. " **Institutional Mechanism** Israel and the Palestinian Authority shall establish within the CAC an Environmental Experts Committee for the coordination of environmental issues, to be convened when the need arises. " Article III(6) and Annex II, Article I(A) 1:Establishment of a Joint Civil Affairs		

TREATY	Summary	Implicit Reference to GW	Explicit Reference to GW
	Coordination and Cooperation Committee (CAC).2.Function:Coordination between the PA and the CA. 3:Equal members, 4:Determines its mode of procedure 5: Matters not settled in the CAC will be passed to the Liaison Committee Annex II, Article II B 31 "**h:** Establishment of a subcommittee to deal with all issues of mutual interest (in the water field)." Annex II, Article II B 31, "h: The Subcommittee shall agree upon the procedures and manners of its meetings and may invite experts or advisers as it sees fit." **Dispute Settlement Mechanism**b Article XVII**:** In cases of dispute, Art. XV of the DOP shall be applied, the same mechanisms as in the DOP		
8. Israeli-Palestinian Interim Agreement on the West Bank and the Gaza Strip, 28 September 28 1995, Washington D.C., 36 ILM 551 (1997), found at http://www.ariga.com/treaties/interim.html> (last visited May 2003).	**Objectives and Scope** Additional Water for the Palestinians, cooperation on the groundwater aquifers shared between them. Institutional mechanism for the implementation of the commitments **General Principles** Good Will Recognition by Israel of Palestinian water rights **Substantive Rules** No Harm Annex III Article 40 3e: Taking all necessary measures to prevent harm to the water resources, including those utilised by the other side." and " h. each side shall take all necessary measures to prevent any harm to the water and sewage systems in their respective areas."		

TREATY	Summary	Implicit Reference to GW	Explicit Reference to GW
	Procedural Rules Cooperation Exchange of Information **Institutional Mechanisms** Establishment of the Joint Water Committee for the implementation of the Agreement. The JWC has wide authorities including the issuance of permits and licenses for well drilling and other infrastructure and development projects implementation. Annex III, Article 40,11: Establishment of a Joint Water Committee, under the auspices of the CAC (Cairo Agreement). Various tasks, listed up in the Article. Annex III, Article 40,13: The JWC is composed of equal members from each side. Annex III, Article 40,14:All decisions of the JWC shall be reached by consensus. Annex III, Schedule 8: Detailed obligations and responsibilities of the JWC, the JWC does not to have to approve the agreed projects in the Treaty, but it has to approve any further development of water resources. In cases of drought or change, the water can be reallocated through the JWC. The establishment of the Joint Supervision and Enforcement Teams whose role is to ensure the enforcement of decisions taken within the JWC: Annex III, Article 40,16: Establishment Joint supervision and Enforcement JSET, detailed tasks in schedule 9, shall report to the JWC. Dispute Settlement Mechanisms Article XXI**:** In cases of dispute, Art. XV if the DOP shall be applied. And Annex III, Article 40,11,f**:** "JWC shall deal with resolution of water and sewage related disputes"		

Appendix IV

Proposed Framework for the Assessment Of Palestinian Legal Entitlement From The Mountain Aquifer

INTRODUCTION

This appendix responds to one of the steps suggested under the progressive approach for solving the Palestinian-Israeli water conflict over the Mountain Aquifer.[1] It aims at providing an illustrative mechanism for assessing the legal

[1] *See supra* Chapter Six pages 207-228.

entitlements of the Palestinians from their shared groundwaters. It builds on the "Legal Assessment Model" that was developed by the Dundee International Water Law Research Institute.[2] Worth mention is the difficulty faced by the researcher in accessing primary sources of data which was discussed in Chapter Three. This framework can be updated and improved once a comprehensive set of data and information is available.

1. FACTORS 1 ("WHAT?")

This factor sets out the physical context, covering the physical or natural characteristics of the watercourse; characteristics of the watercourse including:

1.1 Geography

Palestine is located on the east coast of the Mediterranean Sea, west of Jordan and to the south of Lebanon. The territory of the West Bank covers around 5572 km^2 divided as: 5352 km^2 of land and 220 km^2 *of* water.[3] The water comprises half of the area of the Dead Sea, Huleh Lake, which was dried by the occupation and Tabariyya Lake, which is also known as the Sea of Galilee.[4] The West Bank is mostly composed of limestone hills having an average height of 700 to 900 m. The border countries are Israel 307 km and Jordan 97 km.[5] Israel covers a total, 21,671 km^2 of land (plus 474 km^2 of lakes) reflecting armistice or cease-fire agreements with Lebanon and Syria, and more recent peace treaties with Egypt

---

[2] The project entitled " Transboundary Water Resources Management: Using the Law to Develop Effective National Water Strategy: Poverty Eradication through enforceable Rights to Water", was implemented by the International Water Law Research Institute IWLRI, Department of Law, Dundee University-Scotland. The overall objective of the project is to develop a generic Legal Assessment Model (LAM) capable of being used by any Transboundary Watercourse State. The LAM provides a mechanism by which all relevant factors can be identified and evaluated within a framework consistent with a State's legal rights and obligations. This will provide the Transboundary Watercourse State with a tool, which can be used to determine its "equitable and reasonable use" of shared waters at the national level. Transboundary Groundwater in Palestine was one of the case studies of this project. This researcher was the team leader for that specific case study. The other two case studies are one that deals with upstream case (China) while the other handles the downstream (Mozambique). *See* http://www.dundee.ac.uk/law/iwlri/index.php (*last visited* September 2004).

[3] Valuable Agricultural Land In The West Bank Governorates, Emergency Natural Resources Protection Plan - Thematic Report 03 Directorate. found at http://planning.pna.net/agriculture/agriculture.htm (*last visited* June 2003). Not available any more on the web.

[4] *See generally* http://www.palestine-net.com/geography/ (*last visited* September 2004)

[5] *See generally* http://www.infoplease.com/ipa/A0776421.html (last visited September 2004).

and Jordan. The West Bank is in the upstream of the NEMAB and the WMAB whereas Israel is in the downstream with respect to these same aquifers.[6]

1.2 Hydrography

The sources of water in the WB are those renewable waters of the Mountain aquifer that rises and outcrops in the WB but extends across and below the territories of Israel. The main recharge acceptance area is located in the core of the WB where water originating in altitudes higher than 400 meters feed the major aquifers in the area.[7] The Mountain Aquifer Basin underlying the West Bank highlands is 150 km long between Tamasseh springs [8] in the north and Beer Sheva in the south. Natural outlets are the Al Auja and the Tamaseeh springs with a possible outlet to the Mediterranean Sea. Sub-aquifer Basins constituting the Mountain Aquifer are the WMAB, NEMAB and EMAB. The WMAB is located in the Western part of the WB. The recharge area is 1800 km^2 of which 1400 km^2 lie in the WB. The storage area of 2500 km^2 lies in Israel. Two major natural outlets are Ras El Ein and Tamaseeh Springs.[9] The NEMAB covers the areas of Nablus and Jenin in the WB and the waters flows northwards to their major outlets in Bet Sha'an Springs in Northern Israel. The feeding and storage area of 700 km^2 lies completely in the West Bank (650 km^2).[10] The NEMAB starts near Nablus and flows towards the Gilboa Mountains and Jezreal and Bet Shean Valleys to the northeast.[11] The EMAB is located and recharged almost entirely in the West Bank, with the feeding and storage area spread over 2200 km^2 mostly in the WB. A small part of recharge is located west of the Green Line including West Jerusalem.[12]

1.3 Climate

The amount of available water whether surface or groundwater is highly influenced by the Mediterranean climate, which is characterized in the West

---

[6] Against the water related political, it is obvious that this factor does not have any significance under the *status quo* of absolute Israeli control over the resources.

[7] D. Gross and A. Soffer, International Groundwater and International Rules: The Case of the Middle Easy and the Agreements between Israel and Jordan and between Israel and the Palestinians about the Gaza Strip, Occasional Papers on the Middle East No. 12, University of Haifa, October 1996.

[8] Different terminology is used in Israel and Palestine.

[9] H. Gvirtzman, *Groundwater Allocation in Judea and Samaria.* In J. Issac and H. Shuval (eds.) Water and Peace in the Middle East (Elsevier, Amsterdam: 1994)

[10] Ibid.

[11] Israeli Ministry of Agriculture Annual Report, 1998 (on file with the author).

[12] H.I. Shuval, Towards Resolving Conflicts Over Water Between Israel and Its Neighbours: The Israeli Palestinian Shared Use of the Mountain Aquifer as a Case Study ,found in: J.A. Allan, Water, Peace and the Middle East: Negotiating Resources in the Jordan River Basin, (New York, 1996).

Bank by long, hot, dry summers and short, cool, rainy winters.[13] The great differences in the distribution of temperatures, evaporation and rainfall create a wide variety of climatic conditions within the relatively small area of the country.[14] The climate is strongly affected by the proximity of the desert to the south and east. A dry season with practically no rain prevails from about the beginning of April to the end of October.[15]In summer there are no great fluctuations in average or actual temperature and hence the weather remains stable. The absolute maximum in May-June may be as high as 38° C, and in September-October even 42° C. From July-August the maximum doesn't exceed 32° C. The average daily temperature from June-October is 20-25° C, while in the rainy season temperatures range between 10-11° C.[16] Rainfall is confined to the winter and occurs mainly between November and March. Average annual rainfall ranges from 400 to 800 mm in the northern and western parts of the country and declines sharply toward the south and east. The average annual rainwater supply in the West Bank is approximately 2,800[17] mcm and the average annual rainwater supply in Israel and the occupied Palestinian and other Arab territories does not exceed 10,000 mcm. That means that the rainwater resources in the West Bank alone represent over 25 per cent of the total rainwater resources.[18] The rainy season is distinct occurring mainly from November to May with maximum rainfall in January. There are large deviations from this pattern; and there is also a considerable deviation from the average rainfall from year to year. [19] The western slopes have an annual rainfall of 500-700 mm, the eastern slopes 150-250 mm Precipitation is the most important hydrological variable in arid zones.

1.3.1 Annual Water Balance and Groundwater Recharge

The groundwater recharge in Palestine is the direct infiltration of rainwater through fractured, karstic rocks and porous soils. Of the 2800 million m^3 approximately 1900 is lost through evaporation, 225 flows to the rivers, 50 flow

[13] L. Hosh, Criteria For The Evaluation Of Aquaculture Potentials, 1995, (found on file with the author)

[14]Y. Karmon, *Israel: Regional Geography*, (Wiley-Enterprise, 1971).

[15] D. Sitton, Applied Research Institutes, Ben-Gurion University of the Negev, 1999.

[16] E. Kesler, *Regional Characteristics and Climate of the Judea and Samaria* (in Hebrew), The Civil Administration for Judea and Samaria (Transportation and Meteorological Services) 1982.

[17] Analysis in an unpublished study by the Palestinian Water Authority provides different figures for annual rainfall in the West Bank. This analysis provides that the annual rainfall on the WB is estimated to be 2248 mcm, of which 1529mcm is lost through evaporation, 71mcm as runoff and 648 as natural recharge. The Return Flow has not been assessed in the said analysis.

[18] United Nations Economic and Social Council, Israeli land and water Practices and Policies in the occupied Palestinian and other Arab territories Note by Secretary-General, Document A/46/263 E/1991/8819 June 1991.

[19] Ibid.

as surface runoff and the remaining 625 million m^3 to groundwater basins.[20] The Interim Agreement between Israel and the Palestinian Authority estimated that the EMAB had 172 mcm/yr. recharge, the NEMAB was said to yield 145 mcm/yr. and finally the WMAB is estimated to have a recharge of 362 mcm/yr.[21] The main acceptance for recharge is the core of the WB with altitudes higher than 400 m feeding the major transboundary aquifers in the area.[22] This water drains along the western and eastern slopes. Recharge on the western slopes is much more than that on the eastern slopes since evapotranspiration losses on the latter slopes are very high. In 1963/64 reported values for Mean Annual Recharge (MAR) ranged between 0 mm in the area of the Jordan valley and 325 mm. in the west. As for 1964/65 the range was larger 0-425 mm.[23] The recharge area of the mountainous aquifers lies over 78% of the WB. According to Israeli sources the main storage area of the most significant aquifer the WMAB lies within Israel while 90% of this aquifer's recharge occurs on the WB.

1.3.2 Groundwater Quality

To date there is no comprehensive study for water quality in the WB. Most recent data is collected in the Environmental Profile by ARIJ.[24] By and large geo-chemical processes determine the water quality during contact of water with the rock formation. Since carbonate rocks predominate in the WM aquifers the natural water quality is that of carbonate/low saline waters. The infiltration of water in karstic structures, as exist in most of the carbonate rocks in the WB is through fissures and joints widened by solution weathering. The karstic infiltration is halted by impervious strata.[25] The existence of karst and fissures is an important factor to the vulnerability of groundwater to pollution since water pollutants find their way through these fissures to groundwater.[26]

---

[20] Palestinian Water Authority, *Background Document on Water*, 1999.(found on file with author)

[21] Annex III of Appendix II of the Interim Agreement *supra* note 1, Article 40 Schedule 8. Interim Agreement, Israeli-Palestinian Interim Agreement on the West Bank and Gaza Strip, Washington DC, 28th December 1995. *found at* http://www.miftah.org/Documents/documents/piia/interim.html, (*last visited* June 2003).

[22] Rofe and Raffety Consulting Engineers, 1965, *Nablus and Jerusalem Districts: Water Resources Survey*, Hashemite Kingdom of Jordan, Central Water Authority, Jordan, see also H. Gvirtzman, *Groundwater Allocation in Judea and Samaria* found in J. Issac and H. Shuval (eds.) Water and Peace in the Middle East (Elsevier, Amsterdam: 1994).

[23] *Ibid.*

[24] Applied Research Institute, The Environmental Profile, 1996 (*found on* file with the author), also could be ordered through ARIJ website www.arij.org.

[25] D. Scarpa, The Quality And Sustainability Of The Water Resources Available To Arab Villages To The West Of The Divide In The Southern West Bank, 42 Water Science & Technology, 331-33 (2000).

[26] Ibid.

FACTOR 2 ("WHO?")

The 'who' factor details the population in the area; including present and projected population. This factor must be linked with the next factor(what uses especially planned uses).

2.1 Present and Projected Population Dependence on Groundwater

The population growth rate in Palestine for the last five years ranged from 3.2% to 3.5%. Demographic trends in Palestine have been closely related to the Political developments in the region.[27] The official census of the Palestinian Bureau of Statistics conducted in 1997 shows that the WB population was 1,600,100.[28] Estimates for the current Palestinian population in the West Bank as of the end of 2002 is 2,300,000 at a growth rate of 3.4%.[29] The Estimates of the population in Israel as of the end of 2002 is 6,439,000[30]. Under the current terms of Israeli-Arab conflict in Palestine, Diaspora[31] populations constitute a frequently mentioned potential for immigration, hence a relevant element in the evaluation of possible future population trends in the area.[32] The concept of Vital Human Needs plays here an important role in ensuring equality between all the population without any discrimination.

FACTOR 3 ("WHAT USES?")

This factor identifies the demands on or the uses of the watercourse and the economic and other benefits that flow from the uses; including:

---

[27] Middle East Regional Water Supply and Demand Development, 1998 (*found on* file with the author).

[28] Palestinian Central Bureau of Statistics, PCBS, Population Census 1997 (found on file with the author.

[29] Palestinian Central Bureau of Statistics, PCBS, *Palestinians at the End of 2002.* http://www.pcbs.org/temp/pales002/pal_e.pdf (last visited June 2003)

[30] Based on the results of the population Censuses and on the changes which occurred in the population after the Censuses, as recorded in the Population Register. Last census in Israel was conducted in 1995. Since the end of 1995 and onwards estimates both of population and geographic distribution are based on the 1995 Census (5,612,300inhabitants).

[31] *Ibid.* The Palestinian Bureau of Statistics estimates as of end 2002 for the number of Palestinians in the Diaspora is 4.7 million. One of the most significant aspects of population dynamics in Israel and Palestine was the continuous interaction between trends occurring locally and in the much broader frame of reference of a Jewish and a Palestinian Diaspora.

[32] S. DellaPergola, *Demography in Israel/Palestine: Trends, Prospects, Policy Implications*, The A. Harman Institute of Contemporary Jewry, the Hebrew University, Israel, 2001. Found at http://www.iussp.org/Brazil2001/s60/S64_02_dellapergola.pdf (*last visited* September 2004).

3.1 Existing Uses of Transboundary Groundwater

The total Palestinian utilization from the aquifers according to Interim Agreement figures is 118mcm/yr representing 16% of the total transboundary aquifers yield, while Israel using the remaining. [33] Records show that 22 mcm/yr is being used by the Palestinians from the WMAB, 54mcm/yr from the EMAB and 42mcm/yr from the NEMAB. The additional amount of 12 mcm/yr is imported from Israel through their Mekorot network which makes the total use to be 130 mcm/yr. The total water supply for the West Bank for urban and domestic purposes during 1997 was estimated to be approximately 52.3 mcm. Israeli Use from TGW ranges between 651-676 mcm from the shared aquifer basins) including the use of settlements. 372 mcm/yr from the WMAB, 150mcm/yr from the NEMAB and 129-144mcm/yr from the EMAB. It is also important to note here that illegal Israeli colonies official information on the exact amount of water withdrawn by Israel from within the West Bank to the settlements is in not available. Some estimates indicate that it may be about 35-45 MCM/Yr from the Eastern Aquifer and another 10-15 MCM from the Western and North-Eastern Aquifers for an estimated combined withdrawal of some 45 to 60 MCM/Yr.[34]

The Total Water Consumption in Israel is 2008mcm/yr; 1264 mcm for irrigation; 621 mcm/yr for domestic and 123 mcm/yr for industry. As aquifers are already overused, the only solution to provide enough water for the Palestinians is to reallocate the fresh water re-sources between Israel and Palestine. This reallocation must be fair and equitable, and must simultaneously ensure the sustainability of the fresh water resources. If demands on both sides are not totally fulfilled from the freshwater resources after the reallocation the Parties are recommended to consider the development of new water resources such as desalinated water or the reuse of treated wastewater.[35]

3.2 Potential and Planned Uses

There isn't much information published on the Palestinian plans for the use of water resources. The official standing in the Palestinian Authority emphasizes that a prerequisite to the development of a realistic Palestinian Water Master Plan is a comprehensive understanding of the available resources, their potential and needs for investments to develop them. Up till now the Palestinians are unable to achieve this level of understanding, and the best they can come out

---

[33] Interim Agreement, Israeli-Palestinian Interim Agreement on the West Bank and Gaza Strip, *supra* note 21.

[34] Foundation for Middle East Peace, *The Socio-economic Impact of Settlement on Land, Water and the Palestinian Economy*, Washington DC, 1998, found at http://www.monde-diplomatique.fr/cahier/proche-orient/colonies-2 (*last visited* June 2003).

[35] Middle East Regional Water Supply and Demand Development Study, *supra* note 27.

with is a demand projection scheme for their domestic as well as the other economic sectors demand projections. A complete master plan will only be possible when a new water sharing criteria are applied in the region and the Palestinian start exercising their limited sovereignty over transboundary waters. According to the Framework document on the water issue the Palestinians domestic water demands were based on the principal of equality.[36] The said principle if applied in the case of the Palestinians would imply a dramatic rise in the per capita consumption to be equal with that of the Israelis. The current per capita Israeli consumption for domestic use is estimated at 105 m^3/capita/year. Taking into account a population growth of 3.5% from 1997 results in a projected value of 2,202,641 million Palestinian in the WB for year 2002. To supply the current population with the 105m^3 a total amount of 231.2 mcm/yr of drinkable fresh water is required. Furthermore an additional 15% of the domestic consumption is required for industrial purposes (35 mcm/yr). Finally and assuming a 0.6 m^3 are required to irrigate 0.001 km^2 and the total irrigable land is 200 km^2 then it is estimated that a further 120 mcm/yr is required for irrigation. Assuming an optimistic figure of 15%[37] for water losses provides a total value 444.13 mcm/yr of water demand which is more than triple the amount the Palestinian are receiving now (130mcm/yr). This scenario would require an immediate increase of 314.13 mcm/yr for the fulfillment of the current needs of the Palestinians.

According to a study that was implemented by the GTZ in cooperation with the three core parties in the multilateral talks (Jordan, Israel and the Palestinian Authority) a long-term scenario was given to estimate the future gap between supply and demand. According to this Study the total Palestinian water demand for Y2010 is 376 mcm/yr (divided as follows: 103 for domestic, 22 for industrial and 152 for irrigation). The 2040 water demand is 767 mcm/yr (divided as follows: 418 for domestic, 54 for industrial and 295 for irrigation).[38]

FACTOR 4 ("WHAT IMPACTS?")

This factor aims to set out the consequences of the uses, both within a nation and the effects of use in one nation on others;

- (i) Impacts of existing and potential uses
- (ii) Beneficial and adverse impacts
- (iii) Transboundary and national effects
- (iv) Changes in physical characteristics (quantity, quality)

[36] Palestinian Framework Document on the Final Status Negotiation Related to Water, April 2002.(*found* on file with author).

[37] Awartani 1994 assumes this value is 8%, however since the current losses is more than 30% in many areas the 15% value is proposed here as an alternative value.

[38] Middle East Regional Study on Water Supply and Demand Development, *supra* note 27.

(v) Social and economic impacts

In 1960s the Israelis drilled hundreds of wells all over the mountains on both sides of the Green line[39] and inside the WB, which has greatly influenced the yield of the WMAB. Prior 1960 the natural outlet of this aquifer was through two major springs, Ras El Ain[40] and the Timsah springs.[41] The annual yield of the Ras El Ain spring was 220mcm while that of the Timsah was 100mcm[42].After 1960 both Springs' flow was reduced considerably to the level of drying up the springs of Ras El Ain while the flow of the Timsah springs was reduced from 100 MCM/yr to 20 MCM/yr. It is estimated that 90% of the abstracted water is from the Upper Aquifer and only 10% of the abstraction is from the Lower Aquifer.[43]

After putting into operation, what so called, the national carrier, it was used to inject to the Basin nearly 10% of its total abstraction (see Map 6). In the period from 1971/2 to 1992/3, about 21 MCM/yr on average was injected through the National Carrier to the WMAB Basin. It should be noted that in some years, no water was injected and in other years as much as 55 MCM/yr was injected. The abstraction from this sub-basin was in early 1970's about 12 MCM/yr. This heavy pumping resulted in saltwater intrusion. Saltwater came from the recharge zone at Marj Ibn Amer. This intrusion is noticed when the salinity in some wells increased. As a result, pumping was reduced to 8 MCM/yr in late 1980's and subsequently water levels were raised and salinity decreased. NEMAB: Salinity is low (about 50 to 200 mg/l). High salinity is observed in Majedou and Mahouleh. In the last two years salinity was increased considerably (in some areas in this aquifer).[44] The main abstraction (about 55 MCM/yr) is from the unconfined aquifer. When abstraction started, the water levels dropped about 10 to 20 m. After the rainy winter of 91/92, the water levels were restored to the early levels of early 1970's.[45] The total well abstraction is about 15 MCM/yr. In the mountainous region, salinity of water extracted by wells is about 100 to 200 mg/l. This is a result of heavy irrigation

[39] Between 1949 and 1967 there existed what was called the Green Line boundary, a very artificial boundary between Israel and the West Bank. It had been drawn up in the 1949 Rhodes armistice agreement and then imposed upon the landscape, with a number of changes as it was implemented, as happens when most boundaries are implemented on the ground. For the Palestinian side, the green line is the only boundary. Accepting the notion of a future Palestinian state in the West Bank means for them that they are accepting only 23 percent of the territory of Mandatory Palestine.

[40] Called Yarkoon in the Hebrew terminology.

[41] Called Tanineem in Hebrew terminology.

[42] G. Weinberger , E. Rosenthal and A. Ben Zvi, The Yarkoon Tanineem Groundwater Basin, Israel Hydrogeology: Case Study and Critical Review 161 Journal of Hydrology 227, 1994.

[43] *Ibid.*

[44] *Ibid.*

[45] *Ibid.*

activities and using septic tanks in the region. In 1991/92, water levels rose by 5 to 15 m and spring discharges doubled.[46]

Following the establishment of Israel, Palestinians from the coastal region were forced to migrate towards the West Bank and East Bank of the Jordan. As a result unemployment rose and both banks of the Jordan faced a difficult economic situation in the early 1950s: 50 per cent of the labor force there worked in agriculture, which suffered from underemployment. However, strenuous development efforts enabled agriculture to double its contribution to the gross domestic product (GDP) from 14.2 million Jordanian dinars (JD) in 1954 to JD 27.7 million in 1966. According to the population and housing census of 1961, 35 per cent of the labor force in the West Bank worked in agriculture.[47]

The Jordanian 1964-1970 Development Program, which it was not possible to implement fully owing to the 1967 Israeli occupation of the WB, was aimed at increasing the area of agricultural land in the West Bank in order to raise the level of income from agriculture and improve the standard of living of farmers by increasing agricultural production through both horizontal and vertical expansion. Statistics available indicate that 30 per cent of the agricultural land in the West Bank before 1967 depended on irrigation.[48] The intention was to increase the area of land under irrigation to 40 per cent by the end of the Development Program in 1970. The Program also aimed at increasing the area of land planted with fruit trees and, in particular, at doubling the area planted with olives, almonds, grapes and other fruit trees, as well as various types of vegetables.[49]

A study prepared by UNCTAD on recent economic developments in the occupied Palestinian territory (TD/B/1142) indicated that the policy, practices and measures of the Israeli occupation of the WB and the GS had brought about radical changes in the structure of their economy, which had adversely affected economic growth and development. Agriculture remained the backbone of development in the occupied Arab territories, despite its declining contribution to GDP. That study demonstrated clearly that the total cultivated area had decreased from 36 per cent in 1966 to 27 per cent in 1984 of the total land area of the occupied West Bank, and in the occupied Gaza Strip from 55 per cent of total area in 1966 to 28 per cent in 1985. Confiscation of land had adversely affected agricultural production and income in the West Bank and the Gaza Strip; income from agriculture declined from 1,505 million shekels in 1978 to 1,488 million shekels in 1984 in constant 1980 prices. As a result, according to another study of UNCTAD (UNCTAD/ST/SEU/4), the share of agriculture in

---

[46] *Ibid.*

[47] *Ibid.*

[48] N. Sabri, *The Economic Value of Water,* Palestine 2001 (*found* on file with the author).

[49] *Ibid.*

GDP declined in constant prices from 32 per cent in 1978 to 28 per cent in 1984.[50]

FACTOR 5 ("WHAT OPTIONS?")

The assessment of existing option requires consideration of the comparative efficiency of uses and of alternative uses, both in terms of alternative sources of water and broader alternatives that may yield similar levels of benefits

5.1 Efficiency and Water Conservation

The existing water supply networks in almost all the WB Governorates are old with some dating back to more than 50 years. During the last 34 years of occupation no planned maintenance or rehabilitation have been undertaken due to the limited financial and technical abilities of facilities of the local authorities that were in charge of water services were limited. Additionally, permissions to conduct maintenance work during the years of occupation had to be authorized by Israel. [51] The Palestinian municipality networks are in desperate need of rehabilitation, as up to 50 percent of the water supply is lost due to leaky pipes in the West Bank. Palestinians suffer from water shortages and an irregular water supply, especially in the summer months. Due to the Palestinian water deficiency, it is estimated that Palestinian demand for water is at least twice as much as the present per capita consumption. [52] This situation has lead to high loss percentages through pipe leakage. Unaccounted for water reaches a value of 50% in some of the WB Governorates.[53] Therefore, the inefficiency in those systems is conflict related as it has hindered the Palestinians financially politically and technically to conduct maintenance. After the establishment of the Palestinian Authority and especially after the took- over by the Palestinian Water Authority many projects have been implemented to construct new water networks or to rehabilitate existing ones. In order to reach the efficiency required by the modern water sector management, prodigious efforts have to be exerted by the Palestinians that should be complemented by a real commitment from the international community to continue supporting the build-up of the Palestinian water institutions.

---

[50] United Nations Conference on Trade and Development, *Prospects for Sustained Development of the Palestinian Economy: Strategies and Policies for Reconstruction and Development,* UNCTAD/ECDC/SEU/12, August 1996.

[51] These are the municipalities in the big cities and village councils in the rural areas.

[52] *See* http://www.wws.princeton.edu/~wws401c/agriculture.html.

[53] Excluding the case of the Jerusalem Water Undertaking JWU in the Ramallah region which has records of 15-25% of losses in the network.

As for efficiency on the Israelis side the Israeli Comptroller report of 1990 dropped the veil to reveal the inefficiencies in the Israeli water policy.[54] The report indicates that efficiency is no good in quota allocation. The achievement of fairness among users is claimed to be the ultimate objective of the water allocation policy but still remains the secret of the State. The report concludes that this policy has for long encouraged waste of water. Users, who don't use their total allocations in a specific season, loose the unused portion from their next season quota. Consequently, users prefer to overuse water rather than getting less allocation in the next season. The main guidelines of Israeli water policy have consisted of under pricing, distorted and discriminatory pricing, and gross misallocation among water users. All of this particularly affects farm production and agricultural exports, as the limited water supplies could be used much better to generate farm value and export earnings. In a sense, a large portion of Israel's water resources is "exported," as embodied in farm exports. When water allocation is distorted, Israel ends up exporting its water for too few dollars.[55]

The main problem with Israeli water policy as this report concludes is that it is a politicized system instead of a market system. Administrative and bureaucratic considerations dictate water allocation, not economic considerations. Water distribution is a mechanism for patronage and rent allotment, not a system of rational resource allocation. Political considerations, including the quest for fairness, override all consideration of efficiency in water allotment. The result is not merely enormous waste and misallocation but accelerating ecological disaster. And all of this will no doubt get worse due to the added demands on water resources resulting from the various Israeli-Arab accords. In another report it has been indicated that 71% of the water used for agriculture in Israel is drinking quality water.[56]

A major problem in agriculture is that the cost of water to the user is highly subsidized, especially with regard to water that has been earmarked for agriculture. The true cost of water would reflect all of the pumping, treatment, and delivery costs of that water, most of which are not passed on to the farmers. In Israel alone, 20 per cent of the country's energy is used solely to move water from one place to another [57]. Economic theory argues that only when the price paid for a commodity is a reasonable reflection of the true price can market

---

[54] State Comptroller, *Report on the Administration of Israel's Water Economy* (Jerusalem: State Comptroller,1990), p. 53. [in Hebrew]

[55] *Ibid.*

[56] Z. Grinwald, *Israeli Water Resources Alternatives Administrative and Operative Aspects of Water Allocation Policy (found* on file with the author).

[57] T., Naff and R. Matson (eds.) *Water in the Middle East: Conflict or Cooperation?* (Boulder : Westview Press, 1984).

forces work for efficient distribution of the commodity.[58] In other words, subsidized water leads to waste in agricultural practices, too little incentive for research and development of conservation techniques and practice, and, finally, too much water being allocated to the agricultural sector as opposed to industry. Take away subsidies and allow the price to rise, it is argued, and market incentives are created for both greater efficiency on the farm and a natural shift of water resources from the agricultural sector to industry, where contribution to GNP per unit of water is often much higher. Since in each of the areas discussed between 75 per cent and 95 per cent of water use is allocated for agricultural use, savings could be substantial.[59]

5.2 Available Water Resources Alternatives

Before 1967, Palestinians made use of JR waters through 140 pumping units, which were either destroyed or confiscated by Israeli authorities immediately after the occupation in June 1967. The Palestinians have been denied access to these waters since then. Apart from the water resources in the Jordan River system, which is not accessible by the Palestinians. According to Johnston Plan the Palestinian share from the Jordan River was estimated at 257 mcm/yr.[60] The only surface water available is the runoff in the wadis[61], whose flow is mostly seasonal.[62] The un-exploited potential safe yield available to the Palestinians in the Eastern Aquifer Basin has not been determined accurately but may be somewhere between 50 and 100 mcm/yr.[63] Article 40 of the 1995 Interim Agreement between Israel and the PLO indicated that the potential is 70 mcm/yr.[64] The interim agreement between Israel and Palestine stipulates that future Palestinian water demand might be fulfilled from this reserve in addition to other sources that must be agreed between them.[65] However, it will require a number of surveys and researches in order to assess the quantitative and the qualitative properties. Seeking for other non-conventional water sources is a long term objective for the Palestinians. Presently, priority is given to ensuring that the existing fresh shared resources are utilized in an equitable and

---

[58] D. Wishart, *An economic approach to understanding Jordan Valley water disputes* 21 (4) Middle East Review, 45-53, (1989).

[59] *Ibid.*

[60] Palestinian Water Authority, *supra* note 36.

[61] Wadi is the Synonym for Valley.

[62] An exception is the spring fed Wadis - for instance Wadi Qilt and Wadi Far'ia, but these are already heavily utilized. To a large extent, the Wadis are also overloaded by raw sewerage in the headwater areas.

[63] G. Weinberger, *supra* note 42.

[64] This figure is believed to be an overestimate of the real available quantities for development from the EMAB. Recent studies that were conducted for the benefit of the Palestinian Authority provides lower figures.

[65] Article 40 of the Interim Agreement, *supra* note 21.

reasonable manner. Desalination and recycling wastewater – as carried out by Israel – require more investment than the Palestinian economy can bear. Water importation appears to be impractical and uneconomic at present. The Palestinian National Water Policy[66], Water Resources Management Strategy[67] and the Water Law No. for 2001[68] that have been adopted by the Palestinian Authority encourage conservation and call for the development of non-conventional resources.

In addition to its exploitation of the waters of the transboundary aquifer basins, Israel enjoys the benefit of availability of other fresh water resources. The Israeli revealed quantities by the Israelis on their utilisation from the Jordan River is 680 mcm/yr. Additionally, there are another five groundwater aquifers that lie in Israel and are utilised solely by the Israelis. Israel has gained good experience in developing non-conventional water sources such as reclaiming wastewater effluents, inducing artificial rainfall and desalinating brackish /or Seawater. Treated effluent is estimated at 400 MCM, form the largest water source awaiting intensive development. Currently, about *250 MCM* of effluents, treated to varying degrees, are already utilized for irrigation, constituting about 65% of the generated wastewater. At the end of 1999, an estimated 300 MCM (25%) of the total amount of water supplied for irrigation was in the form of reclaimed sewage effluents. As Israeli sources indicate this is expected to increase to about 600 MCM by the year 2020.[69] In Israel Although surface runoff is sporadic, several regional and local schemes have already been established. The schemes divert storm flow from the rivers into surface reservoirs from which they are pumped into the supply system, or dispersed on spreading grounds and left to percolate into the underground aquifer (mainly along the coastal plain). At present, approximately 40 MCM are intercepted, out of a potential of 135 MCM/year of storm water. [70]

Cloud seeding with silver iodide crystals has been practiced in Israel for last 30 years on a countrywide basis, using ground incinerators and aircraft. Controlled experiments that were conducted between 1960 and 1975 pointed to a positive significant effect of seeding on increasing rainfall. It is assumed that a significant increase of 10-15% in rainfall in the northern part of the country has been materialized. Although a representative of Israel's water authority claims that 15 per cent of Israeli annual rainfall is due to its cloud-seeding

---

[66] Palestine Main Elements of the National Water Policy, adopted 13 September 1993. (*found on* file with the author).

[67] Palestine Water Management Strategy, March 1998 (on file with author).

[68] Water Law in Palestine, No. 3/2002. (on file with author)

[69] The Israeli Ministry of National Infrastructure: The Water Sector: Non-Conventional Water Resources and Conservation, http://www.mni.gov.il/english/units/Water/Non-convetionalWaterResourcesandConservation.shtml, and in Y. Shevah, *Irrigation And Agriculture: Experience And Options In Israel,* Prepared For Thematic Review Iv.2: http://www.damsreport.org/docs/kbase/contrib/opt159.pdf.

[70] Ibid.

programme[71], this has been documented only within the northern Galilee catchment and results seem not to have the consistency necessary for reliable water resources planning. Internal and external sources have cited the Israeli seeding program to statistically show significant success, although the limited cloud occurrence in drought years limits the benefits of cloud seeding when most needed.[72] Israel has many desalination facilities in operation today, and many small and medium desalination plants are used for processing brackish and seawater for domestic water supply in the Arava Valley and the Gulf of Eilat. The largest of these facilities produces 44,000 cum/day of water from brackish groundwater and seawater, thus meeting most of the drinking water requirements of the town of Eilat. Reverse osmosis has been adopted as the leading technique for brackish and seawater desalination. The first sea water desalination plant, with a capacity of 10,000 cum/day was commissioned in 1997 and other modules of various capacity are planned along the Red Sea and the Mediterranean Sea coast. In April 2000 Israel announced its intention to construct the first desalination plant to be built in Ashkelon.[73] The present costs of about US$0.80-1.50/$m^3$ to desalt sea water and about US$0.30/$m^3$ for brackish water do not make this technology an economic water source for most uses.[74] Efforts are being made, however, to lower these costs through multiple-use plants, increased energy efficiency in plant design, and augmenting conventional plant power with solar or other energy sources.[75] In 1985, Israel confirmed the discovery of a large fossil aquifer in the Nubian sandstone underlying the Sinai and Negev deserts. Israel is already exploiting 25 MCM/yr from this source and is investigating the possibility of pumping up to 300 MCM/yr in the twenty-first century.[76] Turkey is considered Israel's best source

---

[71] J. Siegal,. *The World's Best Rainmaker*, Jerusalem Post, 8 April p. 12 (1989).

[72] Ibid.

[73] Yediot Ahronot (Israeli Newspaper), April, 8, 2000. The Israeli government's tenders committee has awarded two Build-Own- Operate (BOO) desalination projects for the production of 30-mil cubic meters (mcm) each, according to an announcement by the Israeli Finance Committee. The contracts to purchase water from the plants by the state are for 25 years. There were six bidders in the tender. The cost of the two projects is put at $140-mil. The cost includes private power plants at the site. In an interview with Global Water Report,, Israel's deputy National Infrastructure Minister Naomi Blumenthal said that all six bidders should be allowed to build their plants because of the severe water shortage. She added this would yield an additional 120 mcm of water by 2005. "We're planning to build a larger plant at the Palmachim site and sell the additional water to private customers," said a senior official in the Via Maris consortium. He estimated that the project would produce 35 to 40 mcm annually. Found at http://www.platts.com/gwr/081202.shtml (*last* visited March 2003). Not found any longer on the web.

[74] L. Awerbuch, *Desalination technology: An overview* In: J. Starr and D. Stoll (eds.), The Politics of Scarcity: Water in the Middle East (Boulder, Colo.: Westview Press 1988)

[75] M. Murakami, *Arid Zone Water Resources Planning for Peace.* Tokyo: United Nations University Press (1995).

[76] D. Lerner, A. Issar, and I. Simmers, *Groundwater Recharge: A Guide to Understanding and Estimating Natural Recharge*, 8 International Association of Hydrogeologists (1990).

of imported water from the mouth of the Manavgat River in southern Turkey to the Ashkelon Port, 600 km away. Water may be transferred using a submerged pipeline, super tankers or floating bags. The Manavgat River has an average flow of 140 cum/sec, of which 60 cum/sec are assumed to be available for export (15% of the unused water on the Turkish Mediterranean Coast). The use of super tankers, with a capacity of 250,000 tons, is currently being considered. They would unload near Ashkelon Port, on an existing floating Single Buoy Mooring (SBM) stationed 3.25 km from the coast. The existing facilities, including the SBM, the floating hose and the submerged pipes connecting the SBM to the coast, have to be cleaned for water use. [77]

---

77 Israeli Ministry of National Infrastructure: The Water Sector: Water Import. http://www.mni.gov.il/english/units/Water/WaterImport.shtml.

Index

IWA Publishing's authorised EU representative for General Product Safety Regulations is Diane D'Arras, 15 rue Duret, 75116 Paris, France, e-mail: safety@iwap.co.uk.

Printed and bound by CPI Group (UK) Ltd, Croydon, CR0 4YY
12/05/2026
02108867-0006